T0271360

Designed cover image: Shuterstock

First edition published 2024
by CRC Press
2385 NW Executive Center Drive, Suite 320, Boca Raton FL 33431

and by CRC Press
4 Park Square, Milton Park, Abingdon, Oxon, OX14 4RN

CRC Press is an imprint of Taylor & Francis Group, LLC

ISBN: 978-1-032-71542-1 (hbk)
ISBN: 978-1-032-71544-5 (pbk)
ISBN: 978-1-032-71543-8 (ebk)

DOI: 10.1201/9781032715438

Typeset in Times
by Apex CoVantage, LLC

Contents

Chapter 4 Energy Conservation and Management: Key to Energy
Sustainability 66

Rao Muhammad Mahtab Mahboob, Kiran Mustafa,
Sara Musaddiq, Nadeem Iqbal, Rao Muhammad Shahbaz
Mahboob, Mueed Ahmed Mirza

Chapter 10   The Push for Renewable Energy Adoption in Africa:
                 Challenges and Prospects... 187

Joan Nyika, Megersa Olumana Dinka

Chapter 11   Energy and the Environment: A Dynamic Partnership 204

Guller Sahin

Chapter 15 Decision-Making in Energy and Environmental
Systems Based on Water-Energy-Food Security Nexus
Principles 283

Abdolvahhab Fetanat, Mohsen Tayebi, Hossein Mofid

Chapter 16 Policy Dynamics for Energy, Environment and Sustainable
Development in the Year 2060 308

Seeme Mallick

Chapter 17 Global Warming and Climate Change: Projections and
Implications 335

Taddeo Rusoke

Chapter 18 Pollution Haven or Pollution Halo? Environmental Effect
of Chinese FDI Regarding the One Belt One Road Project 357

Yahya Algül, Vedat Kaya, Ömer Yalçınkaya

Contributors

Yahya Algül
Erzurum Technical University,
Erzurum, Turkey.

Muhammad Asif
Architectural Engineering &
Construction Management, KFUPM
Dhahran, Saudi Arabia.

Emrah Atar
Recep Tayyip Erdogan University
Recep Tayyip Erdoğan Üniversitesi,
Zihni Derin Yerleşkesi, İktisadi
ve İdari Bilimler/Hukuk Fakültesi
Binası K Fener Mahallesi RİZE,
Turkey.

Syeda Youmnah Batool
Fuel Cell Research Center, Korea
Institute of Energy Research,
Daejeon, Korea.
Department of Energy Engineering,
University of Science and
Technology, Daejeon, Korea.

Evandro Albiach Branco
National Institute for Space Research
(INPE), Av. dos Astronautas,São
José dos Campos, SP, Brazil.

Gisleine Cunha-Zeri
National Institute for Space Research
(INPE), Av. dos Astronautas, Jardim
da Granja, São José dos Campos, SP,
Brazil.

Megersa Olumana Dinka
University of Johannesburg Cnr Kingsway
& University Roads, Auckland Park,
Johannesburg, South Africa.

İlker Yasin Durmaz
Recep Tayyip Erdogan University
Recep Tayyip Erdoğan
Üniversitesi, Zihni Derin
Yerleşkesi, İktisadi ve İdari
Bilimler/Hukuk Fakültesi
Binası K. Fener Mahallesi
RİZE, Turkey.

Estefanía
University of Vigo Faculty of
Economic and Business Sciences
Lagoas-Marcosende University
Campus Vigo, Spain.

Abdolvahhab Fetanat
Department of Electrical
Engineering, Behbahan Branch,
Islamic Azad University,
Behbahan, Iran Behbahan,
Iran.

André Gonçalves
National Institute for Space
Research (INPE), Av. dos
Astronautas, Jardim da Granja,
SP, Brazil.

Nikita Gupta
Babu Banarasi Das University,
Lucknow (U.P) Faizabad Road,
Lucknow (U.P), India.

Muhammad Haseeb Hassan
Fuel Cell Research Center, Korea
Institute of Energy Research,
Daejeon, Korea.
Department of Energy Engineering,
University of Science and
Technology, Daejeon, Korea.

Hafiz Ahmed Ishfaq
Department of Materials Chemistry,
    National Institute of Chemistry,
    Hajdrohova 1900, 1000 Ljubljana,
    Slovinia.
Faculty of Chemistry and Chemical
    Technology, University of Ljubljana,
    Vecna pot 113, 1000 Ljubljana,
    Slovinia.

Nadeem Iqbal
Director Microtech Chemicals
    and minerals, Kasur, Punjab,
    Pakistan.

Nugun P. Jellason
Teesside University International
    Business School Clarendon Building,
    Teesside University, Middlesbrough,
    Tees Valley, UK.

Seethalekshmi K.,
Institute of Engineering and
    Technology, Lucknow (U.P) Sitapur
    Road, Lucknow (U.P), India.

Vedat Kaya
Ataturk University, Erzurum, Turkey.

Muhamamd Khalid
KFUPM, Dhahran, Saudi Arabia.

Fernando León-Mateos
University of Vigo Faculty of Economic
    and Business Sciences Lagoas-
    Marcosende University Campus
    Vigo, Spain.

Lucas López-Manuel
University of Vigo Faculty of Economic
    and Business Sciences Lagoas-
    Marcosende University Campus
    Vigo, Spain.

Rao Muhammad Mahtab Mahboob
University of Agriculture Faisalabad
    (UAF), Pakistan Department
    of Computer Science, Institute
    of Management and Applied
    Sciences, Khanewal, Punjab,
    Pakistan.

Seeme Mallick
Freelance Consultant, Islamabad,
    Pakistan.

Mueed Ahmed Mirza
Department of Computer Science,
    Institute of Management and Applied
    Sciences, Khanewal, Punjab, Pakistan.

Hossein Mofid
Department of Instrumentation and
    Automation, Petroleum University of
    Technology, Ahwaz, Iran.

Sara Musaddiq
Department of Chemistry, The Women
    University Multan, Pakistan Govt.
    Post Graduate College, Pakistan.

Kiran Mustafa
Department of Chemistry, The Women
    University Multan Pakistan Govt.
    Post Graduate College, Pakistan.

Joan Nyika
University of Johannesburg, Cnr
    Kingsway & University Roads,
    Auckland Park, Johannesburg, South
    Africa.

Sérgio Pulice
National Institute for Space Research
    (INPE), Av. dos Astronautas, São
    José dos Campos, SP, Brazil.

Uliana Pysmenna
National Technical University "Kyiv
    Polytechnic Institute" Iskrivska,
    Kyiv, Ukraine.

Saeed-ur Rehman
Research and Development Department,
    Elcogen, OY Niittyvillankuja, Vantaa,
    Finland.

Taddeo Rusoke
Nkumba University, Uganda.

Guller Sahin
Kütahya Health Sciences University
    Evliya Çelebi Campus, Kütahya,
    Turkey.

Antonio Sartal
University of Vigo Faculty of Economic
    and Business Sciences Lagoas-
    Marcosende University Campus,
    Vigo, Spain.

Rao Muhammad Shahbaz Mahboob
University of Agriculture Faisalabad
    (UAF), Pakistan.

Barry Solomon
Michigan Tech, Michigan, USA.

Mohsen Tayebi
Department of Environmental
    Sciences & Engineering, Faculty
    of Agriculture & Natural
    Resources, Ardakan University,
    Ardakan, Iran.

Galyna Trypolska
State Institution "Institute for
    Economics and Forecasting",
    NAS of Ukraine Kyiv, Urlivska,
    Kyiv, Ukraine.

Ömer Yalçınkaya
Ataturk University Erzurum,
    Turkey.

Marcelo Zeri
National Center for Monitoring
    and Early Warning of Natural
    Disasters (CEMADEN) Av.
    Dr Altino Bondensan, São
    José dos Campos, São Paolo,
    Brazil.

Preface

Energy would have hardly been more relevant in human history than it is today. With its pivotal role in every aspect of human life, the technological and socio-economic dimensions of energy are becoming ever more vibrant. At the same time, climate change has already dawned upon the planet with its wide-ranging and intense implications including rising sea levels, seasonal disruptions, and an increase in the frequency and severity of weather-related disasters such as heat waves, wildfires, flooding, and storms. The fossil fuels dominant energy systems have made the energy and environmental scenarios highly interlaced. The energy industry owes to play a leading role in the fight against climate change, as also underlined by the Paris Agreement and the most recent COP 28. Accordingly, the energy sector is embracing a transformation seeking an energy landscape that commensurate with the demands of a sustainable future for the world.

The Handbook of Energy and Environment in the 21st Century aims to advance the knowledge and debate on energy and environmental sustainability in the 21st century. It delves into the intricate interplay between technology and policy dynamics shaping our energy and environmental outlooks. It helps with understanding and navigating the challenges and opportunities that lie ahead. It presents a robust and comprehensive account of the energy and environmental landscape in the 21st century considering the faced challenges and the potential solutions. It discusses the key dimensions of the present energy and environmental scenario as well as the emerging trends. Global response to the challenges is covered taking into account technical, economic, social, and policy perspectives. The handbook is a call to action for the energy and environment stakeholders and the society at large. Together, we can make a lasting impact on our and planet's future.

The book is teamwork and we are grateful to the chapter contributors for helping me accomplish it. We would like to thank the reviewers for their time and efforts in reviewing chapter abstracts and manuscripts. I would also acknowledge the King Fahd University of Petroleum and Minerals (KFUPM) for its appreciative support.

1 Sustainable Energy Transition
Technological Perspective

Muhammad Asif

1.1 INTRODUCTION

Energy is the backbone of modern societies. The 18th-century Industrial Revolution transformed the human–energy relationship. Ever since, extensive and efficient utilization of energy has played an instrumental role in human development. Energy is becoming an increasingly critical commodity on multiple fronts, including technological, socio-economic, and geo-political. Energy has attained the status of a prerequisite for all crucial aspects of societies, like mobility, agriculture, industry, health, education, and trade and commerce [1]. Energy resources exist in a wide range of physical states, which can be harnessed and capitalized upon through various technologies. Energy resources can be broadly classified into two categories: renewables and non-renewables. Renewable energy resources are the ones that are naturally replenished or renewed. Examples of renewable resources include solar energy, wind power, hydropower, and wave and tidal power. Energy resources that are finite and exhaustible are termed non-renewable such as coal, oil, and natural gas.

An important dimension of the human use of energy is its contribution to climate change. Unchecked emissions of greenhouse gases (GHGs) are leading to global warming. Climate change, as a result of global warming, is regarded as the biggest challenge facing the world. Different types of energy resources, especially fossil fuels, contribute to GHG emissions. Fossil fuels are considered the primary reason for the anthropogenic emission of carbon dioxide (CO_2)—the 18th-century Industrial Revolution is considered to have triggered the rapid growth in the release of greenhouse gases [2]. The carbon dioxide (CO_2) concentration in the atmosphere, for example, has increased from the pre-industrial age level of 280 parts per million (ppm) to 415 ppm. The acceleration in the growth of CO_2 concentration can be gauged from the fact that almost 100 ppm of the total 135 ppm increase has occurred since 1960 [3].

Climate change is leading to a wide range of consequences such as seasonal disorder, a pattern of intense and more frequent weather-related events such as floods, droughts, storms, heat waves and wildfires, financial loss, and health problems [4–6]. Climate change is also adversely affecting the water and food supplies around the world. Warmer temperatures are increasing the sea level as a result of the

DOI: 10.1201/9781032715438-1

melting of glaciers. During the 20th century, the global sea level rose by around 20 centimeters. The pace of the rise in sea level is accelerating every year—over the last two decades, it has been almost double that of the last century [7]. As a result of warmer temperatures, glaciers are shrinking across the world, including in the Himalayas, Alps, Alaska, Rockies, and Africa. An extremely alarming dimension of climate change is that it is growing in momentum. Most of the temperature rise since the Industrial Revolution has occurred since the 1960s. Extreme weather conditions and climate abnormalities are becoming so frequent that the situation is already being widely dubbed a climate crisis. With the recorded acceleration in the accumulation of greenhouse gases and consequent increase in atmospheric temperature, climate change–driven weather-related disasters are becoming more intense and recurrent. The seven most recent years have been observed to be the warmest since records began, with 2020 being the hottest year ever [7]. The year 2021 set new records for natural disasters, including heatwaves, wildfires, storms, and floods. Extreme weather events are now considered a new normal as experts predict more intense natural calamities, including wildfires, storms, floods, and droughts.

The global energy scenario also faces a number of other challenges, such as rapid growth in energy demand, depletion of fossil fuel reserves, volatile energy prices, and a lack of universal access to energy. A fast growth in the global energy demand—owing to factors like surging population, economic and infrastructural development, and urbanization—is adding pressure on the energy supply chain. According to the Energy Information Administration (EIA), between 2018 and 2050, world energy requirements are projected to increase by 50% [8]. Access to refined energy resources remains a major challenge for significant proportions of the population in the developing countries.

The energy sector is experiencing a major transformation in order to address the energy and environmental challenges. The primary aim of this transition is to shift the global energy system away from fossil fuels. Renewable and low-carbon technologies are at the heart of this energy transition. This chapter presents an overview of the key dynamics of the ongoing energy transition. It defines the energy transition to have four main dimensions: decarbonization, energy efficiency, decentralization, and digitalization. The chapter discusses the prospects of these four dimensions (4Ds) of the energy transition, especially in terms of relevant technological and policy advancements.

1.2   KEY DIMENSIONS OF THE ENERGY TRANSITION

The use of energy is closely linked to the environment [9–11]. It is estimated that despite the pledges and efforts by the global community to tackle climate change, CO_2 emissions from energy and industry have increased by 60% since the United Nations Framework Convention on Climate Change (UNFCCC) was signed in 1992 [12]. Climate change is already there, with its implications like seasonal disorder; rising sea level; and a trend of more frequent and intense weather-driven disasters such as flooding, droughts, heat waves, wildfires, storms, and associated financial losses

[13, 14]. The situation calls for an urgent paradigm shift across the entire energy sector. Responding to the challenges on hand, and ensuring a supply of energy compatible with the demands of a sustainable future for the planet, the global energy sector is going through a transformation. This energy transition can be defined as "The energy transition is a pathway toward the transformation of the global energy sector from fossil-based to zero-carbon by the second half of this century." At the heart of the ongoing energy transition is the need to reduce energy-related CO_2 emissions to limit climate change [15].

In recent history, humankind has witnessed two major energy transitions. The first energy transition propelled the Industrial Revolution, mainly attributed to coal replacing biomass and wood as a more efficient and effective fuel to drive machines. The second energy transition was a shift from coal to more refined forms of fossil fuels—oil and gas—in the later part of the 20th century. The world is now experiencing the third energy transition. This energy transition is much more vibrant, intriguing, and impactful compared to the earlier ones. It is fundamentally a sustainability-driven energy pathway with a focus on decarbonization of the energy sector by shifting away from fossil fuels. This energy transition, therefore, can also be termed 'sustainable energy transition' or 'low-carbon energy transition'. Holistically, however, the ongoing energy transition is not just about going low carbon or shifting away from fossil fuels. It is rather much more diverse and comprehensive in terms of scope and impact. The 21st-century energy transition is being propelled by unprecedented developments on the fronts of energy resources and their consumption, technological advancements, socio-economic and political response, and evolving policy landscape. This energy transition has four key dimensions: decarbonization, energy efficiency, decentralization, and digitalization [16].

1.3   DECARBONIZATION

Decarbonization of the energy sector is at the heart of the energy transition, as reduction in CO_2 and other GHG emissions is fundamental in the fight against climate change. The energy sector can be decarbonized through a range of technologies and solutions, such as renewable energy, electric vehicles (EVs), hydrogen and fuel cells, carbon capture and storage (CCS), and phasing out of fossil fuels. Renewable energy is the single most critical component of the decarbonization drive.

Through the Paris Agreement, the world has adopted the first-ever universally legally binding global climate deal to avoid dangerous climate change by limiting global warming to well below 2° C. Alarm bells are, however, being frequently rung by concerned circles, including the United Nations Intergovernmental Panel on Climate Change (IPCC), that the world is seriously overshooting this target. In order to be anywhere closer to achieving this target, the world needs to make major changes in four big global systems: energy, land use, cities, and industry. The energy sector is where the greatest challenges and opportunities exist [17]. Energy systems have huge variation in terms of their associated environmental emissions. Table 1.1, for example, shows the carbon dioxide emissions from different types of power generation technologies [18].

TABLE 1.1

Comparison of CO_2 Emissions from Different Energy Systems

Power Plant	Type of Fuel/Energy	CO_2/(kg/kWh)
Steam power	Lingnite	1.04–1.16
Steam power	Hard coal	0.83
Gas turbine	Pit coal	0.79
Thermal power	Fuel oil (heavy)	0.76
Gas turbine	Natural gas	0.58
Nuclear power	Uranium	0.025
Solar thermal	Solar energy	0.1–0.15
Photovoltaic	Solar energy	0.1–0.2
Wind power	Solar/wind energy	0.02
Hydro-electric	Hydropower	0.004

1.3.1 RENEWABLE ENERGY

Renewable energy is the primary pathway for the energy sector's low- or zero-carbon transition. Over the last couple of decades, renewable technologies, especially solar PV and wind turbines, have made great progress in terms of technological developments and economic maturity. The global installed capacity of renewables increased from 2,581 GW in 2019 to 2,838 GW in 2020, exceeding expansion in the previous year by almost 50%. In 2021, renewable energy installed capacity increased globally by 257 GW, reaching 3,064 GW. Over the last few years, renewable energy has increased power generation capacity more than fossil fuels and nuclear power put together. For instance, more than 83% of all new power generation capacity installed globally in 2022 came from renewable sources, as shown in Figure 1.1 [19].

For several years now in a row, renewable energy is adding more power generation capacity compared to the combined addition by fossil fuels and nuclear power. In the year 2020, for example, renewables contributed to more than 80% of all new power generation capacity added worldwide. The growth of the renewable sector is primarily being propelled by solar and wind power, with the two technologies accounting for 91% of the new renewables added during the year. The annual growth in the cumulative installed capacity of solar PV and wind power over the last 10 years has been unprecedented [20]. Renewable energy is already supplying 26% of the global electricity needs. According to IEA, to achieve net zero emissions by 2050, almost 90% of global electricity generation is to be supplied from renewables. There was over US$ 303 billion invested in renewable energy projects during the year [21]. The upward scale of renewable development can be gauged from the fact that China has started developing the first 100-GW phase of massive solar and wind power initiatives. The initiative is likely to be expanded to several hundreds of GW in capacity as China aims to develop 1,200 GW of renewables by 2030 [22]. The renewable growth trends are projected to continue, as the annual capacity addition of solar and wind power is set to grow fourfold between 2020 and 2030 [17]. It is also expected

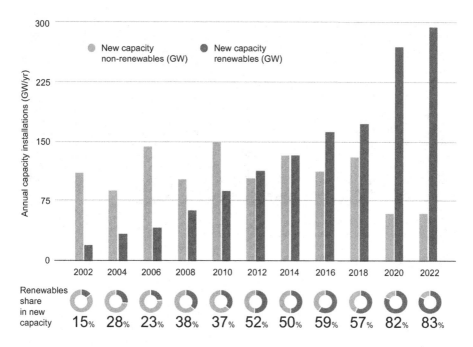

FIGURE 1.1 Growth of renewable energy in the power generation capacity.

that by 2026, solar and wind power will account for around 95% of the total new capacity addition in the power sector.

The success of renewables has been propelled by technological advancements, economy of scale, and supportive policies. Solar and wind power industries are massively benefiting from the scientific and engineering advancements. Solar PV cells, for example, are becoming more efficient and reliable. Concentrated solar PV cells, as shown in Figure 1.2 [23], have achieved efficiency figures of over 40%. The progress renewable technologies, especially solar energy systems, are making is significantly helped by their broadening application domains. The building sector has been a vital area of application for solar PV and solar water heating systems [24–29]. PV systems are also being installed over agricultural farms, termed agrivoltaics. For PV and wind turbines, the issue of low power density is being addressed by offshore applications. The application of PV on water bodies—lakes, canals, rivers, and oceans—termed floating PV, is becoming popular, with the added advantage of higher system efficiency. Wind turbines are witnessing improvements both at the manufacturing and installation ends. Besides improvements in aerodynamic designs, advanced and sophisticated materials are helping develop larger, lighter, and stronger wind blades. These developments have enabled wind turbines to grow rapidly in size, as shown in Figure 1.3. The off-shore application of wind turbines has significantly boosted the capacity factor. Within a couple of decades, larger and more sophisticated wind turbines and better site selection have resulted in the average annual capacity factor increasing from 20–30% to 40–50%. Some offshore wind turbines are now claiming to have a capacity factor of over 60% [1].

FIGURE 1.2   Concentrated solar PV.

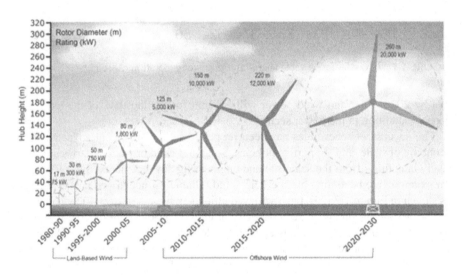

FIGURE 1.3   Growth in size of wind turbines.

1.3.2   DECARBONIZATION IN THE FOSSIL FUEL SECTOR

Fossil fuels have traditionally led the energy sector, presently contributing almost 80% of the total supplies. Oil and gas are the main fuels being used in the transport sector. The industrial sector also heavily relies on fossil fuels. Despite the ground being gained by renewables, in the foreseeable future, fossil fuels are projected to be a major part of the global energy outlook. The emissions associated with fossil fuels, however, need to be curbed in order to avoid irreparable damage to the planet's ecosystem. Among the three main types of fossil fuels, coal is the most polluting one, while natural gas has the lowest environmental burdens. Within the fossil fuel sector, there is a gradual shift away from coal, especially in power generation, which can also be considered a decarbonization trend. There is, however, a need to have more profound decarbonization efforts. There can be two major pathways in this respect, which are carbon capture and storage and the transformation of fossil fuels into hydrogen or hydrogen-rich fuels. The removal of carbon from fossil fuels— which are fundamentally hydrocarbons—allows for the generation of hydrogen as a secondary energy source that does not produce CO_2. Carbon capture and storage, also termed carbon capture and sequestration, involves the removal of CO_2 from the direct combustion of fossil fuels for power generation or industrial processes. The idea is to prevent the release of CO_2 into the atmosphere and instead store it in underground geological formations for the long term [1].

1.3.3 NUCLEAR POWER

There were 442 nuclear power plants in the world as of 2021, with a gross capacity of roughly 394 GWe, and an additional 57 power plants, with a total capacity of 60 GWe, were under construction [30, 13]. As more than 8 GW of reactors were retired in 2021, there was a 3 GW reduction in the world's nuclear power capacity. While the majority of these lengthy shutdowns took place in G7 members Germany, the United Kingdom, and the United States, all of the additional capacity was in emerging market and developing economies. To maintain nuclear power through 2030, the world's nuclear capacity would need to grow by about 10 GW annually. G7 members should give lifetime extensions top priority in order to maximize new nuclear capacity and fortify the current low-emissions infrastructure. The goals embodied in net zero targets have fostered the development of novel nuclear power technologies, including small modular reactors (SMRs), which have a reduced size of under 300 MW per reactor, down to 10 MW. SMRs have the potential to be more affordable, easier, and quicker to build than conventional large reactors. More than 70 designs are currently being worked on. SMRs might be produced in a factory and shipped to the ultimate destination, reducing financial requirements and accelerating project deadlines. With the decarbonization of power systems and the rise in the proportion of solar and wind energy, SMRs could be an important part of providing the increased flexibility required in power generation. They can also be used to produce hydrogen and heat. Expanding nuclear power is necessary to balance the need for fossil fuels and enhanced renewable energy generation in order to realize the net zero scenario. Nuclear power has reduced the need for coal, natural gas, and oil since it has been used to produce electricity for more than 50 years, which has led to a reduction of 66 Gt of CO_2 emissions globally [31].

1.3.4 ELECTRIC VEHICLES

Electric vehicles are leading the decarbonization efforts in the transport sector. Electric vehicles are environment-friendly, require low maintenance due to fewer components, are quiet to operate, and offer convenience in urban use. The growth of electric vehicles is being supported by wide-ranging policies. These include standards (such as requiring a certain share of clean vehicles or setting limits for fleet-wide average emissions intensity), purchase price subsidies (i.e. tax exemptions or tax credits), incentives encouraging the usage of clean vehicles (financial or non-financial incentives, including free parking, zero-low road tax, and bus lane usage), pricing of externalities (such as carbon pricing), and scrappage policies targeting emitting vehicles. Active support for research and development (R&D) and infra-structure development are also instrumental in this respect [3].

Electric vehicles are showing a steep rise in car market share. In 2020, the world-wide sale of EVs, for example, increased by 41% despite the COVID-related eco-nomic downturn and a drop of 6% in the overall sale of vehicles. During the same year, Europe recorded an increase in the registration of new electric cars by 100%, and the number of electric car models available worldwide increased from 260 to 370. The global electric vehicle market saw a 65% YoY growth in 2022 as EV sales climbed over 10.2 million units. EVs accounted for over 14% of the world's pas-senger vehicle sales in 2022, compared to 9% in 2021 [32]. Battery price has been an important cost factor in the economics of EVs. Battery prices are experiencing a rapid decline, recording a drop from $1,191 in 2010 to $137/kWh in 2020, as shown in Figure 1.4. While electric mobility is also paving the way in the aviation and ship industries, the sale of electric cars is expected to increase from around 3.5 million in 2020 to over 23 million by 2030 [33].

1.3.5 ENERGY STORAGE

Energy storage is an important aspect of the energy value chain. The modern elec-tricity infrastructure, especially in the wake of increasing supplies from renewable technologies, is finding energy storage critically important for its smooth operation.

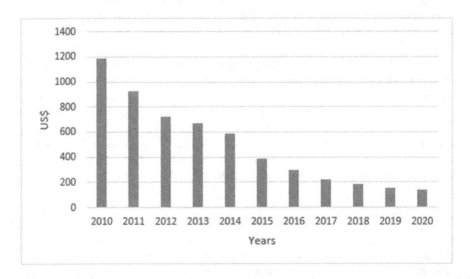

FIGURE 1.4 Declining price trend of lithium batteries.

Energy storage is becoming of particular interest to power sector stakeholders, including utilities, end-users, grid system operators and regulators.

Modern renewables like solar energy and wind power, being dependent on weather conditions, have an intermittency drawback. Solar radiation, for example, is available only during the daytime. The daytime availability of solar radiation can be hindered by multiple weather conditions such as rain, snow, fog, and overcast conditions. Issues like dust storms, smog, haze, and smoke from wild fires also affect the intensity of solar radiation. Similarly, availability of wind is not a constant phenomenon either. Furthermore, even during their spells of availability, solar radiation and wind speed can fluctuate quickly and heavily, accordingly affecting the output from the respective systems. Renewable energy thus needs backup storage to serve as a reliable source of energy.

Pumped storage hydropower projects have traditionally been used as an optimum large-scale energy storage solution. In recent years, battery technology has seen major techno-economic breakthroughs to become another option for large-scale energy storage. The 100-MW lithium battery storage developed by Tesla in Australia in 2017 has been a turning point in the field of large-scale battery storage. Large- and utility-scale battery storage systems, as shown in Figure 1.5 [34], have become a viable option. The USA has planned a 1,500-MW/6,000-MWh lithium-ion battery project, the first phase of which, with a capacity of 300-MW/1,200-MWh, started to operate in December 2020. Australia and the UK are also developing major battery storage projects. Some of these projects include a 1,200-MW project in New South Wales, a 700-MW system by Origin Energy Ltd, a 500-MW system MW in New South Wales, and a 300-MW facility in Victoria. The UK has over 1.1 GW of battery storage capacity in operation, while projects with 600 MW of cumulative capacity are under construction. An overview of the leading battery storage projects currently in operation around the world is provided in Table 1.2.

FIGURE 1.5 Utility-scale lithium ion battery.

Photo: 56318 by Dennis Schroeder, NREL

TABLE 1.2
World's Largest Battery Storage Projects

Project	Capacity (MW)	Battery Technology	Country
Vistra	300	Li-ion	USA
Hornsdale Power	150	Li-ion	UK
Stocking Pelham	50	Li-ion	Australia
Jardelund	48	Li-ion	Germany
Minamisoma substation	40	Li-ion	Japan
Nishi-Sendai substation	40	Li-ion	Japan
Laurel AES	32	Li-ion	USA
Escondido substation	30	Li-ion	USA

1.3.6 HYDROGEN AND FUEL CELLS

Hydrogen as a fuel has unique characteristics. Its use does not release any toxic emissions, and it has the highest calorific value compared to other commonly used fuels. Hydrogen is the simplest and one of the most plentiful elements in the universe. Despite its abundance, however, hydrogen does not occur independently—it exists bonded with other elements and is available in compound forms such as water, hydrocarbons, and carbohydrates [35]. Hydrogen can be produced from fossil fuels as well as renewables. In the former case, hydrogen can be produced through various routes such as reformation of natural gas, partial oxidation of heavy fossil fuels, and coal gasification. Production of hydrogen from fossil fuels—making up over 90% of the current hydrogen supplies—however, leads to GHG emissions and is typically termed 'blue hydrogen'. An environmentally clean option is to produce hydrogen through electrolysis, which involves splitting water into hydrogen and oxygen with the help of electricity. Ideally, the electricity used for electrolysis should be from renewables, making it 'clean hydrogen', which is also termed 'green hydrogen'. Hydrogen can be stored, transported, and used for energy applications through various technologies. Hydrogen in the capacity of an energy vector also has the potential to become an optimum solution for renewables' intermittency by providing an energy storage solution. The vision of building an energy infrastructure that uses hydrogen as a fuel and energy carrier, a concept called hydrogen economy, is the path toward the full commercial application of hydrogen energy technologies [36]. Fuel cells are an important complementary technology. It is a device that converts hydrogen directly into electricity through electrochemical oxidation while generating pure water as its byproduct. Fuel cells come in many types and are classified mainly in terms of the kind of electrolyte they use, which in turn influences other features such as the involved electrochemical reaction, required catalysts, cell operating temperature range, and the needed fuel [37].

1.4 ENERGY EFFICIENCY

The fast growth in the global demand for energy is putting pressure on the entire energy value chain. A one-dimensional approach, of matching the growing energy demands with corresponding capacity addition, is not a sustainable solution, especially when the planet is already overshooting its bio-capacity by almost 70%. A sustainable energy pathway, aiming to satisfy the global energy requirements while protecting the environment, has to begin with decreasing the use of energy through energy efficiency (EE) measures. Energy efficiency is regarded as a better solution to address energy shortages than adding new capacity. To industrial and commercial entities, energy efficiency delivers economic and environmental gains, besides offering a competitive edge.

Energy efficiency is a broad domain in terms of the nature of facilities, types of energy losses and wastes, and range of solutions available to improve efficiency. Given the technological and policy advancements, the field of energy efficiency is continuously evolving. In terms of fundamental approach, energy efficiency solutions can be broadly classified into three types, as also shown in Figure 1.6. An energy efficiency program, typically involving eradication of the unnecessary use of energy and improvement in the efficiency of required energy, starts with an energy audit exercise. The type of energy audit to be carried out primarily depends upon the scope and objectives of the intended energy efficiency program. The energy audit process is also influenced by factors like available resources (funding, human power, and time), type of facility, and provision of data and support. Detailed energy management programs can also include execution of the recommended solutions and post-implementation measurement and verification work to ensure the desired energy-saving goals are achieved.

The use of energy can be reduced across all major sectors, including buildings, industry, and transport. Buildings account for over one third of the global energy consumption [38–41]. Energy use in buildings can be reduced through a range of energy efficiency measures. Energy-efficient solutions for buildings can be broadly classified as active and passive energy saving measures [42]. The choice of energy efficiency solutions depends on factors like the nature of the facility, site condition and local climate, desired levels of comfort and improvement, and financial situation. Through energy efficiency measures, energy demand in existing as well as new

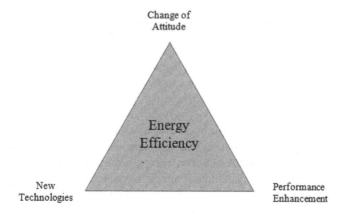

FIGURE 1.6   Classification of energy efficiency approaches.

buildings can be reduced by 30% to 80% [43–46]. Energy efficiency in the transport sector can be improved through measures like incorporating fuel economy standards and eco-driving [47]. Digital technologies can also help save on fuel across road, air, and sea transportation through the optimization of routes. The industrial sector also offers significant potential for energy efficiency, especially in the energy-intensive industry. Improvement in energy efficiency enables industrial entities to enhance their productivity and competitiveness, besides contributing to addressing energy and environmental problems locally, nationally, and globally. The energy efficiency drive in the industrial sector is also being helped by digital energy management technologies. It is estimated that with the help of proven and commercially viable technologies, energy use in the manufacturing industry can be reduced by 18% to 26% [48].

Energy efficiency is beneficial both at the micro and macro levels. While it offers savings on bills to individual consumers, it also helps commercial and industrial entities in terms of economic competitiveness. While energy efficiency helps the utility and energy sector in terms of demand-side management and peak-load shaving, it also helps foster national security. The global economy could increase by $18 trillion by 2035 if energy efficiency is adopted as the "first choice" for new energy supplies, which would also achieve the emission reductions required to limit global warming to 2° C [51, 44]. Besides enabling economic growth and improving energy security, energy efficiency can also play a vital role in the fight against climate change, as it can deliver more than 40% of the reduction in energy-related greenhouse gas emissions over the next 20 years [50].

1.5 DECENTRALIZATION

Decentralized or distributed generation is energy generated close to the point of use, as shown in Figure 1.7. Decentralized generation (DG) avoids/minimizes transmission and distribution setup, thus saving on cost and losses. It offers better efficiency, flexibility, and economy as compared to large and centralized generation systems. There are several energy technologies that can be used in DG systems depending on the application and type of project. Based on the type of energy resource, DG technologies can be classified into two categories: renewables-based systems and non-renewables-based systems. Renewables-based DG systems employ technologies like solar energy, wind power, hydropower, biomass, and geothermal energy. Some of these technologies can be further classified into different types. Solar technologies, for example, can be categorized into solar PV, solar thermal power, and solar water heating. Similarly, biomass can be used to deliver solid fuels, liquid fuels such as biodiesel and bioethanol, and gaseous fuels. Renewables-based DG systems offer several benefits such as reduced greenhouse gas emissions and lower operation and maintenance costs. These systems, however, are typically intermittent and need energy storage to offer reliable solutions. Non-renewable-based DG technologies are also available in a wide range and may include internal combustion (IC) engine, combined heat and power (CHP), gas turbines, micro-turbines, Stirling engines, and fuel cells. These technologies can use different types of fossil fuels.

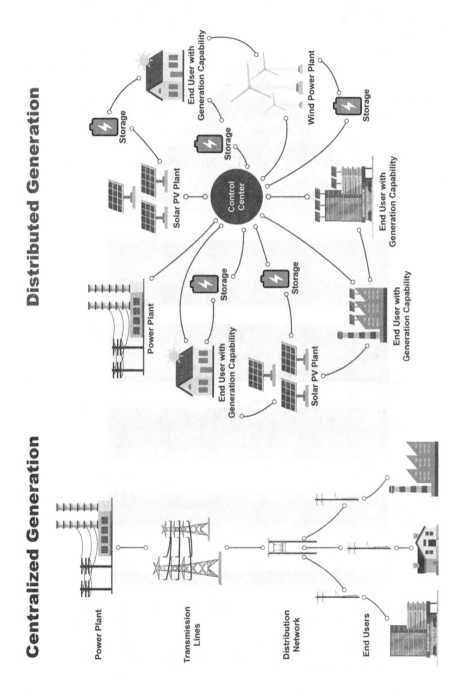

FIGURE 1.7 Overview of central and distributed generation systems.

Renewables like solar and wind power systems are leading the DG landscape. DG is playing an important role in the global electrification efforts and is presenting viable solutions for meeting modern energy needs and enabling the livelihoods of hundreds of millions who still lack access to electricity or clean cooking solutions [51, 46]. Solar PV is one of the most successful DG technologies, especially at small-scale and off-grid levels. The building sector offers tremendous potential for DG PV systems [26–28]. It is estimated that since 2010, over 180 million off-grid solar systems have been installed, including 30 million solar home systems. In 2019, the market for off-grid solar systems grew by 13%, with sales totaling 35 million units. Rooftop PV systems, as shown in Figure 1.8 [52, 47], make up 40% of the total PV installations worldwide. Renewable energy also supplied around half of the 19,000 mini-grids installed worldwide by the end of 2019. Efficient biomass systems such as improved cooking stoves and biogas systems are also helping the global efforts towards clean energy access. In 2020, the installed capacity of off-grid DG systems

FIGURE 1.8   Rooftop PV systems.

grew by 365 MW to reach 10.6 GW. Solar systems alone added 250 MW to have a total installed capacity of 4.3 GW.

1.6 DIGITALIZATION

Digitalization, also referred as the 4th Industrial Revolution, is driving the needed fundamental shift in the energy industry, which is also disrupting traditional market players [53]. Digitalization is a broad term in the context of the energy sector. An important dimension of digitalization is the collection and analysis of energy data to optimize energy demand and supply to achieve system efficiency and cost-effectiveness. While decarbonization, decentralization, and decreasing use of energy are transforming the energy sector, digitalization—through the proliferation of sensors, computing, communication, and predictive and control techniques—is also set to change the way energy services are realized and delivered. This is accomplished through a range of established and emerging technologies, above all artificial intelligence (AI). Digitalization from the perspective of business opportunities created in the energy sector can be regarded as the use of digital technologies to change a business process and enhance efficiency and revenue; it is the process of moving to a digital business. However it is defined, digitalization is having a profound impact on the global energy scenario. While digitalization is leading to new business models, it is also disrupting existing models of generation, consumption, markets, and businesses and employment, potentially pushing some of the established ones on their way out [54]. Digitalization of the energy sector employs technologies like artificial intelligence, machine learning, big data and data analytics, Internet of Things, cloud computing, blockchain, and robotics and automation. These technologies are at various degrees of techno-economic maturity for their application in the energy sector. Digitalization is revolutionizing the energy sector by improving the productivity, safety, accessibility, and overall sustainability of energy systems. New, smarter ways of modeling, monitoring, analyzing, and forecasting energy production and consumption are helping the sustainable energy transition. With the range of advantages it offers, digitalization is also posing several challenges. Most importantly, the digital transformation heavily relies on large data sets, the handling of which is increasingly exposing utilities and the energy industry to cyber security risks.

1.7 POLICY AND INVESTMENT TRENDS

Decarbonization is becoming the central part of the national and international energy policy frameworks across the world. As part of greenhouse gas emission targets, which mandate a reduction in overall emissions and can include net zero and carbon-neutral targets, there have been new emission reduction commitments covering around 47% of total global emissions. Some examples in this respect are China aiming to become carbon neutral by 2060, Japan by 2050, and the Republic of Korea by 2050. A major development is the European Union (EU) committing to reduce carbon emissions by 55% from the 1990 levels by 2030 and to become net zero by 2050. The United Kingdom has planned to reduce carbon emissions by 68% by 2030, compared to the 1990 level to go net zero by 2050. A major decarbonization boost with far-reaching impacts at the global level has been the USA's rejoining of the Paris Agreement in 2021. The USA, after having walked out of the Paris Agreement in 2020, under the Nationally Determined Contribution (NDC), has committed to reduce emissions by 50–52% below 2005 levels by 2030, which is equivalent to a reduction of 40–43% below 1990 levels [55]. Overall, over 30 nations have incorporated climate neutrality by 2050 (or 2060) in laws, proposed legislation, or a national policy document [56]. There have been significant decarbonization efforts on the part of other stakeholders as well, including energy, banking, and the corporate sector. The world's leading corporations are becoming increasingly aware of the threats associated with climate change and the business opportunities in taking action. It will help not only reduce their own emissions but also those of their business associates. For example, the British insurer Prudential is working on an energy transition mechanism (ETM) with the Asian Development Bank on a scheme to buy out coal-fired power plants in Asia in order to shut them down within 15 years. The initiative involves insurance groups, Asian governments, and multilateral banks [57]. As a result of these developments, globally, with the exception of mainly China and India, all other major coal-consuming economies are reducing dependence on coal, with France, Germany, Italy, the UK, and the USA recording an annual decline in coal consumption by 23.2%, 20.7%, 19.3%, 17.4%, and 14.6% respectively, as shown in Figure 1.9 [58].

Despite the fact that coal has been a critical part of their energy mix, several countries have plans in place to phase out coal, while many others have decided to reduce its use. The UK, for example, has ambitious targets towards shifting away from coal. In 2020, it completely avoided the use of coal for more than two months in a row for the first time in history [59]. The country has decided to close down all coal power plants by 2024. This means within a decade, the country will bring down its reliance on coal for power generation from nearly 40% to zero. It is a major step towards the transition away from fossil fuels and decarbonization of the power sector in order to eliminate contributions to climate change by 2050. Germany, one of

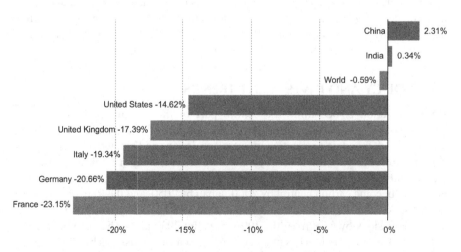

FIGURE 1.9   Annual changes in the coal power capacity.

the leading economies of the world, which has traditionally heavily relied on coal, has also plans to phase it out by 2038. Similarly, France, Canada, and Denmark have plans to go coal free before 2030. While the phasing out of coal is underway in many parts of the world, use of carbon capture and storage is deemed a key incremental technology on the path to net zero emissions. Decarbonization of the power sector in particular is also critical, from the fact that the majority of the energy used by any country is wasted. In the USA, for example, 61% of the total energy goes to waste [60]. While energy waste exists across all sectors, over 90% of the losses are associated with power generation and transport. Currently, both these sectors overwhelmingly rely on fossil fuels; thus hefty GHG emission is an integral part of the process. Switching power generation to renewable energy and transport to electric vehicles can help control the emissions associated with energy waste.

Decarbonization efforts are being crucially helped by vital advancements in the renewable sector on the technical, economic, and policy fronts. There have been significant technological improvements, for example, in terms of efficiency gains in the solar industry and improving capacity factor in the wind power sector. These technological breakthroughs, coupled with economy of scale, are helping renewables become economically competitive with conventional energy options. The renewable energy policy landscape has also steadily improved over the years. From 2010 to 2020, the number of countries with regulatory incentive/mandate policies in the areas of power generation, transport, and heating and cooling has respectively increased from 81 to 145, 35 to 64, and 11 to 24 [21]. Getting to net zero emissions by 2050 is, however, a mammoth task. According to the IPCC, a paradigm shift is needed across four major global systems: energy, industry, land use, and urban development. Besides firm policy commitments and technological advancements, the zero emission target is estimated to require an investment of around $50 trillion by 2050 [61].

Renewable and low-carbon technologies are making significant progress in attracting investment. Global investment in the energy sector is set to rebound in 2021, reversing most of the drop in 2020 caused by the Covid-19 pandemic. Investment in the power sector is also set to rise. Over the last ten years, investment in the power sector has been relatively stable compared with significant fluctuations in the oil and gas industry, mainly due to renewables. Worldwide, since 2010, the annual investment in renewable energy technologies has been over the US$ 200 billion mark. Global new investment in renewable projects (excluding hydropower projects larger than 50 MW) totaled US$ 301.7 billion in 2019, up 5% from 2018. Within the renewable sector, the main focus of investment has been on wind and solar power. In 2019, investment in small-scale solar PV installations (less than 1 MW) increased by 43.5% to US$ 52.1 billion worldwide [21]. Renewables are dominating investments in the power sector, accounting for 70% of the total new investments in 2020. Importantly, investment in renewables is becoming more impactful. Money now goes further than ever in financing clean electricity, with a dollar spent on solar PV deployment today resulting in four times more electricity than ten years ago, thanks to greatly improved technology and falling costs. The International Energy Agency (IEA) warns that not enough is going into clean energy, especially in emerging markets and developing economies. The anticipated investment of US$ 750 billion in renewable and energy efficiency technologies in 2021 remains far below what's required for shifting the energy sector to a sustainable path [62].

1.8 CONCLUSIONS

In the background of the fight against climate change, the global energy sector is experiencing an unprecedented transition. This energy transition is regarded as a pathway of shifting away from fossil fuel–based energy systems by the middle of the century. It is fundamentally a sustainable or low-carbon energy transition, which is having profound impacts across the entire energy value chain. The transition is not just about becoming carbon neutral or zero carbon; it is rather much more vibrant and impactful, thanks to the changes and advancements occurring on the fronts of energy resources and their consumption, technological solutions, socio-economic adjustments, and political and policy responses. The 21st-century energy transition has four main dimensions: decarbonization, energy efficiency, decentralization, and digitalization. Decarbonization of the energy sector is the most important dimension of the energy transition. Reduction in CO_2 and other GHG emissions is fundamental to the fight against climate change. The energy sector can be decarbonized through a range of technologies and solutions, including renewable energy, electric vehicles, hydrogen and fuel cells, carbon capture and storage, and phasing out fossil fuels. The replacement of fossil fuels with renewable energy is the most critical part of the decarbonization drive. Renewable energy is already supplying over 25% of global electricity needs. To achieve net zero emissions by 2050, almost 90% of global electricity generation is to be supplied from renewables. Renewable energy has already become an important stakeholder in the energy sector, accounting for over 80% of global newly added power generation capacity in 2020. Energy efficiency is an imperative part of the energy transition, with massive scope across various sectors, especially in buildings, industry, and transport. The global economy could increase by $18 trillion by 2035 if energy efficiency is prioritized as a solution to addressing energy supply issues. Decentralized or distributed generation offers efficiency, flexibility, and economy and is thus regarded as an integral part of the unfolding energy transition. Digitalization is revolutionizing the energy sector by improving the productivity, safety, accessibility, and overall sustainability of energy systems. New, smarter ways of modeling, monitoring, analyzing, and forecasting energy production and consumption are helping the sustainable energy transition. While advanced countries like EU and OECD member states are relatively well positioned to decarbonize their economies, developing countries critically lack the needed resources, including finance, technical infrastructure, knowledge and awareness, and policy frameworks. There is thus a need for strong international partnerships and support for developing nations to achieve global decarbonization targets.

REFERENCES

1. M. Asif, *Handbook of Energy Transitions*, CRC Press, 2022, ISBN: 978-0-367-68859-2
2. M. Asif and T. Muneer, Energy Supply, Its Demand and Security Issues for Developed and Emerging Economies, *Renewable & Sustainable Energy Reviews*, volume 11, issue 7, September 2007

3. M. Asif, *The 4Ds of Energy Transition: Decarbonization, Decentralization, Decreasing Use, and Digitalization*, Wiley, 2022, ISBN: 978-3-527-34882-4

4. M. Asif, *Energy and Environmental Outlook for South Asia*, CRC Press, 2021, ISBN: 978-0-367-67343-7

5. H. Qudratullah and M. Asif, *Dynamics of Energy, Environment and Economy: A Sustainability Perspective*, Springer, 2020, ISBN: 978-3-030-43578-3

6. M. Asif, *Energy and Environmental Security, Handbook of Environmental Management*, Taylor & Francis, 2019

7. NASA, *Climate Change: How Do We Know, Facts, National Aeronautics and Space Administration, Evidence*, Facts—Climate Change: Vital Signs of the Planet (nasa.gov)

8. EIA, EIA Projects Nearly 50% Increase in World Energy Usage by 2050, Led by Growth in Asia, 24 September 2019, *Today in Energy*, U.S. Energy Information Administration (EIA)

9. M. Asif, Energy and Environmental Security, in *Encyclopaedia of Environmental Management*, Taylor & Francis, 2013, Vol. II, 833–842

10. Kh. Nahiduzaman, A. Al-Dosary, A. Abdallah, M. Asif, H. Kua and A. Alqadhib, Change-Agents Driven Interventions for Energy Conservation at the Saudi Households: Lessons Learnt, *Journal of Cleaner Production*, volume 185, 1 June 2018, pages 998–1014

11. M. Asif, An Empirical Study on Life Cycle Assessment of Double-Glazed Aluminium-Clad Timber Windows, *International Journal of Building Pathology and Adaptation*, volume 37, issue 5, 2019, pages 547–564, https://doi.org/10.1108/IJBPA-01-2019-0001

12. IEA, Net Zero by 2050: A Roadmap for the Global Energy Sector, *Flagship Report*, International Energy Agency, May 2021

13. M. Asif, A. Dehwa, F. Ashraf, M. Shaukat, H. Khan and M. Hassan, Life Cycle Assessment of a Three-Bedroom House in Saudi Arabia, *Environments*, volume 4, issue 3, 2017, page 52, https://doi.org/10.3390/environments4030052

14. M. Khan, M. Asif and E. Stach, Rooftop PV Potential in the Residential Sector of the Kingdom of Saudi Arabia, *Buildings*, volume 7, issue 2, 2017, page 46, https://doi.org/10.3390/buildings7020046

15. IRENA, *Energy Transition*, International Renewable Energy Agency, Energy Transition (irena.org)

16. M. Asif, Role of Energy Conservation and Management in the 4D Sustainable Energy Transition, *Sustainability*, volume 12, 2020, page 10006, https://doi:10.3390/su122310006

17. E. Gillam and R. Asplund. *Will Solar Take the Throne*, Invesco, August 2021

18. M. Asif, *Energy Crisis in Pakistan: Origins, Challenges and Sustainable Solutions*, Oxford University Press, 2011, ISBN: 978-0-19-547876-1

19. IRENA, *World Energy Transitions Outlook 2023: 1.5°C Pathway*, Preview, International Renewable Energy Agency, 2023

20. IRENA, World Adds Record New Renewable Capacity in 2020, *Press Release*, 5 April 2021, International Renewable Energy Agency, World Adds Record New Renewable Energy Capacity in 2020 (irena.org)

21. REN21, *Renewables 2020 Global Status Report*, Renewable Energy Network, 2020

22. J. Scully, China Signals Construction Start of 100GW, First Phase of Desert Renewables Rollout, *PV-Tech*, 12 October 2021 (pv-tech.org)

23. NREL, *Concentrated Solar PV*, Photo by Dennis Schroeder, National Renewable Energy Lab, USA

24. A. Dehwah and M. Asif, Assessment of Net Energy Contribution to Buildings by Rooftop PV Systems in Hot-Humid Climates, *Renewable Energy*, volume 131, February 2019, pages 1288–1299, https://doi.org/10.1016/j.renene.2018.08.031

25. A. Dehwah, M. Asif and M. Tauhidurrahman, Prospects of PV Application in Unregulated Building Rooftops in Developing Countries: A Perspective from Saudi Arabia, *Energy and Buildings*, volume 171, 2018, pages 76–87

26. A. Mahmood, M. Asif, M. Hassanain and M. Babsail, Energy and Economic Evaluation of Green Roof for Residential Buildings in Hot Humid Climates, *Buildings*, volume 7, 2017, page 30, https://doi.org/10.3390/buildings7020030

27. M. Asif, Urban Scale Application of Solar PV to Improve Sustainability in the Building and the Energy Sectors of KSA, *Sustainability*, volume 8, 2016, page 1127, https://doi.org/10.3390/su8111127

28. M. Asif, Growth and Sustainability Trends in the GCC Countries with Particular Reference to KSA and UAE, *Renewable & Sustainable Energy Reviews Journal*, volume 55, 2016, pages 1267–1273

29. M. Asif, *Energy and Environmental Security in Developing Countries*, Springer, 2021, ISBN: 978-3-030-63653-1

30. WNA, *The Nuclear Fuel Report, Global Scenarios for Demand and Supply Availability 2021–2040*, World Nuclear Association, 2021

31. IEA, *Nuclear Power*, International Energy Agency, www.iea.org/reports/nuclear-electricity

32. Global Passenger Electric Vehicle Market Share, Q1 2021–Q4 2022, *Counterpoint* (counterpointresearch.com)

33. K. Adler, Global Electric Vehicle Sales Grew 41% in 2020, More Growth Coming Through Decade: IEA, *HIS Markit*, 3 May 2021, https://ihsmarkit.com/research-analysis/global-electric-vehicle-sales-grew-41-in-2020-more-growth-comi.html

34. NREL, *Utility Scale Lithium Ion Battery*, Photo: 56318 by Dennis Schroeder, National Renewable Energy Lab, USA

35. M. Asif, T. Muneer and J. Kubie, Security Assessment of Importing Solar Electricity for the EU, *Journal of Energy Institute*, issue 1, March 2009

36. T. Muneer, M. Asif and S. Munawwar, Sustainable Production of Solar Electricity with Particular Reference to the Indian Economy, *Renewable & Sustainable Energy Reviews*, volume 9, issue 5, October 2005

37. DOE, *Types of Fuel Cells*, Energy Efficiency and Renewable Energy, Types of Fuel Cells | Department of Energy

38. M. Hamida, W. Ahmed, M. Asif and F. Almaziad, Techno-Economic Assessment of Energy Retrofitting Educational, *Sustainability*, volume 13, issue 1, 2021, page 179, https://doi.org/10.3390/su13010179

39. W. Ahmed and M. Asif, BIM-Based Techno-Economic Assessment of Energy Retrofitting Residential Buildings in Hot Humid Climate, *Energy and Buildings*, volume 227, https://doi.org/10.1016/j.enbuild.2020.110406

40. Rami Alawneh, Farid E. Mohamed Ghazali, Hikmat Ali and Muhammad Asif, A New Index for Assessing the Contribution of Energy Efficiency in LEED Certified Green Buildings to Achieving UN Sustainable Development Goals in Jordan, *International Journal of Green Energy*, volume 6, 2019, pp 490–499, https://doi.org/10.1080/15435075.2019.1584104

41. A. Alazameh and M. Asif, Commercial Building Retrofitting: Assessment of Improvements in Energy Performance and Indoor Air Quality, *Case Studies in Thermal Engineering*, https://doi.org/10.1016/j.csite.2021.100946

42. W. Ahmed and M. Asif, A Critical Review of Energy Retrofitting Trends in Residential Buildings with Particular Focus on the GCC Countries, *Renewable and Sustainable Energy Reviews*, volume 144, July 2021, page 111000, https://doi.org/10.1016/j.rser.2021.111000

43. H. Khan and M. Asif, Impact of Green Roof and Orientation on the Energy Performance of Buildings: A Case Study from Saudi Arabia, *Sustainability*, volume 9, issue 4, 2017, page 640, https://doi.org/10.3390/su9040640

44. H. Khan, M. Asif and M. Mohammed, Case Study of a Nearly Zero Energy Building in Italian Climatic Conditions, *Infrastructures*, volume 2, issue 4, 2017, page 9, https://doi.org/10.3390/infrastructures2040019

45. A. Mahmood, M. Asif, M. Hassanain and M. Babsail, Energy and Economic Evaluation of Green Roof for Residential Buildings in Hot Humid Climates, *Buildings*, volume 7, 2017, page 30, https://doi.org/10.3390/buildings7020030

46. W. Ahmed, M. Asif and F. Alrashed, Application of Building Performance Simulation to Design Energy-Efficient Homes: Case Study from Saudi Arabia, *Sustainability*, volume 11, issue 21, 2019, page 6048, https://doi.org/10.3390/su11216048

47. K. Kojima and L. Ryan, *Transport Energy Efficiency, Energy Efficiency Series*, International Energy Agency, September 2010

48. M. T. Hassan, S. Burek and M. Asif, Barriers to Industrial Energy Efficiency Improvement-Manufacturing SMEs of Pakistan, *Energy Procedia*, volume 113, 2017, pages 135–142

49. UNEP, *Energy Efficiency: The Game Changer*, United Nations Environment, www. unep.org/explore-topics/energy/what-we-do/energy-efficiency

50. IEA, *Energy Efficiency 2020*, Final Report, International energy Agency, Paris, December 2020

51. M. Asif, *Energy and Environmental Outlook for South Asia*, CRC Press, 2021, ISBN: 978-0-367-67343-7

52. NREL, *Rooftop PV Systems*, Photos: 45231 and 45180 by Dennis Schroeder, National Renewable Energy Lab

53. DNV-GL, *Digitalization and the Future of Energy*, DNV-GL, 2019, Digitalization_report_pages.pdf (smartenergycc.org)

54. IEA, *Digitalization and Energy, Technical Report*, International Energy Agency, Paris, November 2017, Digitalization and Energy—Analysis—IEA

55. CAT, *Ambitious US Target Upgrade Reduces the 2030 Global Emissions Gap by 5–10%*, Climate Action Tracker, 23 April 2021

56. Jeffrey D. Sachs, Christian Kroll, Guillaume Lafortune, Grayson Fuller, and Finn Woelm, *Sustainable Development Report 2021*, The Decade of Action for the SDGs, Cambridge University Press, 2021, https://doi.org/10.1017/9781009106559, https://s3.amazonaws.com/sustainabledevelopment.report/2021/2021-sustainable-development-report.pdf

57. Jillian Ambrose, Prudential in Talks to Buy Out and Shut Coal-Fired Plants in Asia, *Guardian*, 3 August 2021

58. OWID, *Annual Percentage Change in Coal Energy Consumption*, Our World in Data, 2019 (ourworldindata.org)

59. Matthew Farmer, What Does Britain's Two Months Without Coal Power Mean? *Power Technology*, 11 June 2020

60. CT, US Wastes 61–86% of Its Energy, *Clean Technica*, 26 August 2013

61. MS, Decarbonization: The Race to Zero Emissions, *Morgan Stanley*, 25 November 2019

62. IEA, Global Energy Investments Set to Recover in 2021 But Remain Far From a Net Zero Pathway, International Energy Agency, *Press Release*, 2 June 2021

2 Energy Sustainability
Dimensions and Prospects

Uliana Pysmenna, Galyna Trypolska

2.1 INTRODUCTION

Energy sustainability and resource efficiency have been recognized as the forefront of sustainable development, being among the Sustainable Development Goals. Climate change mitigation scaled up the essence of sustainable energy. Increasing energy efficiency and the share of renewables became the aims of energy sustainability along with the minimization of carbon intensity of the energy sector, decoupling of economic growth, combating energy poverty, fostering energy justice, and enhancing the security of supply.

Energy worldwide under the transition towards more sustainable and customer-oriented systems faces multiple urgent and deferred challenges. Seeking the balance of interests between global energy-related players and understanding energy trend implications from a long-term perspective is crucial for responsible and informed sustainable policy making with respect to the transitions required for the enhancement of energy supply security and resource efficiency and affordability, as well as for the minimization of energy poverty and ecological footprints.

The mutual influence of sustainable development and economic growth forms multiple direct and feedback links. It is important that the choice of sustainable development as an economic model can accelerate growth. Sustainable energy creates a multiplier effect in economy; it is able to generate new jobs and provide a sustainable basis for economic growth.

2.1.1 METHODOLOGICAL DIMENSIONS OF ENERGY SUSTAINABILITY

The knowledge of sustainable energy's institutional, economic, and political aspects lags far behind the understanding of its technological and cost aspects. A sufficient increase in the efficiency of the resources used by the economies of the world, the search for alternatives to meet energy needs, and the reduction of carbon emissions have given the impetus to resource and energy efficiency policies, which in turn can have a positive impact on all three components: economic, environmental, and social. The shifts in energy balances and technological transitions in the world's economies and changes in the geography of international trade and in the geopolitical balance of interests of net exporters, importers, and transiters of energy resources occur due to socio-technical transitions in the energy sphere. Therefore, energy goals, which are technological in nature, take on an institutional, economic, and geopolitical color, and scenarios need to be analyzed and modeled from the standpoint of the political and economic basis of energy transitions.

DOI: 10.1201/9781032715438-2

At the present stage of economic development, the solution to the threefold task (economic, environmental, and social) has often given preference to the economic component. That has led to the need to move towards the concepts of a "low-carbon", "green", and "circular" economy, which were formed in global socio-economic policy. The increase in resource efficiency of all types of resources and the search for ways to replace carbon-intensive fuels with renewables have given impetus to the "green economy" policy. And it, in turn, can have a positive effect on all three components simultaneously. Shifts in energy balances and technological transitions in the world's economies and changes of the geopolitical balance of interests of net exporters, importers, and transiters of energy resources occur due to the socio-technical transitions towards more sustainable energy.

The methodology of indicative analysis of the International Energy Agency (IEA) is usually used as a tool for assessing the energy balances of the world, the state of their energy security, the level of energy market development, the energy intensity of the economy, and energy-related CO_2 emission intensity to make indicative comparisons between countries. The leading indicators that form the system of energy-economic coordinates in this methodology are the indicators of the energy intensity of GDP and the specific energy consumption per capita (IEA 2005).

At the same time, some researchers consider reliability the determining indicator of energy balance (energy systems) sustainability similar to the reliability of technical systems (Cherp and Jewell 2014), and others consider energy independence together with energy efficiency. Some scholars distinguish energy independence as a separate indicator of energy security from sustainability (Sovacool 2016). However, all researchers believe that it is essential to ensure the optimality of energy balance, and hence sustainability, compliance with the proportions between the primary energy sources, and avoidance of the dominance of a non-renewable energy resource.

Sustainable development is the basis for policy-making in the EU and other developed countries, where national strategies for sustainable development have been adopted. Based on the Energy Sustainability Trilemma, which is ranked and published annually by the World Energy Council, the values of the Energy Sustainability Trilemma Index (WEC 2022) are defined based on energy security, energy availability, and environmental sustainability criteria (Table 2.1, Figure 2.1).

ESMAP also provides Regulatory Indicators for Energy Sustainability (RISE) estimations based on four dimensions of energy use: electricity access, clean cooking, renewable energy, and energy efficiency (Table 2.2, Figure 2.2).

Being infrastructural, the energy sector is of great importance in social development, and its sustainability maintains technologic development and high living standards. It significantly impacts the economic, social, and environmental components of sustainable development. At the same time, economic growth, social behavior, and general environmental policy, in turn, determine the achieved level of energy sustainability.

Among the 17 UNDP Sustainable Development Goals (Global Goals) approved in 2016 (UNDP 2016), nine are directly related to energy sustainability: 1 No Poverty, 6 Clean Water and Sanitation, 7 Affordable and Clean Energy, 8 Decent Work and Economic Growth, 9 Industry, Innovation and Infrastructure, 11 Sustainable Cities and Communities, 12 Responsible Consumption and Production, 13 Climate Action, 17 Partnerships for the Goals.

TABLE 2.1

Energy Sustainability Assessment Criteria in Accordance with World Energy Council Methodology

Criteria	Indicators
Energy security	Diversification of energy supply, security of supply level, level of energy and resource efficiency
Environmental sustainability	Impact on the environment and climate (impact on the level of emissions of pollutants, greenhouse gases, the dynamics of bringing emission levels to the established national emission reduction plans, impact on the water and land resources use, biodiversity, landscape, noise pollution, waste management)
Energy availability	Physical and economic availability of energy resources (level of gasification, supply of meters, level of energy poverty or share of energy expenditures in the structure of household income, impact on the number of recipients of subsidies)

Source: Based on WEC (2022)

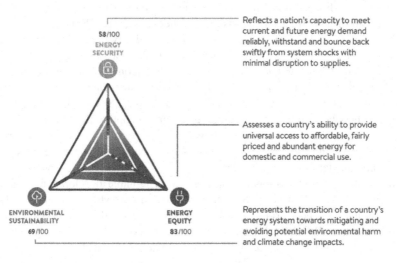

FIGURE 2.1 Energy Sustainability Trilemma Index—2022.

Source: WEC (2022)

TABLE 2.2

Energy Sustainability Assessment Criteria in Accordance with RISE Methodology

Criteria	Indicators
Electricity access	Electrification planning
	Scope of officially approved electrification plan
	Framework for grid electrification
	Framework for minigrids
	Framework for stand-alone systems
	Consumer affordability of electricity
	Utility transparency and monitoring
	Utility creditworthiness

Criteria	Indicators
Clean cooking	Planning
	Scope of planning
	Standards and labelling
	Incentives and attributes
Renewable energy	Legal framework for renewable energy
	Planning for renewable energy expansion
	Incentives and regulatory support for renewable energy
	Attributes of financial and regulatory incentives
	Network connection and use
	Counterparty risk
	Carbon pricing and monitoring
Energy efficiency	National energy efficiency planning
	Energy efficiency entities
	Incentives and mandates: Industrial and commercial end users
	Incentives and mandates: Public sector
	Incentives and mandates: Utilities
	Financing mechanisms for energy efficiency
	Minimum energy efficiency performance standards
	Energy labeling systems
	Building energy codes
	Transport sector
	Carbon pricing and monitoring

Source: Based on RISE (2021)

East Asia & Pacific

Year : 2019

FIGURE 2.2   RISE Energy Sustainability Index.

Source: RISE (2021)

TABLE 2.3
Energy Sustainability Assessment Criteria in Accordance with SEDI Methodology

Group of Indicators	Sub-Indicator (Energy)
Environmental impact	Total CO_2, CO_2/TPES, CO_2/population, CO_2/GDP
Renewable energy	Total energy production from renewable energy/renewable heat consumption, amount of renewable energy in electricity production/ total energy production from renewable energy, TFC Renewable energy consumption in residential/total energy production from renewable energy, TFC renewable energy consumption in commercial/total energy production from renewable energy
Transport	Total TFC in transport, TFC of fossil fuel use in transport/total TFC in transport, TFC of electricity in transport/total TFC in transport, TFC of biofuels and waste consumption/total TFC in transport
Use of energy	Loss/TPES, TFC residential/population, TFC industry/population, TFC commercial/population, TPES/GDP, electricity consumption/ population
Resource access to energy	Total energy production, total fossil fuel production/total energy production, renewable energy production/total energy production
Resilience and safety	Access to electricity (mln population), renewable internal freshwater resources, per capita, electricity consumption/population, CO_2/ population, population/land area
Policy	Energy exports/energy imports, Gini coefficient, GDP

Source: Armin Razmjoo et al. (2019)

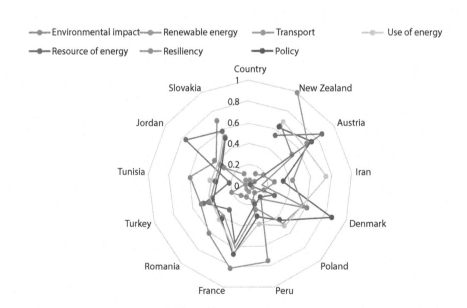

FIGURE 2.3   SEDI Index—2019.

Source: Armin Razmjoo et al. (2019)

2.1.2 ENERGY SUSTAINABILITY AND ECONOMIC GROWTH: INFLUENCE AND TRENDS

The lack of energy sustainability is a source of risks for the economy. The last quarter of the 20th and the beginning of the 21st century evidenced the increasing dependence of the world economy on fuel and energy availability and security of their supply. Reducing the production of primary energy resources, particularly oil and natural gas, without sufficient diversification and replacement significantly complicates the ownership and control over their production, distribution, and transportation. International conflicts, the need to prevent terrorist attacks on major infrastructure facilities for the extraction and transportation of energy resources, and the influence of energy-exporting countries on importing ones enhanced the pressure. In the context of globalization and energy market integration, high price volatility for primary energy resources creates risks for sustainable economic development in many countries with inefficient consumption.

Sorrell (2010) identified five main issues in the interplay between energy sustainability, economic growth, and environmental sustainability:

1. Energy efficiency is needed to ensure decoupling (economic growth without increasing energy consumption)
2. Underestimation of the impact of energy use on productivity
3. Ethics of sufficiency must ensure energy efficiency
4. Sustainability is incompatible with the trend of extensive economic growth
5. A zero-growth economy is incompatible with a banking system of fractional reserves

Energy efficiency is a key to decoupling. Sorrell stressed the ambiguity of the energy efficiency impact on decoupling, referring to the Jevons paradox (a situation where technological progress, while increasing resource efficiency, does not reduce but increases its consumption), as well as the "return effect", redirecting savings from a higher level of energy efficiency to energy-intensive areas of the economy. The result is the reduction in the energy intensity of GDP against the absolute growth in energy consumption. The Jevons paradox shows that changing the focus on energy efficiency does not solve all energy issues. The critical question is whether economic growth is the primary cause of increased energy consumption and/or increased energy efficiency of this consumption, or, conversely, increased energy consumption and/or increased energy efficiency leads to economic growth. It isn't easy to use the typical econometric relationships to describe the relationship between a country's economic and energy dynamics. However, Sorrell emphasizes that this relationship is difficult to establish empirically, and econometric studies do not provide an unambiguous answer. Still, there is a synergistic relationship between these variables with many positive feedback mechanisms. This feedback is provided by the following energy development factors recognized by the IEA (2014): economic growth rate, change in its structure and population growth rate, technological changes and the amount of capital, social relations and consumer behavior (if the income of private households grows, the demand for electrical appliances increases), fossil fuel consumption and external costs, maturity of energy markets (energy markets of many countries are in a state of restructuring, privatization, liberalization, and moving to competitive structures), energy subsidies (in many developing countries), energy prices (electricity, gas, and some petroleum products), environmental and economic policies (limiting emissions of pollutants and greenhouse gases, which are aimed at the III and IV Energy Packages of the EC, the UN Framework Convention on Climate Change, etc.).

The trends that follow the development of energy consumption for the vast majority of regions of the world are the following:

- the increase of potential of final energy consumed, the increase of the electricity share in final energy consumption, and the electrification rate, due to the greater "convenience", versatility, and availability of this form of energy;
- the increase of the secondary energy share in total energy consumption, which characterizes the improvements of efficiency through the deployment of new energy technologies and overcoming the obstacles of secondary energy resource use (insufficient capacity, low environmental friendliness, etc.); and
- the increase in energy consumption per capita and the decrease of it per output unit.

Researchers studying the relationship between energy consumption and economic growth in the scientific literature have formed four main hypotheses (Payne 2010; Omri 2014):

"Neutral"—the lack of econometrically proven dependence. It was formed during the study of countries with a large share of non-energy sectors, with informational and service economies;

"Conservative"—the presence of a multidirectional connection, when, on the one hand, economic activity leads to higher energy consumption, but on the other hand, it reduces energy consumption due to limited resources and the shift of activity to less energy-intensive sectors;

"Growth"—a strong link between energy consumption and economic growth, which is well described by the modified function of aggregated demand depending on the change of GDP rate and changes in energy prices (Cobb-Douglas function);

"Feedback"—energy consumption and economic growth are interdependent. Economic growth requires more energy resource consumption. However, it releases financial resources for technological growth and enhances efficiency, which, in turn, reduces specific energy consumption.

The results of existing studies in favor of one or another hypothesis analyzed in the study (Omri 2014) indicate an approximately equal distribution of results between them: the impact of energy consumption on domestic production confirmed 29% of studies, the impact of the domestic output on energy consumption 27%, GDP impact on energy consumption 23%, and no such impact 21%. This can be explained by the coverage of studies for different countries (both those where economic growth is highly dependent on the energy consumed and those that have achieved decoupling).

The energy sector's contribution through the system of intersectoral relations to the development of the economy is also essential. If energy capacity and end-use energy efficiency enhancement, as well as the upgrade of energy-generating and energy-consuming technologies, are carried out with the use of domestic technologies, equipment, and human and material resources, there would be an increase in production in related sectors, a rise of wages and employment, and total output in the economy due to intersectoral links and multiplicative effect. If not, increasing the capital intensity of energy does not significantly affect economic growth.

Thus, the impact of energy sustainability on economic sustainability through energy consumption (energy efficiency level) has the combined effect of economic growth and energy consumption decoupling, as well as environmental decoupling (environmental growth and environmental impact).

The degree of the economy's independence on fossil fuels as an indicator of energy sustainability should not be neglected. The correlation between the economic growth and the share of energy from RES in energy balance was studied by M. Marinas et al. (2018). They proved a two-way causal relationship between energy consumption with RES and long-term economic growth for all countries surveyed. However, several countries (Romania, Bulgaria) found no link between these indicators in the short term. In contrast, others (Hungary, Lithuania, and Slovenia) increased renewable energy consumption, which enhanced their economic growth.

In addition, economic development based on energy efficiency and the increase in installed capacity of RES improves the quality of economic growth, making it more sustainable in terms of social integration (social justice, reducing inequality) and environmental impact. Again, energy development, based on its technologies and resource base, provides a more significant multiplier effect in the economy and more remarkable growth. It is a two-way link between energy consumption and RES and long-term economic growth. Therefore, in countries where such a connection has not been confirmed, it is advisable to adjust their energy, industrial, and innovation policies to make better use of the domestic technological base, own financial and human capital, domestic resources, and public–private partnership mechanisms and innovation programs for the development of renewable energy technologies to maximize the impact on sustainable growth.

Sorrel (2010), raising the issue of underestimating the impact of energy use on productivity, compares the orthodox and environmental views on energy, productivity, and economic growth (Table 2.3). In this case, the total factor productivity is essentially a reflection of the technological level of the economy. It reflects the overall efficiency of the production factors combination. Innovations, also factors of production, provide qualitative changes in production efficiency.

Ayres and Warr (2005) showed that economic growth in the United States and Japan, since the early twentieth century, can be described by an endogenous production function, which, along with labor and capital factors, takes into account a third factor that measures the production and supply of useful labor to consumers (in terms of thermoeconomics, where useful work is the product of introduced exergy, multiplied by the efficiency of transforming exergy into work, which is

TABLE 2.4

Orthodox and Environmental Views on Energy, Productivity, and Economic Growth

	Orthodox View	Environmental View
The primary source of productivity growth	Exogenous and endogenous technological replacement	Growing availability (technological and financial) of high-quality energy
The marginal productivity of energy supply	Proportional to the share of energy in production costs	More significant than the share of energy in production costs
Alternative supplies for production	Substitute goods according to sector-level elasticity	Substitute goods according to sector-level elasticity
Decoupling of energy consumption and GDP	Decoupling has taken place in OECD countries, as well as the available potential for further decoupling	Traditional energy technologies are slowing down decoupling. There is a strong link between energy efficiency and economic output
Feedback effect on economy	Small	Sufficient

Sources: Sorrel (2010), Cleveland et al. (1984), Ayres and Warr (2005)

essentially the measurement of the technological level). The new theory of growth (endogenous growth) considers energy and energy service supply the key factors of growth: economic growth, in the long run, depends mainly on the historical decline in energy service costs. The same researchers explain why the introduction of carbon taxes, which led to increased energy costs to reduce consumption, is economically ineffective, while promoting energy efficiency is much more efficient. The essential conclusions of this study are the following: (1) rising energy costs through rising carbon tax rates may be relatively inefficient in current market structures and may have a reverse effect on economic growth, and (2) a strategy to reduce greenhouse gas emissions by stimulating economic growth based on new technologies, combined with improved sector regulation and deregulation of the energy market, is seen as effective.

Structuring, maturity of energy markets, and competitive pricing are other sustainable energy issues. The structuring of energy markets is one of the main factors influencing their efficiency. An insufficiently structured market is inefficient. Postponing the correction of distorted market mechanisms often leads to crisis and market failure. An efficient market structure creates the conditions for the balance of the interests of consumers, producers, and energy suppliers. It requires the optimal balance of benefits and costs for each activity in an energy market and the creation of proper incentives for competition.

Each country implements its model of wholesale and retail energy markets, which reflects the general political, geographical, general economic patterns of a country and also the features of the power system (energy mix, location, interconnections) and the energy security, pricing mechanism, and need for centralization.

Competition could increase both the operational and investment efficiency of the energy market. Energy markets are transforming towards technologically diversified, flexible, and liberalized ones in the long run. This takes advantage of growing geopolitical routes and foreign economic relations. Competitive energy markets models may even provide the same degree of liberalization as an unregulated spot market. The competition in energy markets, in essence, can be interpreted as:

- low barriers to market entry for new actors and investors;
- production capacities sufficient to meet demand without limiting supply (production, transmission, supply of energy resources);
- price signals informing about the lack of operational and balance the reliability of power system determine the need for investments in new energy-generating, transmitting, storing, and reserve capacity.

Regulatory instruments that can be used to achieve these competition issues include the requirements for energy companies to generate capacity and to be present in specific segments of the energy market, to create additional market segments, or to develop additional risk-reducing instruments in energy markets, as well as stimulating the cooperation between power exchanges.

2.1.3 NEGATIVE EXTERNALITIES OF CHANGES TOWARDS SUSTAINABILITY

Sustainable energy transitions, for example, the transition to the enhanced use of RES, may have negative externalities, slowing down the implementation of sustainable development practices. These negative externalities are supranational in nature, that is, inherent in some countries or groups of countries at the same time. But, in some cases, they may affect the energy sphere and developing countries by influencing food prices and employment. Since most of the negative externalities are similar for many countries around the world, minimizing them is also universal, considering the specifics of every country. Therefore, to achieve success in reforming energy production, conversion, and consumption, it is better to develop a set of measures to eliminate and minimize negative externalities and use best international practices in dealing with the negative externalities of sustainable transitions.

Simultaneous or separate application of the methods dealing with negative externalities, taking into account the cooperation of socio-technical regimes (e.g., RES and nuclear, RES, and peak natural gas capacity) and international commitments, we can observe the following:

- Introduction of additional markets (market segments);
- Changing the market structure;
- Economic revaluation (e.g., system value together with LCOE);
- Parallel operation of technologies;
- Improved operational strategies (DSM, smart grid, improved scheduling, etc.);
- Classic methods (responsibility of RES operators for imbalances, feed-in tariffs, quotas and emission standards, price caps, etc.).

Negative externalities require detailed attention when applying state energy, environmental, and climate policies, considering the types of externalities. Examples of negative externalities that can significantly degrade the effectiveness of regulatory policy and slow down the sustainable transitions are the following:

- Environmental externalities from the incineration of unsorted waste in positioning this energy activity as the use of secondary energy resources and RES with feed-in tariff application. Insufficient cleaning of waste incineration plants' smoke gases leads to additional air pollution.
- Negative impact on the stability of the power system of charging electric vehicles, not during the night minimum load but the half-peak and peak. The dominance of high-carbon fuels in the energy mix leads to up to four times higher greenhouse gas emissions from electric vehicles than energy systems that use low-carbon and non-carbon energy sources (Schaffartzik et al. 2003). Therefore, improper charging of electric vehicles from the grid generates "shifted" rather than zero greenhouse gas emissions.
- Ecological externalities of HPP, PSP, and small HPP construction include the fragmentation of ecosystems, impact on flora and fauna of water reservoirs, water use regimes, water quality, hydrology of water objects, and their basins. Environmentalist opposition to such facilities is significant in some countries. However, such power units perform a balancing function and allow integration into the power system of more substantial amounts of the intermittent capacity of renewable generation.
- Solar power plants' externalities on land use, impact on flora and fauna, energy and resource consumption of solar panels, and the need for their disposal, despite the rapid movement along learning curves, create significant obstacles to the spread of this energy technology.
- Externalities like "food vs. fuel" for the first-generation biofuels: uncontrolled use of arable land poses a threat to food security and could lead to additional depletion of agricultural land and require the reduction of edible crops or doubling their yields.

Based on the analysis of modern energy transitions globally, we summarized the main negative externalities and correlated them with regulatory measures, as shown in Table 2.4.

The energy crises in 2021 and in 2022 globally have shown the vulnerability of energy systems that rely on both fossil and renewable energy sources. Given that nearly all industries and types of consumers rely on energy, the crisis hit many quite sensitively. It also gave the prerequisites for the countries that major energy carriers' suppliers (mostly natural gas and coal), to dominate. Renewable energy is one of the critical elements of decarbonization and transition towards carbon neutrality. With some negative externalities and being a source of primarily intermittent electricity, it is often subject to criticism or even skepticism. Some of the externalities include the fact that the electricity infrastructure used for transmission of renewable-based electricity becomes overstressed (IEA ETSAP 2014).

TABLE 2.5

Negative Externalities of Sustainable Energy Transitions and Regulatory Policy Measures

Case	Policy Measure	(I)nternalyzing/ (T)echnological
Improper application of the feed-in tariff	Classic methods, introduction of additional markets (segments)	I
Improper EV charging	Economic revaluation	T
	Parallel operation of technologies	
Small HPPS with dams	Parallel operation of technologies	T
Solar power plants	Introduction of externalities market	T, I
Wind power plants	Introduction of additional markets (segments), market structure change, economic revaluation	
	Parallel operation of technologies	
	Improved operational strategies	
First-generation biofuels	Classic, market structure change	I, T

Source: Pysmenna and Trypolska (2020)

2.1.4 CASES OF OVERCOMING BARRIERS AND FOSTERING SUSTAINABLE ENERGY CHANGES

Other externalities are considered in the form of small cases, particular policy measures to overcome them, as follows (but not limited to):

2.1.4.1 Balancing Needs

Intermittent energy sources such as wind and sun require additional efforts to balance the energy system. Over 24 hours, wind and solar electricity is generated as shares. While wind-based electricity hypothetically could be generated close to 24 hours should wind be in place, solar-based electricity is generated a maximum of half of the time (NPC Ukrenergo 2021). This problem is aggravated by the fact that Ukraine has a sharp deficit of maneuvering capacities, as the energy system of Ukraine is one of the least flexible energy systems globally (Wärtsilä 2020). ENTSO-E integration will partially decrease this problem but will have its price. In 2020, intermittent renewable energy sources were even accused by the Security Service of Ukraine of being a threat to the energy security of Ukraine through destabilization of the energy system due to uncontrolled growth of "green" energy producers (EPravda 2020). Back then, relatively cheap nuclear energy was replaced by more expensive green and grey electricity from coal-powered CHPs, which were needed to balance the system. Ukraine witnessed the so-called green-coal paradox. Due to the growing share of solar and wind energy, the dependence on thermal generation to balance the energy system increases. Therefore, the growth of renewables in the energy balance leads to a rise in greenhouse gas emissions.

To tackle the problem of balancing capacities, not less than 2 GW of high maneuvering capacities are needed to integrate RES into the energy system in Ukraine during the last couple of years. The energy storage capacities have to reach 500 MW. RES also must provide a load decrease service, that is, become one of the sources of ensuring the flexibility of the power system. To prevent further accusations of destabilization of the energy system of Ukraine, the transition from feed-in tariff to the auction system has to be made, allowing more reasonable allocation of energy production facilities, preferably closer to energy consumers. Besides, power-to-gas (P2G) technology has to be introduced, allowing the transformation of excess energy into gas, which later produces power to meet the peak load. Together with that, smart grids need to be introduced, coupled with demand management measures involving energy consumers in balancing the grid, for example, charging electric vehicles during the nighttime, when energy demand is lowest (Krynytsky and Aliyeva 2020). The share of energy generation by households has to increase the flexibility of the grid. After the feed-in tariff for home-owned installations expires in 2029, solar-plus-storage incentives should be introduced. Should Ukraine be willing to switch to 100% RES by 2050, according to Wärtsilä forecasting (2020), the energy system of Ukraine would reach 126 GW (compared to 56 GW in October 2021), of which nearly 40 GW should be balancing capacities. Such a composition of the energy system represents the lowest system cost. To ensure that, 9 GW of flexible gas CHPs have to be built by 2030 (should companies find the financing for natural-gas-based projects).

2.1.4.2   Energy Storage Capacities and Batteries

To balance the energy system, energy storage capacities are needed. Storage technologies have negative externalities. In particular, lithium, cobalt, and graphite mining globally is associated with violations of human rights, adverse pollution impacts, and the necessity of handling high volumes of lithium-ion waste (Florin and Dominish 2017), as lithium-ion batteries have only up to 10,000 maximum cycles in a lifetime. Lithium is also widely used for e-vehicle manufacturing, and the high cost of batteries is one of the impediments to e-vehicles expansion (BloombergNEF). Prices for lithium-ion batteries have fallen dramatically, from USD1200/kWh in 2010 to USD137/kWh in 2020 (BloombergNEF) (Figure 2.4).

In the upcoming 10–15 years, some of the already known deposits globally may be exhausted (as in the Democratic Republic of Congo, producing half of the global cobalt supply [Florin and Dominish 2017]). To overcome these externalities mentioned, sustainability criteria for raw materials need to be introduced and strictly obeyed.

2.1.4.3   Green Hydrogen

Hydrogen is a promising technology of the near future. Green hydrogen, that is, produced from renewable energy sources, is especially interesting for implementation. The EU is interested in green hydrogen because it meets the terms of the Green Deal. Some regions ("the Eastern Neighbourhood, in particular Ukraine, and the Southern Neighbourhood countries should be priority partners" [EC 2020]) are included in the European Hydrogen Strategy, in which the EU encourages countries to become the main (primary) partners. As of 2022, the possibility of building up to 10 GW of

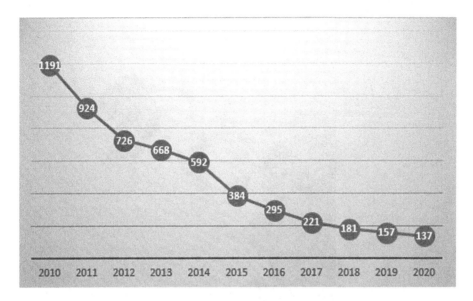

FIGURE 2.4   Volume-weighted average prices for lithium-ion batteries, USD/kWh.

Source: BloombergNEF (2021)

electrolyzers for green hydrogen is considered in Ukraine, but the Russia–Ukraine war hinders the plans. However, water availability for green hydrogen production is an important issue, as Ukraine is a water-deficient country (Snizhko et al. 2021). Water stress is expected to persist in the coming decades and even be aggravated by the projected climate change (Figure 2.5), so Ukraine will experience even higher water stress in the future.

At the current stage of technology development level (i.e. price characteristics), green hydrogen and its products are expected to be exported mainly to the EU. Therefore, there is a significant conflict of interest in place—Ukraine needs to spend a tremendous amount of water on a product not intended for consumption at the domestic market. Overall, a lack of water is a significant obstacle for green hydrogen output, requiring careful planning and care of scarce water use.

2.1.4.4   Renewable End-of-Life Equipment

The growing share of renewable energy in the energy balance usually means growing installed capacities. The equipment for energy generation, like any other, has a lifetime. Used solar and wind power plant equipment is classified into general and industrial waste. Wind power plant equipment has a shorter service life than solar PV, and wind turbines of lower capacity tend to be replaced by more powerful equipment. Waste wind power equipment is more challenging to recycle, especially old blades. The existing applications of old blade processing or reuse are the exception. According to the principles of circular economy, waste energy equipment should be recycled and processed. As of 2022, the EU has the most advanced legislation

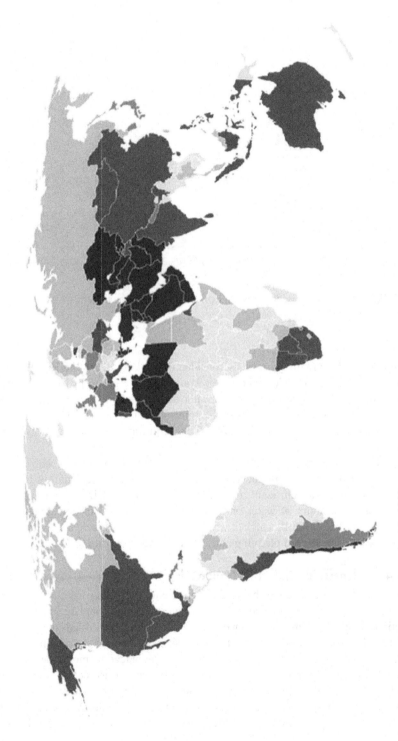

FIGURE 2.5   Projected water stress level in 2040.

Source: WRI (2021)

promoting waste energy equipment recycling, with Directive 2012/19/EU on waste electric and electronic equipment in place and the final Package of the Circular Economy (2019). The directive provides for an extended producer responsibility principle; that is, all equipment manufacturers are responsible for the collection, processing, and control of both electric and electronic equipment sold in the EU, regardless of the country in which it was manufactured. In other words, the manufacturer bears both financial and organizational costs to eliminate its waste product. This directive was adopted to protect people and nature from toxic elements, including used electronics, and to reduce the burden of waste disposal costs, including from municipalities and local governments.

To approach the problem of waste solar and wind equipment, the regulatory document similar to the directive 2012/19/EU has to be developed with the aim to introduce the principle of extended producer responsibility. The cooperation of local governments with extended producer responsibility organizations is also needed.

2.1.4.5   Coal Phase-Out

During COP-26 in Glasgow, more and more countries pledged to carbon neutrality. One of the negative externalities of carbon neutrality is the need for socially just transformation of coal regions and the need to employ the laid-off coal miners. The problem of coal phase-out and just energy transition is extraordinarily complex and long term. There are no simple one-size-fits-all solutions. In some cases, internal social mobility could help, such as in Germany (during their coal-phase out) (Oei et al. 2020) or in the USA. Other countries that already commenced coal phase-out started this process much earlier—for example, Germany started the process in the 1990s, and the role of coal in the energy balance of the United Kingdom began to decline as early as in 1980s, allowing the country to claim coal phase-out by 2024/2025 (Brauers et al. 2020). The coal phase-out process is financially supported: for example, EUR 14 billion is allocated in Germany until 2038 for mining regions. Additionally, another EUR 26 billion will be allocated by the federal government for infrastructure, research, and expansion of existing support policies for coal regions. The cumulative adjustment allowance for laid-off miners will reach EUR 5 billion in 2020–2048 (Niebuhr 2021).

Particular measures that may help solve this negative externality may include, but are not limited to, the following:

- Training programs enabling laid-off miners to get new knowledge and skills during the next decade. The central or regional government should pay for these training programs (i.e., it is one of the forms of subsidies). In some cases, this can be training enabling miners working at the renewable energy facilities, as it was in Canada (Trypolska 2021);
- Stock-taking of jobs and industries available in coal regions;
- If applicable, invitation of large employers to the respective regions. They might include large foreign auto manufacturers or international appliance producers;
- Financial support for the creation of new jobs (subsidies could be provided to companies employing former coal miners);

- Enhanced job-seeking assistance by labor registry offices;
- Simplification of bureaucratic procedures to obtain severance pay;
- Allocation of funds by companies that undertake mining to cover the environmental damages and ensure site restoration;
- Development of programs aimed at switching from coal as the only energy carrier to other energy carriers.

Energy sustainability in different forms requires significant financial resources and an outstanding share of flexibility and adaptableness to develop new policies, courage to pursue communities in the necessity of new energy sources, and use of new technologies. Recent trends such as decarbonization or implementation of new energy technologies impose severe risks and financial burden on less developed countries, making more economically developed countries pioneers in implementing new technologies, driving down the costs. Like any technological paradigm shift, energy transition does not only presume certain job losses (as in the case of coal phase-out). It allows so-called "healthy" or structural unemployment and the creation of new markets and new types of jobs.

Transitions to energy sustainability are often postponed, slowed down, or even rejected to avoid negative externalities that could threaten the system stability (energy and economy security). Plenty of cases are being seen nowadays concerning such issues: the opposition to new energy technologies, the dilemma of "food versus fuel", the stability and cooperation of socio-technical regimes. A deeper understanding of the externalities of energy transitions and the vulnerability of energy systems needs to be reached under the influence of negative externalities caused by sustainable energy transitions. Energy poverty postponing rapid innovations towards energy sustainability is another issue to deal with when maintaining sustainable energy systems.

The prospects of energy sustainability depend in a great way on the success of value-driven energy policies as those which respond to sustainability demands. Such policies should lead efficient energy strategies and implement policy mechanisms towards all the dimensions of the energy sustainability quadrilemma: economic (affordability), political (energy security), environmental (climate change and environmental sustainability), and social (energy justice) simultaneously.

The effective governance of sustainable energy transitions needs to be innovation based to implement a new energy mix and to be flexible enough to deal with the uncertainties of global energy trends and local deficits. Innovations can accelerate economic growth, although the primary reason for their emergence is the restriction of the extensiveness of such growth (inefficient and unsustainable resource use). This paradox determines the viability of economic systems within the sustainable transitions and ensures their further development.

Innovation-based energy policies form a considerable transition of paradigm from cost to value, becoming action plans for achieving sustainable energy systems by creating prerequisites for the emergence, forcing and supporting sustainable transitions, simultaneously minimizing negative external and social factors, and maintaining an optimal energy balance and energy security level.

REFERENCES

Armin Razmjoo, A., Sumper, A., and Davarpanah, A. (2019) Development of Sustainable Energy Indexes by the Utilization of New Indicators: A Comparative Study. *Energy Report*, 5, 375–383.

Ayres, R.U., and Warr, B. (2005) Accounting for Growth: The Role of Physical Work. *Structural Change and Economic Dynamics*, 16, 181–209.

BloombergNEF (2021) *EV Battery Prices Risk Reversing Downward Trend as Metals Surge.* URL: www.bloomberg.com/news/newsletters/2021-09-14/ev-battery-prices-risk-reversing-downward-trend-as-metals-surge

Brauers, H., Oei, P.-Y., and Walk, P. (2020) Comparing Coal Phase-Out Pathways: The United Kingdom's and Germany's Diverging Transitions. *Environmental Innovation and Societal Transitions*, 37, 238–253, ISSN 2210-4224. https://doi.org/10.1016/j.eist.2020.09.001

Cherp, A., and Jewell, J. (2014) The Concept of Energy Security. Beyond the Four As. *Energy Policy*, 75, 415–421. https://doi.org/10.1016/j.enpol.2014.09.005/ (дата звернення: 23.04.2017)

Cleveland, C. J., Costanza, R., Hall, C.A.S., and Kaufmann, R.K. (1984) Energy and the US Economy: A Biophysical Perspective. *Science*, 225, 890–897.

EC (2020). *Communication from the Commission to the European Parliament, the Council, the European Economic and Social Committee and the Committee of the Regions.* A Hydrogen Strategy for a Climate-Neutral Europe. COM (2020) 301 final. URL: https://ec.europa.eu/energy/sites/ener/files/hydrogen_strategy.pdf

EPravda (2020) *СБУ попередила Разумкова щодо загроз енергобезпеки через "зелений тариф".* 23 квітня 2020. URL: www.epravda.com.ua/news/2020/04/23/659703/

Florin, N., and Dominish, E. (2017) *Sustainability Evaluation of Energy Storage Technologies.* Report Prepared by the Institute of Sustainable Futures for the Australian Council of Learned Academies. URL: https://acola.org/wp-content/uploads/2018/08/wp3-sustainability-evaluation-energy-storage-full-report.pdf

IEA (2005) *Energy Indicators for Sustainable Development.* IEA. URL: www-pub.iaea.org/MTCD/publications/PDF/Pub1222_web.pdf

IEA (2014) *More Data, Less Energy.* IEA. 176 p. URL: https://webstore.iea.org/more-data-less-energy/

IEA ETSAP (2014) *Electricity Transmission and Distribution.* IEA ETSAP—Technology Brief E12. URL: https://iea-etsap.org/E-TechDS/PDF/E12_el-t&d_KV_Apr2014_GSOK.pdf

Krynytsky, K., and Aliyeva, O. (2020) *Зелено-вугільний парадокс: зупинити не можна дозволити, де кома?* URL: https://ua.boell.org/uk/2020/06/09/zeleno-vugilniy-paradoks-zupiniti-ne-mozhna-dozvoliti-de-koma

Marinas, M., et al. (2018) Renewable Energy Consumption and Economic Growth. Causality Relationship in Central and Eastern European Countries. *PLoS ONE*, 13(10), e0202951. Published online 2018 October 8. https://doi.org/10.1371/journal.pone.0202951/

Niebuhr, A. (2021) *Just Transition of the Coal Mines Regions.* Webinar 30 November 2021. URL: https://docs.google.com/presentation/d/1wwl34F_lErrUXiIC_mMcjSCrSsRoTw6L/edit#slide=id.p1

NPC Ukrenergo (2021). URL: https://ua.energy/peredacha-i-dyspetcheryzatsiya/dyspetcherska-informatsiya/dobovyj-grafik-vyrobnytstva-spozhyvannya-e-e/

Oei, P.-Y., Brauers, H., and Herpich, P. (2020) Lessons from Germany's Hard Coal Mining Phase-Out: Policies and Transition from 1950 to 2018. *Climate Policy*, 20(8), 963–979. https://doi.org/10.1080/14693062.2019.1688636

Omri, A. (2014) An international literature survey on energy-economic growth nexus: Evidence from country specific studies. *Renewable and Sustainable Energy*, 38, 951–959.

Payne, J. E. (2010) Survey of the International Evidence on the Causal Relationship Between Energy Consumption and Growth. *Journal of Economic Studies*, 37, 53–95.

Pysmenna, U. Y., and Trypolska, G. S. (2020) Sustainable Energy Transitions: Overcoming Negative Externalities. *Energetika Proceedings of CIS Higher Education Institutions and Power Engineering Associations*, 63(4), 312–327. https://doi.org/10.21122/1029-7448-2020-63-4-312-327

RISE (2021) *Regulatory Indicators for Energy Sustainability*. RISE. URL: https://rise.esmap.org/analytics

Schaffartzik, A., Plank, C., and Brad, A. (2003) Ukraine and the Great Biofuel Potential? A Political Material Flow Analysis. *Ecological Economics*, 104, 12–21. https://doi.org/10.1016/j.ecolecon.2014.04.026

Snizhko, S., Shevchenko, O., and Didovets, Y. (2021) Analysis of the Climate Change Impact on the Water Resources of Ukraine. *NGO "Ecoaction"*. URL: https://ecoaction.org.ua/wp-content/uploads/2021/06/analiz-vplyvu-vodni-resursy-full.pdf

Sorrell, S. (2010) Energy, Economic Growth and Environmental Sustainability: Five Propositions. *Sustainability*, 2, 1784–1809. https://doi.org/10.3390/su2061784/

Sovacool, B. (2016) *Fact and Fiction in Global Energy Policy: Fifteen Contentious Questions*. With Brown, M. A. and Valentine, S. Baltimore: Johns Hopkins University Press.

Trypolska, G. (2021) Prospects for Employment in Renewable Energy in Ukraine, 2014–2035. *International Journal of Global Energy Issues*, 43(5/6), 436–457.

UNDP (2016) *Sustainable Development Goals*. URL: www.undp.org/sustainable-development-goals

Wärtsilä (2020) *The Optimal Path Forward for Ukraine's Power System*. White Paper on Power System Optimization. URL: www.finnishenergyhub.com/post/оптимальний-шлях-розвитку-енергетичної-системи-україни

WEC (2022) *World Energy Trilemma Index—2022*. WEC. URL: https://www.worldenergy.org/assets/downloads/World_Energy_Trilemma_Index_2022.pdf?v=1669839605

WRI (2021) *Under Pressure. The Economic Costs of Water Stress and Mismanagement*. World Resources Institute via the Economist Intelligence Unit. URL: https://bluepeaceindex.eiu.com/pdf/EIU_Under%20Pressure_Economic%20Costs%20of%20Water%20Stress_2021.pdf

3 Renewable Energy
Technologies, Applications and Trends

Muhammad Asif

3.1 INTRODUCTION

The sustainability of the global energy landscape is one of the major concerns the world faces today. Fossil fuels dominate the current energy scenario, meeting almost 80% of the total energy requirements [1]. Fossil fuels—coal, oil, and natural gas—have their own set of problems such as depletion of reserves, volatile and surging prices, security of supplies, and above all environmental effects. These fuels are extracted, transported, and burned, releasing greenhouse gases that contribute to climate change. In addition, they cause habitat degradation, air and water pollution, and negative health effects [2, 3]. The use of renewable energy is an essential response to the serious problems the world faces with energy production and consumption.

Renewable energy is leading the global energy landscape. It is the energy derived from renewable natural resources. Renewable energy can be classified into six main categories: solar energy, wind energy, hydropower, biomass, geothermal energy, and wave and tidal power. Renewable energy has been in use since the beginning of human civilisation. Biomass, hydro energy and wind energy have been in use for thousands of years. Utilising natural resources, renewable energy systems offer energy services with virtually no emissions of greenhouse gases or air pollutants. Currently, over 28% of the world's overall electricity needs and more than 13% of its primary energy requirements are being satisfied by renewable energy [4]. Renewable resources and abundant and can supply much more energy than the world needs. They can increase market diversity for energy supply, guarantee long-term sustainable energy sources, and lower emissions into local and global atmospheres. They can also provide new employment opportunities, offer prospects for local equipment production, and offer commercially appealing choices to address unique needs for energy services especially in developing countries and rural areas [5–7].

Renewable energy is the main foundational block of the unfolding sustainable energy transition as the world targets to move away from fossil fuels. Increasing the proportion of renewable energy in the energy mix is another need of Sustainable Development Goal 7 (SDG7) [8]. Renewable energy sources can deliver energy services with nearly zero emissions of greenhouse gases and air pollutants. These resources have the potential to supply all of the global energy needed for today and in the future. The use of renewable energy sources can increase market diversity for energy supplies, help secure long-term sustainable energy supplies, lessen negative

DOI: 10.1201/9781032715438-3

environmental effects locally and globally, and offer commercially appealing options to meet particular energy service needs, particularly in developing nations and rural areas, leading to the creation of new job opportunities. With their declining cost trends, financially, renewable energy is also becoming more viable than fossil fuels. Solar and wind energy systems have grown rapidly over the past 20 years as a result of policy and technological developments, as well as economies of scale [9, 10].

3.2   RENEWABLE ENERGY SYSTEMS: TECHNOLOGIES AND APPLICATIONS

3.2.1   SOLAR ENERGY

The planet Earth revolves in an elliptical orbit around the Sun at an average annual distance of nearly 150 million km. The Sun's nuclear fusion process generates 3.8×10^{23} kW of electricity in its core. A tiny portion of the Sun's energy, around 1.73×10^{16} kW, travels to Earth in the form of energy particles known as photons. Even this portion is enormous when viewed from Earth's perspective; the planet receives more solar energy in a single hour than what the entire human race needs [11]. As a result, the sun is the world's primary source of energy because most other energy sources, including biomass, hydropower, wind power, and fossil fuels, are direct and indirect derivatives of solar energy.

3.2.1.1   Solar PV

Solar photovoltaic (PV), often known as a solar cell, is the most prominent of the solar technologies. PV uses the photoelectric process to convert solar energy directly into electricity. When exposed to light, solar cells turn light's photons into electricity. There are many different types of solar cells, which are typically composed of semiconductor materials, depending on the material and manufacturing process. The most common material used in PV manufacture is silicon. Figure 3.1 depicts a classification of typical PV cell types in terms of photovoltaic material. Crystalline, thin-film, and multi-junction PV cells commonly fall under the categories of first-generation, second-generation, and third-generation PV cells.

Solar PV systems can be broadly classified into two main types: grid-connected and off-grid or stand-alone systems. Sometimes PV systems' additional types are also considered, such as utility-scale and hybrid systems. Batteries are often needed for energy storage in off-grid systems. These systems ought to be built to handle peak loads as well as requirements for load during times when solar radiation is not available. Grid-connected systems have a two-way exchange with the grid rather than needing a battery bank to store energy. A grid can provide electricity when a PV system is not producing electricity, and excess/unused energy can be exported to the grid. Building rooftops are where solar PV systems are most frequently installed, as shown in Figure 3.2 [12].

3.2.1.2   Solar Thermal Energy

The harnessing of solar radiation to generate heat that may be used for a variety of purposes is the second primary method for using solar energy. Utilising various heat transfer qualities, including absorption, insulation, and the "greenhouse effect,"

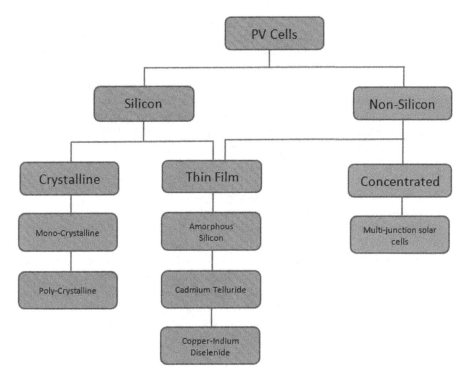

FIGURE 3.1 Classification of PV cells by material type.

FIGURE 3.2 Rooftop PV on residential buildings.

solar energy is converted into heat. The quantity of heat radiation that is absorbed by various surfaces varies on a number of variables, including the incident surface area, radiation intensity, and radiation angle. Additionally, it relies on the surface's composition, colour, and texture. Solar thermal energy, often known as solar energy captured as heat, is used through a variety of processes. In addition to power generation, solar thermal energy can also be applied for heating and cooling purposes. Regarding applications, scientific development, and financial sustainability, solar thermal technologies display a large variation. According to their operating temperature, solar thermal technologies can be divided into the three categories listed as follows.

- Low-temperature (below 70° C) technologies
- High-temperature technologies (above 200° C)
- Medium-temperature technologies (between 70° C and 200° C)

Solar thermal power generation is the use of solar thermal energy to produce electricity. It is one of the most advanced applications of solar thermal energy. The first and fundamental feature of a solar thermal power system is to capture heat from solar radiation. Different types of concentration techniques are used to improvise medium- to high-temperature heat from solar radiation. The concentrated solar power (CSP) systems have the following four common types:

- Parabolic trough systems
- Linear Fresnel systems
- Parabolic dish systems
- Tower systems

These four types of CSP systems can be broadly classified into two groups, depending on whether the solar collectors concentrate the sun rays along a focal line or on a single focal point (Figure 3.3) [13], the latter providing much higher concentration factors.

FIGURE 3.3   Concentrated solar power (solar tower) system.

3.2.2 WIND POWER

As a result of the unequal heating of the Earth's surface by solar radiation, wind is the movement of air from areas of high pressure to those of low pressure. One of the first sources of energy used by mankind was wind power. Sailing was the first application of wind power, with boats along the Nile River being propelled as early as 5,000 BC. By 200 BC, the Middle East and Persia were using wind mills—the forerunners of today's wind turbines—to pump water and grind grain. Electricity production from wind turbines, the current form of wind power, first appeared in the 1880s. The first wind farm in the world, with 20 wind turbines and a combined generation capacity of 600 kW, was built in New Hampshire, United States, in 1980. The use of wind turbines offshore is a more recent development. The first offshore wind farm, which had 11 wind turbines with 450 kW capacity each, was built in Vindeby, Denmark, in 1991. Although a wind turbine of the megawatt (MW) size was constructed in the United States in 1941, the commercial market for wind turbines did not expand over the kilowatt (kW) level until the 1980s. The size of wind turbines has rapidly increased during the 1990s. Since the wind industry wants to increase the capacity to 20 MW, wind turbines with a 14 MW capacity have already been created [11].

With the help of a number of parts, including rotor blades, shafts, gearboxes, and electric generators, wind turbines convert the kinetic energy of blowing air into electrical energy. The two main categories of wind turbines are the horizontal axis (Figure 3.4) [14] and the vertical axis. Three-bladed horizontal-axis wind turbines have become the norm in the current wind industry. The size of wind turbines varies

FIGURE 3.4 On-shore wind turbines.

greatly, and they can be utilised both off-grid and in grid-connected configurations. In terms of big and utility-scale projects, wind power is one of the leading modern renewables. The main difficulty with wind energy is its intermittent nature, as it can not be relied on to meet base load on its own.

3.2.3 HYDROPOWER

One of the most affordable, dependable, and flexible sources of producing clean electricity is hydropower. When water is flowing, its kinetic energy is converted into mechanical energy, which is then used to power a generator and a hydro turbine to produce electricity. The head height or water drop and water flow affect power production. Currently, hydropower projects account for 16% of the world's electrical supply. Compared to other types of power projects, hydropower projects last longer. The capacity of hydropower plants can vary greatly and might be anything from a few watts to tens of gigawatts. Hydropower projects are mainly classified into the following three types.

- Dam-based
- Run-of-river
- Pumped-storage

Dam-based hydropower projects employ dams to raise the storage of water in reservoirs. By discharging water from the reservoir to power a turbine connected to a generator, electricity is generated. Using high operational flexibility, dam-based hydropower projects, as depicted in Figure 3.5 [15], are employed to meet base load. Projects based on dams can aid with water management as well. Contrary to other types of hydropower projects, these are more expensive and take longer to develop. Run-of-river hydropower projects generate energy by using flowing water to turn a turbine through a canal or penstock. Most run-of-river projects do not need water

FIGURE 3.5 Dam-based hydropower project.

storage, although occasionally a small reservoir may be used to help the water flow through the penstock.

Pumped-storage projects consist of two reservoirs at different heights. These are often used to help with the peak-load management. As with traditional hydropower projects, electricity is produced when demand is at its highest by releasing water from the higher reservoir. After generating electricity, the water is stored in the lower reservoir. Water from the lower reservoir is transferred to the upper reservoir during times of low demand. Dam-based and pumped storage hydropower projects are excellent in balancing demand and supply as they offer a great deal of operational flexibility due to daily swings in demand.

3.2.4 Bioenergy

Organic material formed from recently extinct or alive creatures is called biomass. The first energy source from which humans have profited is biomass. Biomass was the predominant fuel up until the Industrial Revolution in the 18th century. The share of biomass gradually decreased as the use of fossil fuels increased. Nevertheless, it continues to contribute significantly to the world's primary energy resources. For cooking and heating, almost 3 billion people still rely on raw biomass. The main categories of biomass include:

- Wood and forestry
- Agricultural residues
- Animal waste
- Energy crops (Figure 3.6)
- Food waste
- Industrial waste

Through a variety of technical conversion possibilities, biomass can be used for different energy uses. Energy can be delivered directly from biomass conversion

FIGURE 3.6 High-yield energy crops.

technology in the form of heat or electricity, or it can be transformed into different forms like biogas and biofuels. A variety of energy outputs, including heat, electricity, biogas, and biofuels, can be produced from biomass. Techniques for converting biomass and their associated products can be generally categorised into three categories [11].

- Direct combustion is the most popular method of using biomass for heating in residential, commercial, and industrial applications.
- Thermochemical conversion: Pyrolysis, gasification, carbonisation, and direct liquefaction are a few examples of the several types of thermochemical conversion. Heat and biofuels are two possible outputs.
- Biochemical conversion: Anaerobic digestion and hydrolysis fermentation are two methods for biochemical conversion that can be used to create biogas/methane and biofuel, respectively.

3.2.5 GEOTHERMAL ENERGY

The Earth's interior is extremely hot, with temperatures that can exceed 6000° C. The heat energy that is drawn from under the earth's surface and converted into hot water or steam is known as geothermal energy. Geothermal energy can be used for heating and cooling in a variety of settings [16]. Medium- to high-temperature geothermal resources are required for power generation. The technological viability of geothermal technology varies. Technologies for the direct use of geothermal energy, such as district heating, geothermal heat pumps as shown in Figure 3.7 [17], and the production of electricity from hydrothermal reservoirs with naturally high permeability, are well developed. Geothermal fields with medium temperatures can be exploited for combined heat and electricity [18].

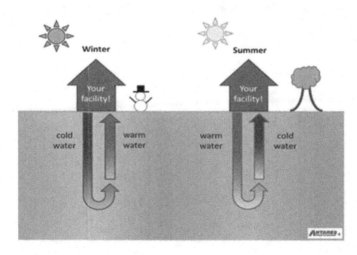

FIGURE 3.7   Ground source heat pump.

3.2.6 WAVE POWER

The energy carried by ocean waves and tides is used to generate wave and tidal power, often referred to as marine energy, ocean energy, and hydrokinetic energy. Large bodies of water experience waves as a result of wind and water surface contact brought on by temperature differences. Compared to wind, waves have a lot more energy. The theoretical yearly energy potential of waves off the shores of the United States is estimated by the US Energy Information Administration to be as high as 2.64 trillion kWh, or the equivalent of roughly 64% of the country's electricity production in 2019 [19]. However, the marine environment presents a number of difficulties, and systems to harness wave power are still in the research and development stage. Various designs for wave power technologies are being developed, as seen in Figure 3.8 [19], depending on the installation location and the desired wave characteristics. Installation of wave power technologies on or below the water's surface can generate electricity as waves drive the turbine-generator system.

Ocean tides are produced by the gravitational attraction of the moon and sun. Water levels close to the shore can fluctuate up to 40 feet due to tides. By stopping the high tide behind a barrage constructed across an inlet of an ocean bay or lagoon that creates a tidal basin, tidal power generates electricity. To allow the tidal basin to fill during high tides and empty through an electrical turbine system during ebb tides, sluice gates on the barrage manage water levels and flow rates. Both incoming and outgoing tides can be used to produce electricity using two-way tidal power systems [20]. Although tidal power projects have been identified as viable around the globe, there have only been three noteworthy tidal projects to date. The first one was 240 MW in La Rance, France, built in 1966, followed by a 20 MW project in Nova Scotia, Canada, finished in 1984, and a 254 MW project in South Korea, finished in 2011 [11].

FIGURE 3.8 Wave power system.

3.3   ADVANTAGES OF RENEWABLE ENERGY

3.3.1   COST-EFFECTIVENESS

One of the key strengths of renewable energy is its increasing cost-effectiveness. Renewable energy sources are developing as a possible replacement for traditional fossil fuels in the changing global energy landscape. The decreasing cost of renewable energy technology has been one of the major forces behind this transition. Over the past decade, the cost of solar photovoltaic energy has significantly decreased. The International Renewable Energy Agency (IRENA) reports that between 2010 and 2020, utility-scale solar PV projects' global weighted-average levelised cost of electricity (LCOE) decreased by almost 80%. This trend has been significantly influenced by technological improvements, economies of scale, and falling solar panel prices. The costs for onshore and offshore wind energy have significantly dropped. According to IRENA, the LCOE for onshore wind projects dropped by almost 40% between 2010 and 2020. Offshore wind has shown even greater cost reductions than onshore wind, with LCOEs falling by over 70% over the same time span, despite traditionally being more expensive [4]. These cost reductions have been made possible by improved turbine design, increased capacity factors, and simplified installation procedures.

The price of traditional fossil fuels has shown more volatility and, in some circumstances, price hikes than that of renewable energy. While there have been moments of fluctuation, geopolitical events, supply disruptions, and market speculation have all had an impact on the overall trend in fossil fuel prices. Exploration and extraction costs have increased as a result of the ongoing transition away from fossil fuels, which could push up prices.

In terms of levelised cost of energy renewable energy is becoming competitive with conventional energy systems. The LCOE of utility-scale solar PV and onshore wind in the United States is frequently equal to or lower than that of natural gas and coal, according to the U.S. Energy Information Administration (EIA). Numerous other locations throughout the world are seeing same patterns. Because of this, renewable energy is a desirable alternative from an economic as well as environmental perspective.

Energy industry investment patterns support the cost-competitiveness of renewable energy even more. Global investments in renewable energy have regularly surpassed investments in fossil fuel-based projects in recent years. It shows that investors are aware of renewable energy's long-term economic potential. The integration of renewable energy into the grid depends on energy storage. Over the past ten years, the price of energy storage technology like lithium-ion batteries has fallen dramatically. Between 2010 and 2020, the average price of lithium-ion batteries decreased by about 89%. By lowering costs, it is now more feasible to store extra renewable energy for use during times of high demand or low supply, improving grid stability.

The renewable energy job market is also growing across the world. Investments in the infrastructure needed for renewable energy, such as wind farms, solar panels, and bioenergy plants, create a wide range of employment possibilities. Manufacturing, setup, upkeep, and research and development are all included in this. These jobs frequently benefit local communities by boosting economic activity in places that may have previously struggled to find employment.

3.3.2 ENVIRONMENTAL PROTECTION AND CLIMATE CHANGE MITIGATION

One of the leading advantages renewable energy offers over conventional energy resources is its environmental friendliness. Renewable resources produce no or very little emissions during their operational phase. Habitat degradation, deforestation, and ecological disruption are frequently side effects of conventional energy extraction and production techniques. Wind and solar farms, for example, leave a reduced ecological footprint and may cohabit more peacefully with natural ecosystems. Ecosystem protection is also essential for preserving biodiversity. Renewable energy contributes to climate change mitigation. Renewable energy sources contribute to limiting the increase in global temperatures, which is crucial for maintaining the planet's climate balance by considerably reducing greenhouse gas emissions. As a result, ecosystems are protected, extreme weather occurrences are avoided, and the welfare of future generations is ensured.

3.3.3 ENERGY SECURITY

By varying the energy mix and lowering dependency on imported fossil fuels, renewable energy sources help to advance energy independence. This lessens a country's susceptibility to changes in the world energy markets and geopolitical conflicts over resource access. Additionally, it aids in price stabilisation, lowering consumer energy costs. The price volatility of renewable energy sources is lower than that of fossil fuels. Renewable resources' costs are generally steady because they are abundant and naturally arise. Consumers and businesses are protected by this stability from the erratic price swings that frequently follow the markets for fossil fuels. Grid resilience is improved through distributed renewable energy systems, including rooftop solar panels. These systems lessen transmission and distribution losses and the susceptibility of centralised power grids to outages and severe weather occurrences by producing electricity closer to the point of use. This adaptability ensures a source of energy that is secure.

3.3.4 RESOURCE SUSTAINABILITY

Renewable energy sources are essentially unbounded, unlike finite fossil fuel supplies. As long as the sun is shining, solar energy is available. Wind and hydropower depend on the natural cycles of the wind and water, respectively. Future generations will have access to clean and sustainable energy sources thanks to this long-term availability. The use of renewable energy reduces the need for limited resources like coal, oil, and natural gas. We can increase the availability of these resources for crucial uses and lessen the detrimental effects of their extraction on the environment by limiting their extraction and consumption.

3.4 TECHNOLOGY ADVANCEMENTS

One of the main causes of the success of renewable energy over the past few decades has been strong technological improvements. These advancements have occurred on a number of fronts, including a better comprehension of resource data and site selection, the creation of novel designs and materials, effective manufacturing procedures, and precise system deployment. The scientific and engineering developments in wind turbine production and installation have a significant positive impact on the wind power sector. In addition to advancements in aerodynamic designs, sophisticated and advanced materials are assisting in the development of bigger, lighter, and stronger wind blades. As seen in Figure 3.12, these advancements have allowed wind turbines to quickly increase in size in recent years. The wind industry is positioned itself to increase the size of offshore wind turbines to 20 GW by 2030, with 14 MW wind turbines already deployed as shown in Figure 3.9. The capacity factor has also been significantly enhanced by improvements in wind turbine technology and better site selection. The capacity factor of General Electric's new 12–14 MW offshore wind turbines with 220 m of rotor diameter is said to be in the range of 60%–64%, an increase of about 150% in just two decades. Wind farms in Europe and the USA produced capacity factors in the mid-20s in 2000 [21].

Low power density is one of the main issues with renewable energy. With popular metrics like W/m^2 and W/kg, power density refers to the energy flow that may be captured from a given unit of area, volume, or mass. Renewable energy sources have a very low power density when compared to non-renewable energy sources. Like wind turbines, solar PV has also gone large-scale, with hundreds of MW scale projects becoming common around the world, as shown in Figure 3.10 [22]. Going offshore is one way to address the problem, in addition to technological developments like improved wind turbine technology and higher PV cell efficiencies that increase the capacity factor of wind farms. In comparison to onshore wind farms, offshore wind applications have various advantages. In addition to resolving the land rights issue,

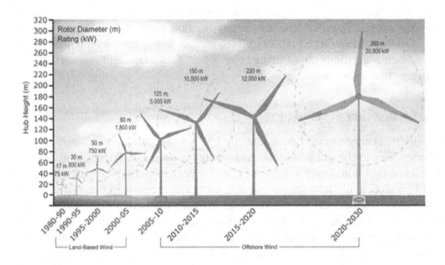

FIGURE 3.9   Growth of wind turbines in size and capacity.

FIGURE 3.10   Utility-scale solar farm.

FIGURE 3.11   Floating solar PV.

offshore wind power offers improved efficiency and a larger capacity factor because of stronger and more consistent wind. As depicted in Figure 3.11 [23], floating photovoltaic (FPV) systems are also being constructed on water bodies, such as lakes, the ocean, rivers, canals, hydropower dams, irrigation ponds, and waste water treatment plants. Particularly helpful are floating PV systems in land-constrained nations and islands. Reduced water evaporation from freshwater reservoirs is another advantage floating PV systems can provide, particularly to areas where water is scarce. The cooling impact of the water underneath floating PV systems increases their efficiency. According to estimates, 25% of the world's electricity needs might be satisfied by the application of FPV on just 1% of the natural basin surfaces [11].

Another challenge large-scale application of renewables, including solar and wind farms and energy crops, is the energy vs food competition. The issue has been well resolved in the case of PV farms through the concept of agrivoltaics, an approach that shows the potential for combining agricultural and renewable energy. Agrivoltaics is a solution that tackles energy sustainability, agricultural production, and environmental conservation by placing solar PV arrays alongside agricultural land, as shown in Figure 3.12 [24]. Agrivoltaics is positioned to play a significant part in our search for a more sustainable and resilient future as technology develops and lessons learned from successful deployments continue to accumulate.

FIGURE 3.12   Agrivoltaics.

3.4.1 Energy Storage

One of the main disadvantages of renewable energy is its impermanence. For instance, solar radiation is only available during the day. The availability of solar radiation during the day can be hampered by a variety of meteorological conditions, including rain, snow, fog, and gloomy skies. The intensity of solar radiation is also impacted by things like dust storms, smog, haze, and smoke from wildfires. Wind availability is not a constant occurrence either. Additionally, solar radiation and wind speed can fluctuate quickly and dramatically even when they are present, which will have an impact on the output from each system. Therefore, in order to be a dependable source of energy, renewable energy needs backup energy storage. The most recent advancements in battery technology have aided renewable energy in terms of energy storage. Large-scale battery storage has changed dramatically as a result of Tesla's development of 100 MW of lithium batteries in Australia in 2017. Figures 3.13a and 3.13b show a 100 MWh battery storage system supporting a 28 MW PV farm [25].

The United States currently boasts the largest battery storage system in the world, with a 300 MW/1200 MWh system, whereas the United Kingdom has a 150 MW system. The largest battery storage system in the world is expected to be built in California, USA, by Vistra, and will have a capacity of 1,500 MW/6,000 MWh. In December 2020, the 300 MW/1,200 MWh first phase of the lithium-ion battery plant began to supply the California unified grid. Australia is working hard to create big battery storage systems for renewable energy. In order to support the expanding renewable generation, the nation has announced the creation of a 1,200 MW battery bank in the state of New South Wales. By 2023, the project is expected to be finished.

FIGURE 3.13A Utility-scale lithium ion battery energy storage system

FIGURE 3.13B   Utility-scale lithium ion battery energy storage system

Other proposals include plans for a 300-megawatt facility in Victoria, a 500-mega-watt system in New South Wales from France's Neoen SA, and a 700-megawatt system by Origin Energy Ltd. Over 1.1 GW of battery storage capacity is now in use in the UK, and 0.6 GW of additional projects are in the planning stages. Over 16 GW of battery storage capacity is reportedly operational, being built, or being planned across 729 projects in the UK [11].

3.5   RENEWABLE POLICY ADVANCEMENTS

Conducive policy support has been instrumental in the strong growth of renewable energy in recent years. Governments all over the world have launched a variety of programmes and policies in recent decades to encourage the creation and uptake of renewable energy technology. Governments are adopting policies that subtly encour-age the use of renewable energy, such as targets for net-zero greenhouse gas emissions, economic and environmental recovery plans, and regulations addressing climate change. Such measures may have an effect on the economy's supply and demand sides. Government initiatives, financial incentives, and market forces have all been instrumental in driving the cost drop of renewable energy sources. By encouraging investment and innovation in the renewable sector, competitive bidding procedures, feed-in tariffs, and tax incentives have helped to further drive down costs.

Feed-in-tariff (FiT) policies have played a key role in the promotion of renew-able technologies, especially small-scale systems, around the world [26–28]. In FiT policy, a set, premium price is ensured by the financial incentive programme for the power produced from these renewable sources. A FiT policy's main objective is to give renewable energy project developers a steady and predictable cash stream to promote investment and the quick deployment of clean energy projects. The performance of a FiT policy can be affected by elements like tariff rates, regulatory stability, and grid capacity. Some nations have recently switched to alternative procedures, such as competitive auctions, to choose the incentives for renewable energy projects. However, feed-in tariffs continue to be an important tool in the larger policy toolbox for promoting the uptake of renewable energy. With solar panels or other renewable energy sources, net metering policies enable homeowners and business owners to resell any excess electricity to the grid. This supports investment in home and business renewable energy systems and fosters distributed energy generation. Grid access regulations make certain that producers of renewable energy have equitable and non-discriminatory access to the electrical grid, fostering market entrance and competition. Some regions have put in place carbon pricing mechanisms, including carbon taxes or cap-and-trade systems, to internalise the external costs of greenhouse gas emissions. These laws give businesses and utilities a financial incentive to cut emissions and switch to renewable and other low-carbon energy sources.

The level of ambition that nations have to quicken the energy transition is demonstrated by their renewable energy objectives, laws, and committed funds for implementation. Although only 31 nations declared plans for 100% renewable energy, most of them for the year 2050, 128 countries have economy-wide targets for renewable energy by the end of 2022, as shown in Figure 3.14 [4].

In 2022, 13 nations revised their renewable energy targets or made new ones public. New targets were set in seven of these nations (Azerbaijan, Bhutan, Egypt, Jamaica, New Zealand, the Federated States of Micronesia, and Vietnam), while five other nations had their targets revised. Power was the primary goal of sector-specific renewable energy targets, which were made public in 133 nations and 41 subnational authorities.

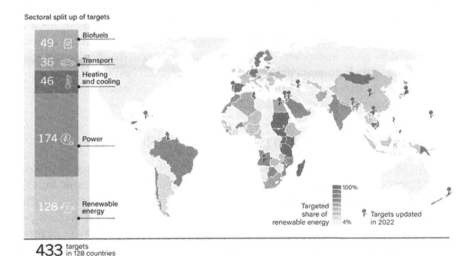

FIGURE 3.14 Account of policies targeting various renewable technologies.

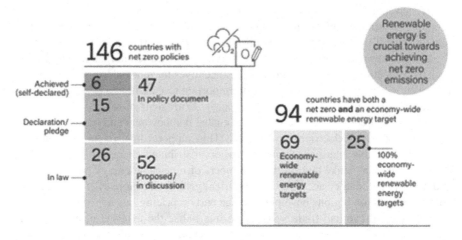

FIGURE 3.15   Status of renewable polices in the world.

Targets for reducing greenhouse gas emissions, commitments to climate neutrality, and policies for net zero emissions are some examples of policies aimed at decarbonisation that also indirectly promote the use of renewable energy. One hundred forty-six nations had declared or implemented a net zero aim as of May 2023, as shown in Figure 3.15 [4]. The net-zero goal was mentioned in legislation in 26 nations, including the EU, and in 47 policy documents, 52 discussions or proposals, and 15 declarations or pledges. In order to achieve new zero emissions, renewable energy is essential, and 94 nations have both a net zero and an economy-wide renewable energy objective. It is also noteworthy that energy efficiency policies are instrumental in driving the energy transition in all sectors (including buildings, heating, and cooling) and complement renewable energy policies. Policies for carbon pricing place a cost on carbon emissions to promote the adoption of low-carbon technologies and consequently lower greenhouse gas emissions. The policies, which can be implemented by governments and/or regional groups, can take the shape of carbon taxes or cap-and-trade programmes. By offering incentives for businesses to cut their carbon footprints, they can assist in generating income for investments in renewable energy and other low-carbon technology.

3.6   MARKET GROWTH AND ECONOMIC TRENDS

In recent years, renewable energy has been making headlines in the energy world. In 2022, renewable energy received over USD 495.4 billion of investment worldwide. Renewable energy sources, particularly solar and wind power, have taken huge strides in terms of technical sophistication, commercial viability, and popular acceptance. In terms of growth rate, installed capacity, and investment, renewable energy has been driving the global energy landscape for the past few years. The global total final energy consumption (TFEC) increased 16% between 2011 and 2021. Between 2011 and 2021, the amount of contemporary renewable energy in TFEC increased from 30 exajoules (EJ) to 50 EJ. The proportion of fossil fuels in TFEC decreased from 81.2% in 2011 to 78.9% in 2021 as the share of renewables increased; however, despite the lower proportion of fossil fuels in TFEC, the overall consumption of fossil fuels increased by 35 EJ over this time. Refer to Figure 3.16. The growth of renewable energy across the power sector is highlighted in Figure 3.17 [4].

Contrary to surging fossil fuel prices, renewables are facing a declining cost trend. The average cost of silicon PV has decreased from around $77/kWh to $0.3/kWh since 1977. The levelised cost of electricity for utility-scale solar PV for newly commissioned

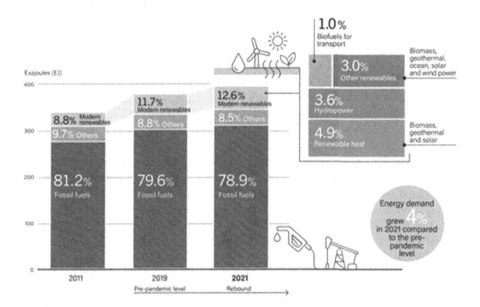

FIGURE 3.16   Trends in total final energy consumption by sources.

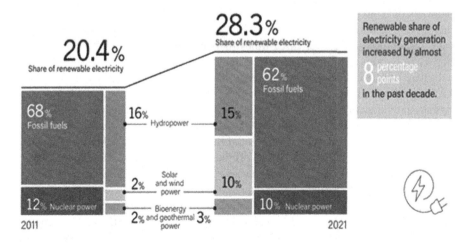

FIGURE 3.17   Growth in share of renewable power generation between 2011 and 2021.

plants has reportedly decreased by 85% over the past ten years, from $0.381/kWh to $0.057/kWh. Particularly in the past two years, PV pricing has gained attention in the energy sector. For example, Dubai obtained a tariff offer of US cents 1.69/kWh for its 900 MW project, while Brazil had a project awarded at US cents 1.69/kWh. Portugal was able to negotiate a solar PV contract with the lowest price in the world in 2019 at US cents 1.65/kWh. The installation cost of PV systems has decreased throughout this time from $4,731/kW to $883/kW. With such a drop in price, PV has moved from being more than twice as expensive as the most expensive fossil fuel-fired power generating choice to being the least expensive alternative for new fossil fuel-fired capacity. Residential PV systems' LCOE sharply decreased during that time as well. Between 2010 and 2020, the LCOE of residential PV systems in Australia, Germany, Italy, Japan, and the United States decreased by between 49% and 82%, falling from between $0.304/kWh and $ 0.460/kWh in 2010 to between $0.055/kWh and $0.236/kWh [29]. Globally, the LCOE of onshore wind projects decreased from $0.089/kWh to $0.039/kWh between 2010 and 2020, a 56% decrease [11].

As a result of historic growth in solar PV and wind power, the addition of renewable power capacity increased 17% in 2021 to reach a new high of more than 314 GW. The total installed capacity of renewable energy increased 11% globally to around 3,146 GW. Solar photovoltaic and wind power technologies are significantly responsible for the success of renewable energy sources. For instance, solar PV contributed most to the increase in total renewable capacity in 2021. The industry for solar photovoltaics continued to set records, with new capacity additions totalling 175 gigawatts in 2021, an increase of 36 GW from the previous year. The total global solar PV capacity increased by the greatest annual rate ever seen, reaching 942 GW. Despite disturbances along the whole solar value chain, mostly brought on by sharp rises in the price of raw materials and transportation, the market kept expanding steadily [7].

Based on installed capacity, solar PV is now the fourth-largest source of energy generation, trailing only coal, gas, and hydropower, and is expected to become the leading source by 2027, as shown in Figure 3.18. [30]. Globally, an estimated 102 gigawatts of wind power capacity were installed in 2021; this figure includes more than 83 GW of onshore capacity and about 19 GW of offshore capacity. Annual offshore installations were over three times their previous peak, with total additions increasing by almost 7% compared to 2020 to the greatest level to date. By year's end, there were 845 GW of wind power installed worldwide (791 GW onshore and the remainder offshore), an increase of 13.5% over 2020. In operation, wind power capacity provided an estimated 7% of the world's electricity production in 2021. As on-shore installations fell in China and the US, global additions onshore decreased compared to 2020; offshore, the substantial increase in capacity added was mostly caused by a dramatic policy-driven rise off the coast of China. In 2021, annual additions were record-breaking in almost every region. Global installations, excluding China, increased by more than 14% from 2020, with at least 55 new wind farms being fully operational, up from 49 in 2020, with Saudi Arabia establishing its first commercial wind farm [7]. Figures 3.19 and 3.20, respectively, display the yearly increase and total installed capacity of solar PV and wind power during the past 10 years [7].

The building sector is experiencing a rapid deployment of solar PV [31, 32]. All sectors of the construction industry, including residential, commercial, and industrial

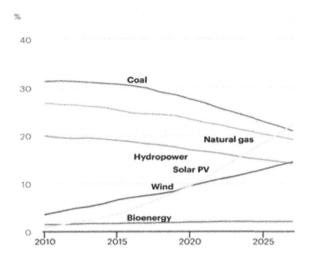

FIGURE 3.18 Projection of solar PV growth.

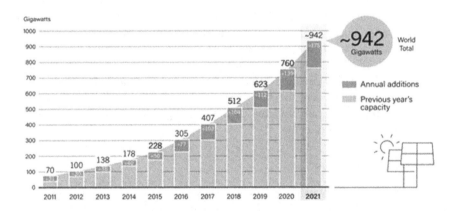

FIGURE 3.19 Solar PV growth trend, 2011–2021.

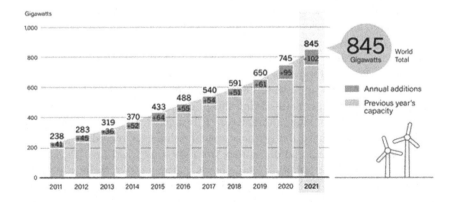

FIGURE 3.20 Wind power growth trend, 2011–2021.

structures, are embracing PV systems [33–35]. PV panels can be put on facades and roofs in the construction industry. However, rooftop PV applications are more common since they provide higher solar radiation while also making use of available space. Rooftop PV system installation in India has reached about 7 GW [36]. The installed rooftop solar PV capacity in the European Union has surpassed 80 GW. Rooftop PV could theoretically generate 700 TWh of electricity, which would satisfy 25% of the world's current electricity needs [37]. The National Renewable Energy Laboratory (NREL) estimates that rooftop PV systems have a potential installed capacity of approximately 1,100 GW. Rooftop PV is also thought to be able to provide close to 39% of the nation's overall electricity needs [38]. According to recent research assessing the global potential for solar PV, rooftop solar PV can supply 25% to 49% of the world's total electricity requirements. Currently, 40% of the PV installed capacity worldwide is used for rooftop applications. Nearly a quarter of all renewable energy capacity additions in 2018 were made by rooftop PV. The cost to generate 27 PWh annually is anticipated to be between US\$40 and US\$280/MWh, with Asia, North America, and Europe having the highest potential. With total capacity estimates of 1,170 GW for hydropower and 145 GW for bioenergy, respectively, hydropower capacity climbed by 20 GW, up from 12 GW added in 2019, and bioenergy capacity increased by 8 GW. The share of renewables in the growth of overall power production capacity reached 82% in 2020, up from 72% in 2019, as a result of this growth in new capacity additions not being observed for fossil fuels or nuclear. Over 50% of all new capacity increases since 2015 have come from renewable sources [39]. The International Energy Agency projects that by 2022, renewable energy sources will account for 30% of global power generation, up from 24% in 2016 and doubling their rate of increase from that of gas. These trends are expected to continue in the years to come.

3.7 CONCLUSIONS

The urgent need for a switch to renewable energy sources is highlighted by energy sustainability concerns such as environmental impact, resource depletion, energy security, economic stability, air quality, and climate change mitigation. While tackling these crucial global issues, renewable energy offers a sustainable and responsible approach to meet our energy needs. It offers a way towards a future where energy is cleaner, safer, and more environmentally friendly.

Fossil fuel prices in 2022 experienced unprecedented hikes in the backdrop of the Russia–Ukraine war, since there was increased volatility, especially as the amount of Russian fossil gas that was sent to Europe was decreased sharply. Various regulatory measures were put in place to protect consumers from fuel poverty and rising prices, maintaining the price performance of fossil fuels in comparison to low-emission alternatives. Consequently, global subsidies for the use of fossil fuels have the greatest annual value ever recorded, surpassing USD 1 trillion. To protect end consumers from growing fossil fuel prices, many nations have put in place temporary safeguards, particularly in the transportation and energy industries [REN21, Renewables 2023 Global Status Report, Global Overview, IRENA].

The benefits of renewable energy over traditional energy sources are convincing and obvious. Renewable energy offers a route to a cleaner, more sustainable future, with benefits for the environment like decreased greenhouse gas emissions and improved air quality as well as economic advantages like job development and energy independence. Furthermore, it is an essential part of an all-encompassing plan to deal with the world's energy and environmental concerns due to its long-term sustainability, resilience, and significant contribution to reducing climate change. The world makes major strides towards a more sustainable, secure, and fair energy future for everybody as it embraces renewable energy sources.

The cost of renewable energy is on a steady falling track, making it more and more competitive with traditional energy sources. Technological breakthroughs, economies of scale, and supportive regulations are the main drivers of this cost trend. As renewable energy sources continue to drop in price and become more widely available, they present a strong economic argument for moving away from fossil fuel dependence and towards a cleaner, more sustainable energy future.

Due to their reliance on energy imports, many countries are vulnerable to changes in the world energy market and geopolitical unrest. This reliance on importing fossil fuels can be reduced by switching to renewable energy. Instead, nations can make use of their own renewable energy sources to improve energy security and lessen the effects of interruptions to the external energy market.

The development of renewable energy policies has been essential in accelerating the switch to cleaner and more sustainable energy sources. These regulations cover a wide range of enforcement techniques, including mandates, rewards, market reforms, and research funding. A commitment to expediting the transition away from fossil fuels and towards a more sustainable and climate-friendly energy future is reflected in the current developments in renewable energy policies.

Innovative solutions that tackle several problems at once are gaining steam as the globe struggles to solve urgent crises like climate change and food security. Such a solution is exemplified by a novel technique called agrivoltaics, which combines the use of agricultural land with solar photovoltaic installations. In addition to promoting the generation of renewable energy, this synergy also improves environmental preservation and agricultural output.

REFERENCES

1. IEA, *World Energy Outlook 2022*, International Energy Agency, 2022
2. M. Asif, *Energy and Environmental Security in Developing Countries*, Springer, 2021, ISBN: 978-3-030-63653-1
3. H. Qudratullah and M. Asif, *Dynamics of Energy, Environment and Economy: A Sustainability Perspective*, Springer, 2020, ISBN: 978-3-030-43578-3
4. REN21, *Renewables 2023 Global Status Report, Global Overview*, International Renewable Energy Agency, 2023
5. M. Asif, *Handbook of Energy and Environmental Security*, Elsevier, 2022, ISBN: 978-0-128-24084-7
6. M. Asif, *Energy and Environmental Outlook for South Asia*, CRC Press, 2021, ISBN: 978-0-367-67343-7

7. REN21, *Renewables 2022 Global Status Report, Global Overview*, International Renewable Energy Agency, 2022

8. A. Pandey and M. Asif, Assessment of energy and environmental sustainability in South Asia in the perspective of the Sustainable Development Goals, *Renewable and Sustainable Energy Reviews*, Volume 165, September 2022, Page 112492

9. M. Asif, *Handbook of Energy Transitions*, CRC Press, 2022, ISBN: 978-0-367-68859-2

10. T. Nadeem, M. Siddiqui, M. Khalid and M. Asif, Distributed energy systems: A review of classification, technologies, applications, and policies, *Energy Strategy Reviews*, Volume 48, 2023, Page 101096, ISSN 2211-467X, https://doi.org/10.1016/j.esr.2023.101096

11. M. Asif, *The 4Ds of Energy Transition: Decarbonization, Decreasing Use, Decentralization, and Digitalization*, Wiley, 2022, ISBN: 978-3-527-34882-4

12. W. Slocum, Photo: 66354, National Renewable Energy Laboratory, USA

13. NREL, *Solar Thermal Power*, National Renewable Energy Laboratory, www.nrel.gov/csp/

14. D. Schroeder a, Photo: 47331, National Renewable Energy Laboratory, USA

15. T. Harder, *Hydropower-What Are the Key Conclusions*, https://mysolarperks.com/benefits-drawbacks-hydropower/

16. GM, *Ground Source Heat Pumps*, Green Match, www.greenmatch.co.uk/heat-pump/ground-source-heat-pumps-in-the-uk

17. PS, *Ground Source Heat Pumps*, Plastic Surge, www.swplasticsurg.com/ground-source-heat-pumps-advantages/

18. IRENA, *Geothermal Energy*, International Renewable Energy Agency, www.irena.org/geothermal

19. EIA, *Hydropower Explained: Wave Power*, Energy Information Administration, Wave Power—U.S. Energy Information Administration (EIA)

20. EIA, *Hydropower Explained, Tidal Power*, Energy Information Administration, Tidal Power—U.S. Energy Information Administration (EIA)

21. GE, *Driving Efficiency and Decreasing the Cost of Offshore Wind Energy*, General Electric, www.ge.com/renewableenergy/wind-energy/offshore-wind/haliade-x-offshore-turbine

22. D. Schroeder b, Photo: 58016, National Renewable Energy Laboratory, USA

23. S. Rai-Roche, Floating solar on the rise globally, significant synergies with pumped hydro storage set to emerge, *PV-Tech*, 31 March 2022, www.pv-tech.org/floating-solar-on-the-rise-globally-significant-synergies-with-pumped-hydro-storage-set-to-emerge/

24. W. Slocum, Photo: 65582, NREL

25. D. Schroeder c, Photo: 56318, National Renewable Energy Laboratory, USA

26. M. Asif, H. Swalha, M Hassanain and K Nahiduzzaman, Techno-economic assessment of application of solar PV in building sector-A case study from Saudi Arabia, *Smart and Sustainable Built Environment*, Volume 8, Issue 1, 2019, Pages 34–52

27. A. Dehwah and M. Asif, Assessment of net energy contribution to buildings by rooftop PV systems in hot-humid climates, *Renewable Energy*, Volume 131, February 2019, Pages 1288–1299, https://doi.org/10.1016/j.renene.2018.08.031

28. A. Dehwah, M. Asif and M. Tauhidurrahman, Prospects of PV application in unregulated building rooftops in developing countries: A perspective from Saudi Arabia, *Energy and Buildings*, Volume 171, 2018, Pages 76–87

29. IRENA, *Renewable Power Generation Costs 2020*, IRENA, 2020

30. IEA, *Solar PV*, International Energy Agency, www.iea.org/energy-system/renewables/solar-pv

31. B. Ghaleb and M. Asif, Assessment of solar PV potential in commercial buildings, *Renewable Energy*, 2022, https://doi.org/10.1016/j.renene.2022.01.013

32. B. Ghaleb and M. Asif, Application of solar PV in commercial buildings: Utilizability of rooftops, *Energy and Buildings*, Volume 257, 15 February 2022, Page 111774, https://doi.org/10.1016/j.enbuild.2021.111774

33. M. Asif, Urban scale application of solar PV to improve sustainability in the building and the energy sectors of KSA, *Sustainability*, Volume 8, 2016, Page 1127, https://doi.org/10.3390/su8111127

34. A. Alazazmeh, A. Ahmed, M. Siddiqui and M. Asif, Real-time data-based performance analysis of a large-scale building applied PV system, *Energy Reports*, Volume 8, November 2022, Pages 15408–15420

35. W. Ahmed, A. Alazazmeh and M. Asif, Energy and water saving potential in commercial buildings: A retrofit case study, *Sustainability*, Volume 15, Issue 1, 2023, Page 518, https://doi.org/10.3390/su15010518

36. GUPTA U, *PV Magazine –Photovoltaics Markets and Technology*. India Closing in on 7 GW of Rooftop Solar, 2021, www.pv-magazine-australia.com/2021/04/13/india-closing-in-on-7-gw-of-rooftop-solar/

37. Solar Power Europe. *EU Market Outlook for Solar Power 2019–2023*, Solar Power Europe, Brussels, Belgium, 2019, www.solarpowereurope.org/wp-content/uploads/2019/12/SolarPower-Europe_EU-Market-Outlook-for-Solar-Power-2019-2023_.pdf?cf_id=7181

38. P. Gagnon, R. Margolis, J. Melius, C. Philips and R. Elmore, Rooftop solar photovoltaic techncial potwential in the United States: A detailed assessment, *Techncial Report: NREL/TP-6A20-65298*, NREL, 2016

39. S. Joshi, S. Mittal, P. Holloway, P. Shukla, B. O'Gallachoir and J. Glynn, High resolution global spatiotemporal assessment of rooftop solar photovoltaics potential for renewable electricity generation, *Nature Communication*, Volume 12, 2021, Article number 5738

4 Energy Conservation and Management

Key to Energy Sustainability

Rao Muhammad Mahtab Mahboob,
Kiran Mustafa, Sara Musaddiq, Nadeem
Iqbal, Rao Muhammad Shahbaz
Mahboob, Mueed Ahmed Mirza

4.1 INTRODUCTION

Our common future defines sustainable development as meeting the needs of the present without compromising the ability of future generations to meet their own needs (World Commission on Environment and Development, 1987). One of the most significant problems in protecting global sustainability is reducing the ever-increasing energy demand. A review of the literature suggested that changing or improving user behavior could be the solution to reduce energy demand. The following section elaborates on the threats to energy sustainability and how energy conservation can aid in achieving a sustainable energy future.

The four primary types of energy risks are economic, security, environment, and social. The first category of global energy challenges, energy-economic, is inextricably connected to the sharp rise in energy trade prices. High energy prices result in greater manufacturing costs, which have a huge impact on global and local socio-economics. Malaysia, although one of the world's top oil exporters, has not been immune to the world's continually rising oil costs. When the world oil price reached a new historic high of US$147.29 (RM470.87) per barrel in 2008, the Malaysian government changed the petrol price. The energy-security danger is the second threat. It's about the unpredictability of future energy supply, a scenario in which more energy resources are extracted than discovered. Fossil fuel reserves will be depleted in the next 40 to 200 years, with oil reserves lasting just 40 years, natural gas 70 years, and coal 200 years. The energy-environment nexus, which includes climate change and environmental degradation, is the third. The cause is the combustion of fossil fuels. The current energy-related emissions trajectory is expected to raise the average world temperature by $6°$ C in the long term, with global energy-related CO_2 emissions peaking in 2025.

Another energy-diminishing hazard is social-energy difficulties, such as the gradually increasing global population and food price stress. According to the International Energy Outlook 2009, the world population is expected to expand at a

DOI: 10.1201/9781032715438-4

rate of 3.9 to 6% per five years between 2005 and 2030. It is necessary to feed, clothe, and house people. As the world's population grows, so does the need for more food and energy. The clash of multiple energy challenges involving economics, security, the environment, and social issues has put the world's energy sustainability in jeopardy. The world must utilize less energy to achieve a sustainable society (Gyberg and Palm, 2009).

As a result, sufficient measures must be made to ensure that threats are resolved as soon as possible before they have a negative influence on the ecosystem. Only then will we be able to weather the storm of an uncertain energy future. In light of the multiple threats to energy sustainability, a quick solution to minimize the constantly rising energy demand is required. One of the urgent remedies to these overarching energy risks is energy management. Perhaps it can lessen the strain on our environment and give us time to create other energy resources, even though it is not the ultimate solution. "Energy conservation is the need of the hour" (Nandi and Basu, 2008).

Energy conservation is defined by the Dictionary of Energy as "a collective term for operations that minimize end-use energy demand by minimizing the service demanded". In general, energy conservation entails using less energy in order to reduce overall energy consumption. Due to the knowledge that burning fossil fuels is one of the main contributors to global warming, energy conservation has recently become a major concern and a major problem (Ting et al., 2011). One of the most important, indigenous, and sustainable energy choices is energy conservation. The rationale is that reducing energy use has always needed no cost or a modest cost. Energy conservation activities provide a diverse set of advantages. The greatest benefit is financial savings from reduced energy costs. They also have nonmonetary benefits, such as promoting environmental preservation, creating a positive company image, and fulfilling social obligations (Fenerty-McKibbon and Khare, 2005).

This chapter will provide a complete overview of the current predominance of energy consumption in specific fields such as construction, agriculture, and industry, as well as alternative energy conservation options for those fields. Different research initiatives in various parts of the world will be used to address energy conservation and management.

4.2 ENERGY CONSERVATION IN THE CONSTRUCTION SECTOR

Historically, most building designs were environmentally mindful and had distinct regional characteristics. This is because their designs were based on the idea of creating harmony with the environment and using locally accessible resources. Building designs have become independent of climatic and material limits since the introduction of contemporary materials and construction processes, as well as the availability of inexpensive energy. The so-called modern international style of today's architecture has completely neglected the effects of climate change on buildings. Buildings appear the same in cold and temperate climates, as well as warm humid and hot dry climates. These structures are designed to be completely reliant on mechanical and electrical systems to function, that is, to provide light and air and maintain comfortable conditions. As a result, our buildings are consuming too much energy. We could afford it when energy was inexpensive, but not any longer. Every country has been pushed to find solutions to its energy and related problems since the oil crisis a decade ago. In recent years, there has been a strong global awareness of the need to reduce energy waste in all industries, including buildings. Building energy conservation is becoming a necessity for all countries, regardless of their geographical location or climatic zone.

This new circumstance has prompted architects and designers to reconsider building designs and architectural styles in light of climate change. This trend is more visible in industrialized countries such as the United States, Europe, and Japan. The majority of these countries have moderate or frigid climates. Though the importance of developing energy-efficient building designs is frequently emphasized in warm areas, rigorous studies on design factors and their impact on building energy use are limited. The prospects for energy conservation in warm climates through building design and passive cooling systems are discussed in this chapter.

Buildings use energy for a variety of purposes, including heating and cooling, lighting, vertical transportation (in skyscrapers and many other tall buildings), and power for appliances and other services. The rate and pattern of energy consumption in buildings vary depending on the climate and building type. In a given climate, the energy consumption of any given building type varies depending on its architectural design, such as size, platform, orientation, space planning and envelope design, air conditioning and lighting system selection and its design, operational and maintenance timeframes, consumer behaviors, and so on (Iwaro and Mwasha, 2010).

Some type of space heating and hot water supply is a frequent characteristic for all buildings in countries with cold and temperate climates that are also developed and have higher living standards. Residential buildings account for a significant portion of the building sector's total energy consumption, which accounts for 35 to 40% of the country's primary energy consumption (Sarkar, 2011). Residential structures in warm regions, which are primarily developing, use far less energy than their counterparts in cold and temperate climates because the bulk of them are not air-conditioned. However, huge commercial and office buildings in warm-climate cities are air-conditioned and consume a lot of energy. Therefore, high-rise buildings in warm regions should be given special consideration.

The comparison of climatic data and thermal comfort requirements serves as the foundation for energy-efficient building designs. The building design includes making decisions on several aspects, such as building form, orientation, building materials for walls and roof, window-to-wall ratios (glass area), external shading devices, and so on, that are appropriate for the location's climate (Rao, 1984).

Given the relevance and repercussions of energy consumption, it is the responsibility of civil engineers to reduce the energy consumption of buildings through good planning, the use of energy-efficient materials and processes, and construction methods. In this context, passive architecture plays a significant role in reducing energy consumption for space heating and cooling. We plan the components of the building, such as walls, windows, and roofs, in such a way that the temperature inside the building is reduced during the summer and increased during the winter compared to the outside, and we use natural lighting instead of artificial electric lights and natural ventilation instead of air conditioning. Sometimes a hybrid system is used as a backup (Hofmann, 1983).

For this goal, adequate orientation, ventilation, and shading, as well as proper materials for the building's body, a wind tower, thermal mass, and other novel methods, should be used. By conducting research, new low-cost materials and technologies will be developed. It's important to remember that one unit of energy saved equals one unit of energy created. The impacts of global warming and climate change have recently been seen in the Chennai floods in Tamil Nadu, India. In a single day, Chennai had enough rain to last a month, making life miserable for everyone. If we look for the root cause, we will find ourselves. So we must conserve the environment and pass it on to future generations in a secure state, remembering that we are merely trustees of the cosmos, not owners (Molykutty, 2015).

4.2.1 SIMULATION OF ENERGY EFFECTIVENESS

The performance of a structure can now be studied by modeling the energy transfer between it and its surroundings. It allows you to evaluate a building's design effectiveness and help the development of better designs for energy-efficient structures with comfortable indoor environments. It's made up of mathematical and thermodynamic algorithms that use the simulation engine's model to calculate energy performance.

A variety of computer modeling techniques are now available to test a building's thermal and daylighting performance quickly and accurately. These tools assess the performance of various building designs for a particular environmental situation, allowing a designer to select the one that uses the least amount of energy. Thermal calculations aid not only new designs but also the selection of energy-efficient retrofits for existing structures. Thus, an energy-efficient building can be achieved by combining the simulation of a structure's thermal performance with its architectural design.

One of the tools is DOE-2.1E, which anticipates a building's hourly energy use and cost. EnergyPlus is a structured, modular software tool for predicting temperature and comfort. Another tool for energy simulation modeling is eQUEST. Another commercially accessible simulation tool is TRNSYS. Simulators include Ecotect and Revit architecture. A number of simulation tools are accessible (www.building-energysoftwaretools.com) in alphabetical order. The quality of the input determines the outcome of the simulation. As a result, providing reliable data for simulation is critical. Simulation tools, it is believed, will play a significant role in future energy conservation in buildings and, as a result, pollution reduction (Ayompe et al., 2011; Ke et al., 2013; Thuesen et al., 2010).

4.3 ENERGY CONSERVATION IN THE AGRICULTURAL SECTOR

Major agricultural development and conversion of natural ecosystems to agro-ecosystems have resulted from a greater demand for food security. Increased use of external inputs and conversion of marginal land to cropland may jeopardize agricultural ecosystem services, including natural resource conservation, soil health, and biodiversity. Increased food production at the expense of ecosystem services (ESs) can jeopardize the long-term viability of agro-ecosystems, including crop output. There is a need to boost ecosystem services by implementing resource conservation measures on farms in order to increase food and nutritional security. One of the most pressing concerns of the 21st century is how to strike a balance between the need to feed a growing population while still preserving healthy ecosystems and thriving habitats. Conservation agriculture (CA) is a resource-conserving agricultural crop production strategy focused on the enhancement of natural and biological processes above and below ground. Concerns about soil erosion, deterioration of soil quality, and chemical hazards have prompted researchers to look back in time to develop conservation agriculture-based practices that aim for higher productivity and profitability over the long term by making rational and sustainable use of available resources (Das et al., 2016).

CA is defined by three interconnected principles, as well as other good agricultural practices: continuous zero or minimal mechanical soil disturbance (implemented by no-till seeding or broadcasting of crop seeds and direct placing of planting material into untilled soil, causing minimum soil disturbance from any cultural operation, harvest operation, or farm traffic); maintenance of a permanent biomass soil mulch cover on the ground surficial layer; and crop species diversification, implemented through the use of a cropping system that includes crop rotations, sequences, and/or associations incorporating annuals and perennial crops, as well as a balanced mix of legume and non-legume crops (Wong, 2011).

Another principle that has been discussed in recent years is controlled traffic, which reduces soil compaction. CA is a potential method for maximizing the utilization of available resources and ensuring long-term production. Conservation agriculture-based management strategies are realistic solutions for long-term agricultural sustainability and useful instruments for halting land deterioration. CA was used on roughly 180 Mha of cropland globally in 2015–16, accounting for around 12.5% of total global cropland, a significant increase of 74 Mha from 2008–09 (Bhattacharyya et al., 2013; Das et al., 2013). Several practices are employed for energy conservation in agriculture.

4.3.1 WATER USE AND EFFICIENCY

Agroecosystems require a sufficient supply of clean water, and agriculture accounts for over 70% of global water consumption. In agroecosystems, water availability is determined not only by infiltration and flow but also by soil moisture storage, which is another sort of ecosystem service. Rainfall provides around 80% of agricultural water, which is held as soil moisture (green water). In various places, irrigation is required to provide surface water and groundwater (blue water) inputs to crops. Increased rainfall unpredictability is expected to increase the danger of drought and flood as a result of climate change, while higher temperatures will increase water demand. Water scarcity projections can be drastically altered by on-farm management strategies that target green water (Power, 2010).

CA boosts water productivity by boosting infiltration, lowering soil evaporation, and increasing soil water storage capacity for stomatal transpiration. In a study, it was found that the plots under no-tillage (NT) had a higher ultimate infiltration rate due to residue retention on the surface, less disruption to the continuity of water-conducting pores, and increased aggregate stability. In a study, it was found that in a CA-based maize-wheat cropping system, a permanent broad bed with residue and permanent narrow bed with residue resulted in higher water-use efficiency and more carbon accumulation in soil with higher sequestration potential, in addition to providing long-term production. In a study, it was found that after 12 years, the steady-state infiltration rate with no tillage was 60% higher than for conventional tillage in northwestern Canada. Groundwater table depletion has been steadily increasing in the northwestern Indo-Gangetic Plains (IGPs) since the early 1970s, and it has accelerated dangerously in recent years (He et al., 2009; Humphreys et al., 2010).

4.3.2 Crop/Food Production

The most significant feature of CA is conservation tillage, which is intended to protect soil health, plant growth, and the environment. Conservation tillage improves crop root growth, water and nutrient use efficiency, and, ultimately, agronomic production. It has been discovered that no-till wheat is more resistant to drought and high temperatures than traditional wheat. CA-based management strategies such as dry direct-seeded rice (DSR), zero tillage, and residue retention have the potential to boost yields, cut costs, and increase farmers' profitability in rice-maize systems (RMSs). When comparing zero-tilled DSR followed by zero-tilled maize with residue retention to traditional RMSs in different soil layers to 60 cm depth, they found that root mass density was 6 to 49% higher in rice and 21 to 53% higher in maize. Many other studies have found that the zero-tillage wheat seeding method can save inputs, reduce turnaround time, reduce energy consumption and pollutants, improve productivity, and increase farmers' revenue (Kumar et al., 2013; Singh et al., 2016).

4.3.3 Carbon Sequestration

Carbon sequestration involves converting atmospheric CO_2 into long-lived pools and securely storing it so that it is not re-emitted right away. When compared to typical tillage-based crop production systems, one of the most favorable elements of conservation agriculture is its potential to raise soil carbon. In general, soil organic carbon (SOC) content rises as the amount of residue returned to soil grows, and returning residues to soil has transformed many soils from sources to sinks of atmospheric CO_2 by increasing soil productivity.

In conventional agriculture, intensive tillage operations enhance the rate of oxidation of organic matter, resulting in CO_2 emission into the atmosphere and the greenhouse effect. One of the benefits of no-tillage has been high carbon sequestration. According to several studies, no-tillage can achieve carbon sequestration of 367–3667 kg CO_2/ha/year. Increased soil C sequestration enhances soil aggregation, which increases accessible soil moisture storage capacity, which can help plants grow and develop more effectively. Several research works have found that implementing no-tillage systems improves soil nutrient recycling, particularly in terms of increased organic C near the soil surface. In the case of a conservation tillage system, a higher organic C content increases aggregate stability as well as soil health (Ghosh et al., 2019).

4.3.4 ENERGY USE AND EFFICIENCY

The implementation of conservation tillage can help reduce energy consumption by lowering the number of tillage operations. Conservation tillage (CT) can offset up to 16% of global fossil fuel emissions and can limit or avoid the loss of organic carbon in the soil. Farmers could save 36 liters of diesel per hectare on average by using zero tillage (Amundson et al., 2015) for land preparation and crop establishment in the IGP rice-wheat system. A group of researchers have investigated energy auditing in a CA-based maize wheat-mungbean system and discovered that ZT bed planting with wheat and maize residue retention could be a substitute for the conventional agricultural system for adoption in maize wheat-green gram cropping systems. In comparison to CT, ZT increased operating field capacity by 81%, specific energy by 17%, and energy usage efficiency by 13%. According to another study, ZT and PB plots used less energy (7 years average) in land preparation (49.7–51.5%) and irrigation (16.8–22.9%) than CT plots. They also suggested that CA methods with a varied maize-based rotation (maize-wheat-mungbean) could be a viable option for achieving high energy efficiency, biomass yield, and bio-energetics (Kour et al., 2011; Kumar et al., 2013; Saad et al., 2016).

4.3.5 CLIMATE REGULATION

Agricultural activities and land utilization changes account for nearly a third of overall GHG emissions and are the leading source of N_2O emissions. Agricultural GHG emissions can be lowered by reducing fossil fuel usage in agricultural activities, enhancing soil carbon absorption, and reducing N_2O emissions from soil (Mosier et al., 2005). Some possible solutions include switching from intensive tillage to zero or minimum tillage, with at least 30% crop residue remaining after harvest. Intensive soil tillage speeds up the oxidation of organic matter and transforms agricultural waste into CO_2, which is released into the atmosphere and contributes to the greenhouse effect and global warming. Increased carbon sequestration through residue retention could be an important strategy for addressing climate change. Conservation tillage strategies reduce unmineralized organic material's exposure to microbial activities, lowering SOM degradation and CO_2 emissions. Flooded rice production is responsible for almost 15% of worldwide CH_4 emissions. Anoxic conditions develop in rice fields as a result of frequent flooding, resulting in the emission of CH_4 gas, which is known as methanogenesis. According to (Yan et al., 2003), rice fields with alternate wetting and drying can efficiently minimize CH_4 emissions. However, this can raise N_2O emissions at the same time, which may serve as a partial offset for its broader advice. The potential for direct-seeded rice to minimize CH_4 emissions is significant. DSR is a labor-saving, fuel-saving, time-saving, and water-saving alternative to puddled transplanting.

4.3.6 Nutrient Accumulation and Cycling

The efficient use of nutrients is critical to boosting agricultural production. It has been established that the chemical qualities of the surface layer of no-till soils are generally better than those of tilled soils. Conservation rotations (e.g., cover crops, leguminous crops, deep-rooted crops) can improve system nitrogenous efficiency, following crop yields and recovering nitrate from the lower soil profile that had been leached from previous shallow-rooted crops.

After 37–40 years of tillage treatments, minimum tillage improved aggregate stability and raised the concentrations of soil organic carbon and nitrogen within the aggregates in the upper 5–8 cm soil depth. A CA-based maize-wheat system in India's IGP improved SOC and N status. In a study done in Poland, higher levels of soil organic carbon, total N, accessible K, and Mg were found in the 0–5 cm soil layer under reduced tillage and no-tillage circumstances than under CT after a total of 7 years of tillage. In another study, it was discovered that higher organic matter (organic C and total N) and exchangeable K under NT than CT and RT. When stubble is burned or removed, no tillage may cause an increase in NO_3-N levels in the soil. As a result, it was suggested that stubble retention be used in conjunction with NT to limit the risk of NO_3-N leaching into the soil (Delgado et al., 2007; Malecka et al., 2012; Thomas et al., 2007).

CA-based systems are essential for long-term agricultural production. These systems provide a wide range of supplying, regulating, and supporting ecosystem services that are critical for increasing the efficiency of natural resources (soil, water, air, and fuel) and meeting the UNDP Sustainable Development Goals for environmental and food security goals. They have the potential to influence a variety of ecosystem services in a variety of contexts, as well as promoting agricultural sustainability by increasing food production, improving soil health through carbon sequestration, reducing GHG emissions, and preserving biodiversity. Studies should concentrate on CA-based agricultural techniques that are specific to the area, cropping system, and cropping season, as well as how they affect ecosystem services. To better measure CA, there should be explicit comparisons of ecosystem services supplied by conservation and conventional agriculture across a wide variety of soil and climatic conditions. This will aid CA's widespread adoption and long-term sustainability of natural resources and productivity (Ghosh et al., 2019).

4.4 ENERGY CONSERVATION IN THE INDUSTRIAL SECTOR

In the future, technological innovation and socio-societal development will be the criterion for evaluating long-term growth. So far, the progress of sophisticated electronic components, transportation systems, communications, and energy technologies has taken precedence over the actualization of a stable techno-economic framework of operation in all things and in the years ahead. Appropriate energy management, according to research, will result in enhanced social, economic, and environmental performance over time (Abdelaziz et al., 2011). To meet the fast-changing trends of the future predicted for efficient and effective output in all aspects of the industrial processes, medium and large industrial production and service delivery will transition to the use of robotic mechanisms and autonomous systems for manufacturing operations. Furthermore, the information technology sector is undergoing a rapid transition, which is equated to bridging the gap between data management difficulties and the execution of complicated technology. As a result, energy production and consumption, which play a crucial role in socioeconomic transformation, have become a harmful management concern in light of energy conservation and climatic deterioration control and sustainable development objectives (Lee and Cheng, 2016).

The importance of the foundation for energy saving cannot be overstated. The amount of work put into considering the processes and methods employed is dependent on the human desire for the achievement of contemporary technology and how the goals are reached. Energy management and environmental security must be carefully examined while technological growth occurs. Otherwise, our current inventions may contribute to the horror that threatens our life in the future (anthropogenic climatic situation). As a result, the Sustainable Development Goals (SDGs) were born, attracting intergovernmental partnerships and anticipating support for the 17 SDG agendas. Some researchers have already embraced the simple definition of sustainable development as "development that meets the current generation's requirements without jeopardizing future generations' ability to meet their own needs" (Baumgartner and Rauter, 2017).

From a philosophical standpoint, there is a clear insight into the current technicalities and methods that enable energy consumption proliferation in manufacturing as it affects economic, social, and ecological growth. Concerning energy utilization throughout industrial production processes, the authors propose changes in strategy about the nature and types of services and goods to conserve energy.

4.4.1 SUSTAINABILITY FROM THE PERSPECTIVE OF MANUFACTURING

Sustainable manufacturing is defined by the United States Department of Commerce as "the creation of manufactured products that have minimal negative environmental impacts, conserve energy and natural resources, are safe for employees, communities, and consumers, and are economically sound". Manufacturing has become a major source of worry in terms of raw material conversion and product creation, owing to its large share of energy consumption and socio-environmental implications. The concept of manufacturing sustainability currently entails determining the best procedures for efficiently utilizing resources through competitive conventional and developing approaches while ensuring that the operations are not damaging to the environment. To achieve sustainability, various aspects of manufacturing have been examined to address energy usage of products, processes, and system performance. Researchers looked into the prospect of combining sustainability concepts with manufacturing, specifically in terms of design options. Various studies on energy management at various stages of manufacturing operations have yielded ground-breaking conclusions on the most effective routine and strategy for lowering energy usage. Baumgartner and Rauter (2017) identify viable challenges involving sustainable development and technological transfer, capturing developmental capabilities and relational acceptance of technological systems, which influence decision makers' willingness to embrace low-emission initiatives and technologies. Researchers need to get a deeper understanding of the major difficulty of assessing the progress of sustainability and the criterion for measuring it (Baumgartner and Rauter, 2017; Bond and Morrison-Saunders, 2011; Deiab, 2014; Menzel et al., 2010; United, 2016).

Other studies have used a holistic approach to urban settlement to foster sustainable development as a means of encouraging technology and establishing the road for manufacturing innovation to thrive. The big picture is that climate change mitigation will require a collective and collaborative response.

4.4.2 STRATEGY FOR INDUSTRIAL SUSTAINABILITY AND ENERGY CONSERVATION

According to Bellard et al. (2012), future energy policies can be deemed effective if they achieve security, equity, and sustainability goals in terms of protecting the planet in the face of rapidly changing climatic conditions. As a result, the pursuit of renewable energy becomes critical.

Various studies suggest that the ability to achieve the goals outlined by the three-dimensional model of sustainable development depends on the capability and capacity of the stakeholders involved, as well as the effectiveness of policies designed to facilitate the accomplishment of these goals.

Within the context of eco-innovation, the newly evolving global requirement that calls for nascent sustainable manufacturing initiatives, considering rising pollution indices and the need to ensure cleaner production initiatives for habitability through the establishment of closed-loop production, can be viewed. To fulfill industrial energy demand, technological advancements in power generation, renewable energy exploitation, and alternative energy technology development are on the rise. The hunt for high-capacity storage systems, sustainable material selection, and durable product designs to consume less energy invariably leads to a reduction in harmful gaseous emissions at the site of power generation. Initiatives in R&D, demonstration, and deployment, as well as eco-innovation collaborations, are critical to ensuring sustainable production. Manufacturing companies "have the potential to be a driving force for the realization of a sustainable society by implementing efficient production techniques and providing products and services that assist reduce negative impacts", according to the report (Jenkins et al., 2016; Machiba, 2011; Song et al., 2017).

4.4.3 MANUFACTURING PROCESSES THAT MAXIMIZE ENERGY

4.4.3.1 Thermal Processes

Thermal heat operations in the steel, mining, pulp and paper, and cement (building) sectors, among others, have consumed a substantial amount of energy in recent years, resulting in increased carbon emissions due to the burning involved, and this trend is continuing. There is a pressing need to investigate new dimensions in terms of heat energy conservation and conversion to replace the prevalent and traditional way of burning (woods and fossil fuels as fuels) and to minimize the production cost of using electricity (Jenkins et al., 2016; Song et al., 2017).

4.4.3.2 Waste and Material Processes

High-performance thermo-electric, thermo-galvanic, and thermochromic materials for waste heat re-utilization and energy maximization have sparked interest in research and development, particularly for industrial applications, and continue to produce outstanding results in terms of effective electricity conservation. Other manufacturing aspects, such as final product configuration with the ability to conserve energy, are critical during the design process when using innovative materials. The onus must be on zero-waste creation solutions to eliminate energy and financial losses, particularly in design. The creation of biofuels from ethanol-producing crops will save money on waste disposal all year long, as well as the burden on land space management. As a result, bioengineering and quantum leap breakthroughs in biofuels production are critical to maintaining manufacturing as fledgling energy-intensive bio-based crops gradually replace fossil fuels utilized in land, sea, and air transportation (Peter and Mbohwa, 2019).

4.4.3.3 Energy Systems and Efficient Processes

Process efficiency and energy system performance optimization, for example, in furnaces, boilers, compressor equipment, motors, and driving mechanisms, is expected to increase operational frequency. Apart from system operational losses, the total available energy consumed by a system for energy management considerations must not exceed that required for manufacturing (net energy demand). A more effective and efficient system will run consuming even less energy and reducing the number of emissions produced during operation (Peter and Mbohwa, 2019).

4.4.3.4 Energy Conservation and Carbon Dioxide Reduction through a Resilient Approach

While carbon capture techniques and technological innovations that can be integrated into existing factory layouts are receiving scientific attention, the initiative that is aimed at multibillion-dollar corporations and well-known energy-intensive industries must also be scaled down to small workshops, mini-factories, and power plants. Carbon capture techniques, among other advanced technological options, require chemical looping procedures and retrofitting of high carbon dioxide emission-based plants (biofixation). For industrial energy management plans, when grid electric power systems are evaluated for industrial use, the development of battery storage technologies continues to see steady improvement. Nuclear and hydel power resources are the most promising for future electricity generation to ameliorate climatic degradation, notably in the case of third-generation nuclear power plants, due to modern safety precautions and advancement in operational techniques. The energy systems indicated previously are carbon free and the most efficient for meeting the substantial power demands of enterprises as they grow in size (Bhutto et al., 2012; Langlois, 2013).

4.4.3.5 Advanced Technological Adoption and Information Technology Integration

One question that looms over today's world of innovation and technological transformation is: how important is new? It's difficult to choose between current hazards connected with energy-intensive systems, cost, and societal acceptability and being accountable for the uncertainty that may arise from the negative repercussions of future technological advancement. Progress in additive manufacturing has demonstrated the ability to handle a wide range of material resources and the capacity to improve product development in a timely manner, making it a critical technological option for transforming the manufacturing industry through conservative designs and appropriate material utilization while consuming the least amount of energy. As a result, it is a cost-effective procedure for the environment and a long-term energy source for future industries (Peter and Mbohwa, 2019).

4.5 ENERGY CONSERVATION IN TRANSPORTATION

Logistics and transportation are one of the most important cornerstones of global economic progress. Over the previous decade, transportation has continued to provide a major contribution to the gross domestic product (GDP). The average yearly rise in transportation-related demand from 1999 to 2016 was roughly 10%, according to the United States Bureau of Transportation Statistics. This demand accounted for 5.6% of GDP in 2016. Similarly, transportation contributes around 5% of Europe's GDP. Both freight transportation and passenger mobility activities have increased around the world. Freight transportation activity in the United States increased by 17.6% between 1995 and 2015, while passenger transportation activities climbed by 18.3% during the same period. The increase in freight and passenger transportation in the EU-28 was roughly 24% apiece. However, this economic growth is accompanied by a number of negative externalities. Transportation is one of the most significant contributors to world energy consumption, accounting for over 29% of total global energy consumption and up to 24% of global CO_2 emissions. Furthermore, according to the UN Department of Economic and Social Affairs, these figures will only worsen by 2100, when the world's population is expected to reach 17.6 billion (Chen et al., 2019; Corlu et al., 2020).

The main source of air pollution in passenger and freight transportation is fuel burning, which has a direct influence on the environment. It also contributes to the development of contaminants that are potentially dangerous to humans. Sustainability is gaining traction, and energy efficiency is attracting a lot of attention from practitioners and researchers as social awareness about the environmental risks of our modern economies grows. Technology, fuel, vehicle operation, demand, demography, and road design all play a role in energy consumption and emissions. Available optimization measures, according to Juan et al. and Fan et al., can be divided into three categories: (i) energy efficiency enhancement, (ii) renewable energy and electrification, and (iii) optimal mode configuration. Operations research (OR) is also making a significant contribution to pollution reduction and sustainable transportation management. Bektas et al. gave a comprehensive evaluation of OR approaches for green freight transportation, with a focus on marine and overland transportation. In the transportation sector, mathematical programming optimization, metaheuristic algorithms, and simulation techniques are among the most common OR methodologies (Bektaş et al., 2019; Corlu et al., 2020; Dekker et al., 2012; Faulin et al., 2019; Juan et al., 2016; Van Fan et al., 2019).

There are two types of optimization methods; one is exact and the other is approximate being utilized to address the previously mentioned difficulties. Exact approaches, such as linear and integer programming, are designed to formulate and solve a problem with an objective function (such as cost minimization or revenue maximization) within a set of restrictions. These methods are capable of locating global best-fit solutions. When a problem is computationally NP-hard, however, they can only address limited instances or require extraordinarily extensive computer durations to find a solution. Metaheuristic algorithms, on the other hand, can deliver near-optimal solutions to difficult transportation issues in a short amount of time (Neumann and Witt, 2010).

Because of the rising complexity and magnitude of real-world transportation networks, these algorithms are becoming more important in the transportation literature. Energy-related transportation challenges in express road, passenger train, maritime, and air transportation are solved using heuristic-based techniques. The focus is on issues that have an explicit energy aspect, such as lowering gasoline costs or pollution levels.

4.5.1 TECHNIQUES BASED ON GREENING THE OBJECTIVE FUNCTION

The energy minimization VRP and the pollution routing problem are two seminal papers that propose mathematical formulations for these two energy-related VRPs. The former was proposed by Kara et al., and the goal is to reduce energy consumption by minimizing a distance-weighted load function. Bektas and Laporte suggested a pollution routing issue that takes into account the driver, fuel usage, and CO_2 emission costs all at once. Once academics and practitioners accepted the prior issues, the scientific literature evolved in that direction, providing several variations and solutions. For the energy minimization problem, Fukasawa et al., and for the pollutant routing problem, Eshtehadi et al. have presented solutions. Two mixed integer linear programming models were proposed by the previous authors: a set partitioning formulation and an arc-load formulation. A branch-and-cut algorithm is used to solve the first, while a branch-cut-and-price method is used to solve the second. Finally, they demonstrated how their proposed techniques outperformed several previously published findings. The energy minimization problem was also extended by the latter authors by including demand and travel time uncertainty in the original problem (Bektaş and Laporte, 2011; Eshtehadi et al., 2017; Fukasawa et al., 2016; Kara et al., 2007).

When it comes to large-scale energy-related road transportation concerns, exact techniques are unusual. This is because the VRP is an NP-hard combinatorial optimization problem. As a result, as the complexity of the problem rises, accurate approaches become increasingly difficult to deliver an ideal solution in a reasonable amount of time (Juan et al., 2014). Recently research has concentrated on constructing metaheuristics for resolving realistic versions of the aforementioned issues.

To solve a VRP with alternative fuel cars, Koç and Karaoglan presented a simulated annealing approach. The focus here is on the vehicles' restricted range as well as the limited availability of recharging facilities. Simulated annealing is also used in a recent study by Hooshmand and MirHassani to solve a very similar problem with electric vehicles. Their heuristic algorithm is split into two parts. The problem is first broken down into clustering and routing steps. The answer found in the first stage is then improved via simulated annealing. For a situation identical to the one proposed by Kara et al., Huang et al. merged the prior approach with a tabu search metaheuristic. To improve the simulated annealing outcomes, the solution technique uses k-means clustering, local search, and tabu list-guided searching. Kirci used a tabu search as well. The problem is similar to the pollution routing problem but with a few additions for modeling CO_2 emissions in a time-dependent vehicle routing context (Bektaş and Laporte, 2011; Hooshmand and MirHassani, 2019; Huang et al., 2017; Kirci, 2019).

4.5.2 Strategies Based on Improving Load Factors

The majority of the research on VRPs with backhauls focuses on heuristic-based solutions. The most common solution approaches are tabu search algorithms or methods that combine them with other heuristics. Küçükoglu and Ztürk devised an innovative hybrid metaheuristic method for solving the VRP with back-hauls and time windows. A tabu search technique is used with a simulated annealing procedure. The resulting strategy is implemented in three steps. In the first stage, a nearest-neighbor heuristic is used to find an initial solution. The tabu search generates diverse neighborhoods in the second step. Finally, simulated annealing selects a solution from these neighborhoods based on an acceptance criterion. Reil et al. assumed three-dimensional loading limitations in a twofold approach and considered time frames. The first step, which focuses on product packing, uses a tabu search heuristic to solve a 3D strip packing problem for each customer. This produces a VRP with back-hauls, which is solved using the tabu search in the second stage. Similarly, Lai et al. provided an integer linear programming formulation and an adaptive guidance metaheuristic for a real-world scenario of an Italian carrier (Chuang et al., 2009; Granada-Echeverri et al., 2019; Küçükoğlu and Öztürk, 2015; Lai et al., 2015; Reil et al., 2018). The adaptive-guidance approach was developed by combining a tabu search with the well-known savings heuristic presented by Clarke and Wright. In the literature, population-based heuristics for dealing with the VRP with back-hauls have also been offered. Paraphantakul et al., for example, proposed an ant colony system for a real-life instance in Thailand. The unique characteristics of that real-world scenario prompted the deployment of this adaptable algorithm. In addition, Küçükoglu and Ztürk created an evolutionary algorithm for a real-world situation. Their proposed methodology was put to the test with a variety of benchmarks, demonstrating its efficacy (Klibi et al., 2010; Küçükoğlu and Öztürk, 2014; Paraphantakul et al., 2012). For the most part, the method outperforms some of the most well-known solutions.

4.5.3 Strategies Based on Horizontal Cooperation

Companies create a coalition for completing their logistics activities in horizontal collaboration. They can gain from mutual cooperation in this way. The companies can get the following benefits:(i) a significant decrease in transportation costs, (ii) a noteworthy improvement in transportation reliability, and (iii) a clear reduction in CO_2 emissions. Problems involving multiple companies are difficult to tackle using accurate procedures alone, similar to the strategy for improving load factors. Nonetheless, the multi-depot VRP ideas are used in some horizontal cooperation systems (i.e., the classical VRP considers more than one depot). Each depot is believed to be owned by a firm in these methods. As a result, a multi-depot VRP setup can be created if the horizontal cooperation agreement successfully establishes the allocation criteria and/or compensation mechanisms. However, multi-depot VRP ideas are used in various horizontal collaboration systems (i.e., the classical VRP considers more than one depot). Each depot is believed to be owned by a business in these approaches. As a result, if the horizontal cooperation agreement establishes the allocation criteria and/or compensation mechanisms successfully, it can be changed to a multi-depot VRP setting. A vehicle-flow and set-partitioning formulation was proposed by Contardo and Martinelli. For a multi-depot VRP, Bektas proposed Benders decomposition, but Fernández et al. employed a traditional branch-and-cut approach. There is a wealth of literature on the use of metaheuristics in horizontal cooperation. Pérez-Bernabeu et al. concentrated on the energy savings that can be realized when horizontal cooperation strategies are implemented. To generate high-quality solutions for the cooperative scenario, they used an iterated local search algorithm. In fact, in a favorable distributed topology, energy consumption can be reduced by up to 92% (Contardo and Martinelli, 2014; Fernández et al., 2018; Pérez-Bernabeu et al., 2015).

Quintero-Araujo et al. presented simheuristics, an extension of metaheuristics for stochastic combinatorial optimization problems. The former examined a collaborative with a non-collaborative setting. The findings demonstrate the advantages of horizontal cooperation in realistic urban settings. The latter put a similar method to the test in a real-world scenario involving the distribution of items in convenience stores (Quintero-Araujo et al., 2017).

4.6 TRENDS AND BEST PRACTICES

Electricity and energy resources are more than just commodities; they are blessings that provide comfort and raise living standards. Appropriately smart management and controls result in large savings as direct benefits and long-term sustainability. Nature's balance and controls always have greater, long-lasting reactions. Humankind has been driven to this point of natural reaction; otherwise our survival will be jeopardized. Others recognize when worldwide standards are formed and entrepreneurs begin to implement those standards, with a special focus on a holistic approach to determine recommendations that would aid in the management of energy crises.

The industrial sector accounts for almost one-quarter of total energy consumption worldwide, and it is concentrated in a small number of industrial facilities in many nations, making it relatively straightforward to identify large energy consumers. As a result, several governments are prioritizing improving energy efficiency in the industrial sector. Industrial energy efficiency improvements can result in significant energy savings, increased production, and reduced pollution. However, in many circumstances, information, financial, and regulatory impediments continue to obstruct businesses from fully realizing the potential benefits of energy efficiency improvements. To help overcome these barriers, a variety of policies and programs can be implemented.

Sharing best practices, particularly different approaches to policymaking and success stories, can help countries improve energy efficiency more quickly, and governments are eager to learn from foreign experiences in these areas. This chapter aims to assist governments in developing industrial energy efficiency policies by sharing international experiences. It examines the prerequisites for successful policy and program implementation and provides real examples. The main purpose is to encourage businesses, governments, and civil society to work together to achieve the SE4All target of doubling global energy efficiency improvement by 2030.

4.7 CONCLUSION

Over the last few decades, the world's electrical and fossil fuel energy consumption has steadily climbed. The adoption of an energy conservation and management system is thought critical for the modern human world's long-term development. Different sustainable fuel alternatives, such as solar energy and biodiesel, are used in energy conversion measures. Thermal insulation and building envelopes are key energy conservation measures in the construction industry. In the same way, holistic frameworks are used in smart homes, and the concept is expanded to include smart cities. Mobile measuring technology has been introduced in communications centers to conserve energy and space. These and other energy-saving measures have already been implemented in a variety of sectors, and energy management and conservation policies have been implemented in several countries over the last decade. However, current measures are insufficient; the urgent need is to develop more energy-efficient systems and favorable policies in a variety of sectors. This chapter provides a complete overview of the current predominance of energy consumption in specific fields such as construction, agriculture, and industry, as well as alternative energy conservation options for those fields. Energy conservation and management are examined through case studies in various parts of the world, as well as prospects, public opinion, policy support, and research initiatives. These and other energy-saving measures have been implemented in a variety of sectors, as well as energy management and conservation policies, in several countries over the last decade. However, current measures are insufficient; it is necessary to develop more energy-efficient systems and favorable policies in a variety of sectors. This chapter provides a complete overview of the current predominance of energy consumption in specific fields such as construction, agriculture, and industry, as well as alternative energy conservation options for those fields.

REFERENCES

ABDELAZIZ, E. A., SAIDUR, R. & MEKHILEF, S. 2011. A review on energy saving strategies in industrial sector. *Renewable and Sustainable Energy Reviews*, 15, 150–168.

AMUNDSON, R., BERHE, A. A., HOPMANS, J. W., OLSON, C., SZTEIN, A. E. & SPARKS, D. L. 2015. Soil and human security in the 21st century. *Science*, 348.

AYOMPE, L. M., DUFFY, A., MCCORMACK, S. J. & CONLON, M. 2011. Validated TRN-SYS model for forced circulation solar water heating systems with flat plate and heat pipe evacuated tube collectors. *Applied Thermal Engineering*, 31, 1536–1542.

BAUMGARTNER, R. J. & RAUTER, R. 2017. Strategic perspectives of corporate sustainability management to develop a sustainable organization. *Journal of Cleaner Production*, 140, 81–92.

BEKTAŞ, T., EHMKE, J. F., PSARAFTIS, H. N. & PUCHINGER, J. 2019. The role of operational research in green freight transportation. *European Journal of Operational Research*, 274, 807–823.

BEKTAŞ, T. & LAPORTE, G. 2011. The pollution-routing problem. *Transportation Research Part B: Methodological*, 45, 1232–1250.

BELLARD, C., BERTELSMEIER, C., LEADLEY, P., THUILLER, W. & COURCHAMP, F. 2012. Impacts of climate change on the future of biodiversity. *Ecology Letters*, 15, 365–377.

BHATTACHARYYA, R., DAS, T. K., PRAMANIK, P., GANESHAN, V., SAAD, A. A. & SHARMA, A. R. 2013. Impacts of conservation agriculture on soil aggregation and aggregate-associated N under an irrigated agroecosystem of the Indo-Gangetic Plains. *Nutrient Cycling in Agroecosystems*, 96, 185–202.

BHUTTO, A. W., BAZMI, A. A. & ZAHEDI, G. 2012. Greener energy: Issues and challenges for Pakistan-hydel power prospective. *Renewable and Sustainable Energy Reviews*, 16, 2732–2746.

BOND, A. J. & MORRISON-SAUNDERS, A. 2011. Re-evaluating sustainability assessment: Aligning the vision and the practice. *Environmental Impact Assessment Review*, 31, 1–7.

CHEN, G. Q., WU, X. D., GUO, J., MENG, J. & LI, C. 2019. Global overview for energy use of the world economy: Household-consumption-based accounting based on the world input-output database (WIOD). *Energy Economics*, 81, 835–847.

CHUANG, L.-Y., YANG, C.-H. & YANG, C.-H. 2009. Tabu search and binary particle swarm optimization for feature selection using microarray data. *Journal of Computational Biology*, 16, 1689–1703.

CONTARDO, C. & MARTINELLI, R. 2014. A new exact algorithm for the multi-depot vehicle routing problem under capacity and route length constraints. *Discrete Optimization*, 12, 129–146.

CORLU, C. G., DE LA TORRE, R., SERRANO-HERNANDEZ, A., JUAN, A. A. & FAULIN, J. 2020. Optimizing energy consumption in transportation: Literature review, insights, and research opportunities. *Energies*, 13, 1115.

DAS, T. K., BANDYOPADHYAY, K. K., BHATTACHARYYA, R., SUDHISHRI, S., SHARMA, A. R., BEHERA, U. K., SAHARAWAT, Y. S., SAHOO, P. K., PATHAK, H., VYAS, A. K., BHAR, L. M., GUPTA, H. S., GUPTA, R. K. & JAT, M. L. 2016. Effects of conservation agriculture on crop productivity and water-use efficiency under an irrigated pigeonpea–wheat cropping system in the western Indo-Gangetic Plains. *The Journal of Agricultural Science*, 154, 1327–1342.

DAS, T. K., BHATTACHARYYA, R., SHARMA, A. R., DAS, S., SAAD, A. A. & PATHAK, H. 2013. Impacts of conservation agriculture on total soil organic carbon retention potential under an irrigated agro-ecosystem of the western Indo-Gangetic Plains. *European Journal of Agronomy*, 51, 34–42.

DEIAB, I. 2014. On energy efficient and sustainable machining through hybrid processes. *Materials and Manufacturing Processes*, 29, 1338–1345.

DEKKER, R., BLOEMHOF, J. & MALLIDIS, I. 2012. Operations research for green logistics–An overview of aspects, issues, contributions and challenges. *European Journal of Operational Research*, 219, 671–679.

DELGADO, J. A., DILLON, M. A., SPARKS, R. T. & ESSAH, S. Y. C. 2007. A decade of advances in cover crops. *Journal of Soil and Water Conservation*, 62, 110A–117A.

ESHTEHADI, R., FATHIAN, M. & DEMIR, E. 2017. Robust solutions to the pollution-routing problem with demand and travel time uncertainty. *Transportation Research Part D: Transport and Environment*, 51, 351–363.

FAULIN, J., GRASMAN, S. E., JUAN, A. A. & HIRSCH, P. 2019. Sustainable transportation: Concepts and current practices. In *Sustainable Transportation and Smart Logistics*. Elsevier.

FENERTY-MCKIBBON, B. & KHARE, A. 2005. Canada post delivers energy conservation. *Energy and Buildings*, 37, 221–234.

FERNÁNDEZ, E., LAPORTE, G. & RODRÍGUEZ-PEREIRA, J. 2018. A branch-and-cut algorithm for the multidepot rural postman problem. *Transportation Science*, 52, 353–369.

FUKASAWA, R., HE, Q. & SONG, Y. 2016. A branch-cut-and-price algorithm for the energy minimization vehicle routing problem. *Transportation Science*, 50, 23–34.

GHOSH, S., DAS, T. K., SHARMA, D. & GUPTA, K. 2019. Potential of conservation agriculture for ecosystem services: A review. *Indian Journal of Agricultural Sciences*, 89, 1572–1579.

GRANADA-ECHEVERRI, M., TORO, E. & SANTA, J. 2019. A mixed integer linear programming formulation for the vehicle routing problem with backhauls. *International Journal of Industrial Engineering Computations*, 10, 295–308.

GYBERG, P. & PALM, J. 2009. Influencing households' energy behaviour—how is this done and on what premises? *Energy Policy*, 37, 2807–2813.

HE, J., WANG, Q., LI, H., TULLBERG, J. N., MCHUGH, A. D., BAI, Y., ZHANG, X., MCLAUGHLIN, N. & GAO, H. 2009. Soil physical properties and infiltration after long-term no-tillage and ploughing on the Chinese Loess Plateau. *New Zealand Journal of Crop and Horticultural Science*, 37, 157–166.

HOFMANN, W. M. 1983. Energy conservation in building construction. *Technische Rundsch (Switzerland)*, 65.

HOOSHMAND, F. & MIRHASSANI, S. A. 2019. Time dependent green VRP with alternative fuel powered vehicles. *Energy Systems*, 10, 721–756.

HUANG, Y., ZHAO, L., VAN WOENSEL, T. & GROSS, J.-P. 2017. Time-dependent vehicle routing problem with path flexibility. *Transportation Research Part B: Methodological*, 95, 169–195.

HUMPHREYS, E., KUKAL, S. S., CHRISTEN, E. W., HIRA, G. S. & SHARMA, R. K. 2010. Halting the groundwater decline in north-west India—which crop technologies will be winners? In *Advances in Agronomy*. Elsevier.

IWARO, J. & MWASHA, A. 2010. A review of building energy regulation and policy for energy conservation in developing countries. *Energy Policy*, 38, 7744–7755.

JENKINS, K., MCCAULEY, D., HEFFRON, R., STEPHAN, H. & REHNER, R. 2016. Energy justice: A conceptual review. *Energy Research & Social Science*, 11, 174–182.

JUAN, A. A., GOENTZEL, J. & BEKTAŞ, T. 2014. Routing fleets with multiple driving ranges: Is it possible to use greener fleet configurations? *Applied Soft Computing*, 21, 84–94.

JUAN, A. A., MENDEZ, C. A., FAULIN, J., DE ARMAS, J. & GRASMAN, S. E. 2016. Electric vehicles in logistics and transportation: A survey on emerging environmental, strategic, and operational challenges. *Energies*, 9, 86.

KARA, I., KARA, B. Y. & YETIS, M. K. 2007. *Energy minimizing vehicle routing problem*. Springer, 62–71.

KE, M.-T., YEH, C.-H. & JIAN, J.-T. 2013. Analysis of building energy consumption parameters and energy savings measurement and verification by applying eQUEST software. *Energy and Buildings*, 61, 100–107.

KIRCI, P. 2019. *A novel model for vehicle routing problem with minimizing CO2 emissions*. IEEE, 241–243.

KLIBI, W., LASALLE, F., MARTEL, A. & ICHOUA, S. 2010. The stochastic multiperiod location transportation problem. *Transportation Science*, 44, 221–237.

KOÇ, Ç. & KARAOGLAN, I. 2016. The green vehicle routing problem: A heuristic based exact solution approach. *Applied Soft Computing*, 39, 154–164.

KOUR, S., ARORA, S. & GUPTA, M. 2011. Conservation tillage a gateway to sustainable soil and crop management: An overview. *Journal of Soil and Water Conservation*, 10, 242–247.

KÜÇÜKOĞLU, İ. & ÖZTÜRK, N. 2014. A differential evolution approach for the vehicle routing problem with backhauls and time windows. *Journal of Advanced Transportation*, 48, 942–956.

KÜÇÜKOĞLU, İ. & ÖZTÜRK, N. 2015. An advanced hybrid meta-heuristic algorithm for the vehicle routing problem with backhauls and time windows. *Computers & Industrial Engineering*, 86, 60–68.

KUMAR, V., SAHARAWAT, Y. S., GATHALA, M. K., JAT, A. S., SINGH, S. K., CHAUD-HARY, N. & JAT, M. L. 2013. Effect of different tillage and seeding methods on energy use efficiency and productivity of wheat in the Indo-Gangetic Plains. *Field Crops Research*, 142, 1–8.

LAI, M., BATTARRA, M., DI FRANCESCO, M. & ZUDDAS, P. 2015. An adaptive guidance meta-heuristic for the vehicle routing problem with splits and clustered backhauls. *Journal of the Operational Research Society*, 66, 1222–1235.

LANGLOIS, L. 2013. IAEA action plan on nuclear safety. *Energy Strategy Reviews*, 1, 302–306.

LEE, D. & CHENG, C.-C. 2016. Energy savings by energy management systems: A review. *Renewable and Sustainable Energy Reviews*, 56, 760–777.

MACHIBA, T. 2011. Eco-innovation for enabling resource efficiency and green growth: Development of an analytical framework and preliminary analysis of industry and policy practices. In *International Economics of Resource Efficiency*. Springer.

MALECKA, I., BLECHARCZYK, A., SAWINSKA, Z. & DOBRZENIECKI, T. 2012. The effect of various long-term tillage systems on soil properties and spring barley yield. *Turkish Journal of Agriculture and Forestry*, 36, 217–226.

MENZEL, V., SMAGIN, J. & DAVID, F. 2010. Can companies profit from greener manufacturing? *Measuring Business Excellence*, 14(2), 22–31.

MOLYKUTTY, M. V. 2015. Energy conservation in building construction. *International Journal of Construction Engineering and Planning*, 1, 20–22.

MOSIER, A. R., HALVORSON, A. D., PETERSON, G. A., ROBERTSON, G. P. & SHER-ROD, L. 2005. Measurement of net global warming potential in three agroecosystems. *Nutrient Cycling in Agroecosystems*, 72, 67–76.

NANDI, P. & BASU, S. 2008. A review of energy conservation initiatives by the Government of India. *Renewable and Sustainable Energy Reviews*, 12, 518–530.

NEUMANN, F. & WITT, C. 2010. Combinatorial optimization and computational complexity. In *Bioinspired Computation in Combinatorial Optimization*. Springer.

PARAPHANTAKUL, C., MILLER-HOOKS, E. & OPASANON, S. 2012. Scheduling deliveries with backhauls in Thailand's cement industry. *Transportation Research Record*, 2269, 73–82.

PÉREZ-BERNABEU, E., JUAN, A. A., FAULIN, J. & BARRIOS, B. B. 2015. Horizontal cooperation in road transportation: A case illustrating savings in distances and greenhouse gas emissions. *International Transactions in Operational Research*, 22, 585–606.

PETER, O. & MBOHWA, C. 2019. Industrial energy conservation initiative and prospect for sustainable manufacturing. *Procedia Manufacturing*, 35, 546–551.

POWER, A. G. 2010. Ecosystem services and agriculture: Tradeoffs and synergies. *Philosophical Transactions of the Royal Society B: Biological Sciences*, 365, 2959–2971.

QUINTERO-ARAUJO, C. L., GRULER, A., JUAN, A. A., DE ARMAS, J. & RAMAL-HINHO, H. 2017. Using simheuristics to promote horizontal collaboration in stochastic city logistics. *Progress in Artificial Intelligence*, 6, 275–284.

RAO, K. R. 1984. Energy conservation possibilities through building design and passive cooling techniques in warm climates. In *Energy Developments: New Forms, Renewables, Conservation*. Elsevier.

REIL, S., BORTFELDT, A. & MÖNCH, L. 2018. Heuristics for vehicle routing problems with backhauls, time windows, and 3D loading constraints. *European Journal of Operational Research*, 266, 877–894.

SAAD, A. A., DAS, T. K., RANA, D. S., SHARMA, A. R., BHATTACHARYYA, R. & LAL, K. 2016. Energy auditing of a maize–wheat–greengram cropping system under conventional and conservation agriculture in irrigated north-western Indo-Gangetic Plains. *Energy*, 116, 293–305.

SARKAR, A. A. 2011. Adaptive climate responsive vernacular construction in high altitude. *International Journal of Architectural and Environmental Engineering*, 5, 761–765.

SINGH, V. K., DWIVEDI, B. S., SINGH, S. K., MAJUMDAR, K., JAT, M. L., MISHRA, R. P. & RANI, M. 2016. Soil physical properties, yield trends and economics after five years of conservation agriculture based rice-maize system in north-western India. *Soil and Tillage Research*, 155, 133–148.

SONG, L., FU, Y., ZHOU, P. & LAI, K. K. 2017. Measuring national energy performance via energy trilemma index: A stochastic multicriteria acceptability analysis. *Energy Economics*, 66, 313–319.

THOMAS, G. A., DALAL, R. C. & STANDLEY, J. 2007. No-till effects on organic matter, pH, cation exchange capacity and nutrient distribution in a Luvisol in the semi-arid subtropics. *Soil and Tillage Research*, 94, 295–304.

THUESEN, N., KIRKEGAARD, P. H. & JENSEN, R. L. 2010. *Evalution of BIM and Ecotect for Conceptual Architectural Design Analysis*. University of Nottingham.

TING, L. S., MOHAMMED, A. H. B. & WAI, C. W. 2011. Promoting energy conservation behaviour: A plausible solution to energy sustainability threats. In *International Conference on Social Science and Humanity*, 5(1), 372–376.

UNITED, N. 2016. *The Sustainable Development Goals Report*. United Nations.

VAN FAN, Y., KLEMEŠ, J. J., WALMSLEY, T. G. & PERRY, S. 2019. Minimising energy consumption and environmental burden of freight transport using a novel graphical decision-making tool. *Renewable and Sustainable Energy Reviews*, 114, 109335.

WONG, K. K. 2011. Towards a light-green society for Hong Kong, China: Citizen perceptions. *International Journal of Environmental Studies*, 68, 209–227.

WORLD COMMISSION ON ENVIRONMENT AND DEVELOPMENT. 1987. *Our common future*. Oxford University Press.

YAN, X., OHARA, T. & AKIMOTO, H. 2003. Development of region-specific emission factors and estimation of methane emission from rice fields in the East, Southeast and South Asian countries. *Global Change Biology*, 9, 237–254.

5 Technology and Architecture of Smart Grids

Muhamamd Khalid

5.1 INTRODUCTION

The current global energy sector is experiencing a significant shift due to the growing need for electricity and the fast process of urbanization. However, there are numerous challenges associated with the existent conventional power systems that lack efficacy. The global challenges include the growing demand for energy, the deterioration of existing infrastructure, the integration of renewable energy sources, and environmental considerations. Additionally, the local grid also has multiple challenges, such as energy inefficiency, peak load management, power grid resilience, and the transition from a consumer-based grid to prosumer-based grids. Furthermore, with the increased level of global urbanization, the reliability of the grid is further reduced with the ageing infrastructure.

Furthermore, the incorporation of unpredictable, uncontrollable, and transient sources of renewable energy such as solar and wind into the electricity grid presents challenges in terms of managing the swings in power production, negatively impacting its practical significance, and hindering its possibilities of establishing sustainable energy systems. Concurrently, the issue of climate change necessitates the shift towards a low-carbon energy framework. A smart grid (SG) represents a fundamental transformation at distribution and administration levels of the electrical sector. It presents a wide array of advantages in comparison to conventional grid systems. The integration of renewable energy sources in an effective manner is a significant factor contributing to the widespread adoption of SGs.

The inevitability of a shift in grid dynamics is accompanied by the need for further auxiliary support for the generally selected renewable energy technologies. To mitigate these problems, the use of advanced and interoperable control methods, deregulation, modern monitoring systems, and robust communication networks is essential. Ideally, SG enables a seamless integration of intermittent renewable energy sources, such as solar and wind, by offering improved grid balance and flexibility. This advancement has the potential to contribute to a more environmentally friendly and sustainable future. Moreover, SGs effectively tackle the urgent issue of managing peak loads and mitigating electrical demand surges. In addition, they enable users to effectively manage their energy usage during peak hours by using demand response techniques and real-time monitoring.

DOI: 10.1201/9781032715438-5

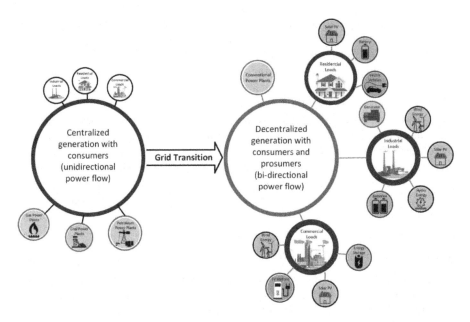

FIGURE 5.1 Transition of centralized conventional grid to decentralized grid with bidirectional power flow.

This capability not only reduces the vulnerability of the system but also mitigates the potential for catastrophic events, such as blackouts. The communication systems have seen significant technical developments, resulting in enhanced monitoring and coordination capabilities. These advancements contribute to improved grid monitoring, controllability, and flexibility, ultimately leading to decreased operating costs. This aligns with the current trend of integrating renewable energy sources. Implementing SG enables the use of information and communication technologies to upgrade the power network system, hence enhancing accessibility in this regard [1]–[4].

However, the extensive network of current power systems necessitates the implementation of an optimized SG. This is justified by the various requirements of the grid, including communication, sustainability, interoperability, and power quality. These factors are crucial in maintaining the technological and economic importance of the entire network [5], [6]. Subsequently, the concept of SG, in recent years, had been further developed to possess better adaptability or interoperability through the concept of smart grid 2.0, second-generation smart grid, or internet of energy (IoE). While both SG and IoE share the same goals, the latter utilizes internet-connect SG to introduce an accelerated and interoperable energy market.

Hence, the concept of SGs provides the means of augmenting the dependability and resilience of the grid. With modernization and urbanization comes heightened susceptibility to cybersecurity breaches, hence expediting the implementation of effective security recovery procedures, and enhancing the overall strength and resilience of the energy infrastructure is important, especially considering critical loads associated with healthcare and public-welfare organizations. Similarly, with the increasing use of electric vehicles, the implementation of SGs becomes imperative to effectively manage the growing demand for electricity in the transportation sector. SGs facilitate the establishment of intelligent charging infrastructure and enable customers to assume the role of energy "prosumers," allowing them to produce, store, and potentially trade energy. Therefore, SG equips its users to tackle multi-faceted challenges faced during the decarbonization of the grid by developing a more dynamic and decentralized energy system framework. Therefore, in line with the multi-dimensional technologies and process associated with smart grids, the intent of this chapter is to outline the concept of smart grids and introduce their multi-disciplinary technologies and terminology while highlighting their components.

5.2 UNDERSTANDING THE NEED

The interconnection and intersection of network links give rise to a collective formation known as a grid. The concept of a SG involves the use of both analog and digital information, together with sophisticated communication technology. Within the realm of renewable energy (RE), the concept of a smart grid pertains to a distinct domain or a communication framework that facilitates the integration of RE sources with the existing power system. However, the integration of RE production into SG presents many challenges, including stability issues, complex operational protocols, and the need for remote control capabilities. The electric power grid plays a significant role in sustaining the infrastructure of our modern civilization. The electricity distribution network is extensive, including residential, commercial, industrial, and many other types of loads. In this way, the electrical power grid is a network of interrelated and intricate systems. As such, a single organization or authority does not have complete control over the monitoring, regulation, planning, and expansion aspects of these systems. Furthermore, it will be increasingly challenging to conduct real-time analysis, monitoring, and maintenance of these energy systems. The grid covers not only the transmission of energy from power producing sources to substations but also the distribution of electricity from substations to individual end-users.

Hence, the conventional grid has several procedural and technical problems, particularly in the areas of coordinated monitoring, automation, analysis, and control. The network becomes more complex with the introduction of variable RE sources at each voltage level. In contrast to conventional centralized power systems that rely on fossil fuels, electricity from renewable energy sources fluctuates due to their dependency on the location and weather. The use of forecasting and monitoring technology combined with expeditious and adaptable grid operations becomes an operational need. The smooth integration of numerous RE sources scattered over a vast geographic region is facilitated by interconnected systems, which are considered a core element of smart grids.

These interdependent networks facilitate the flow of excess energy from one region to other regions with energy deficiencies, hence enhancing the efficient utilization of RE while maintaining the power quality standards and decreasing reliance on traditional power plants. Consequently, the acquisition of data via the use of phasor measuring units, digital frequency recorders, and dynamic swing recordings necessitates the presence of better communication and information systems. Moreover, the implementation and functionality of global positioning systems are crucial, particularly in urban regions, to facilitate the establishment and functioning of wide area management systems. Correspondingly, the use of such technologies in combination with complex bulk energy systems with big data requires protective and safety measures in response to security concerns. This demands the development of an integrated energy-information system that is interoperable across all domains.

Accordingly, the rise of prosumers in the electricity market proves an unprecedented transformation to the grid dynamics. As a result, there is an increasing demand for a virtual inertia market due to the decrease in traditional power production systems that traditionally supplied and supported the overall inertia of the system. The introduction of prosumers will lead to the introduction of RE sources that negligibly contribute towards inertia. Pre-emptively, many protocols associated with voltage, frequency, and black-start process will have to be upgraded and revised. Such significant transformation requires advanced metering infrastructures (AMIs), effective user interfaces, integration of devices related to the Internet of Things (IoT), sensors, and long-term as well as short-term energy management strategies.

While the introduction of prosumers proves beneficial considering long-term sustainable energy establishment, in the early stages of deregulated market competitiveness, it might also introduce many cybersecurity threats due to the introduction of decision-making devices that have comparatively low levels of security such as smart phones. As a result of deregulation of the grid and the addition of power production markets ranging from a few kilowatts to thousands of megawatts, which include consumers, businesses, and utilities, the energy market needs to be able to price energy in real time. This is very important to do in order to deal with the complicated aspects of real-time supply and demand in a market that is changing and becoming more diverse. Thus, not only will a bi-directional power flow exist but the complexity of functionalities also requires bi-directional information flow. This demand-response mechanism fosters energy efficiency and load balancing while reducing the need for peak power generation from fossil fuel plants.

5.3 CHALLENGES IN CONVENTIONAL POWER GRIDS

When electricity networks throughout the globe expanded, they were not built to accommodate occasional fluctuations. The establishment of grid connections first focused on linking point-to-point loads and generating stations. Typically, the allocation of loads may be restricted to significant consumers, such as industrial facilities and substations catering to residential communities. The network architecture plays a significant role in determining the optimal placement of both customers' power plants and customers themselves. Distinct hierarchies exist for each of the connections, wherein the magnitude of the voltage, the length of the distance, and the amount of power being exchanged are positively correlated with the extent of the distance.

Over time, the interconnection linkages inside networks have undergone changes and development, resulting in the formation of star topologies. In these topologies, the central component of the star comprises substations, while the upper levels of the hierarchy include redundant links. In the past, the control network that was co-located with the grid transmission network used a mix of automated switches and human control. This included the construction and management of automatic circuit breakers at crucial locations to address grid disturbances. The proliferation of technological advancements has led to the availability of a diverse range of solutions aimed at mitigating the issues faced by the grid.

Deteriorating infrastructure is among the major challenges with conventional power networks due to them being constructed decades ago without considering the exponential rise in the demand of electricity. As a result, the ageing components and equipment are prone to failures, leading to frequent outages and reduced grid reliability. Considering the ageing infrastructure aspect, a significant number of power system networks worldwide, including the transmission and distribution networks, as well as generating systems, have reached their anticipated lifespan and hence require replacement. The expense associated with refurbishing and reinstalling these items to their original technical standard might be substantial. Furthermore, there is a shortage of qualified personnel to carry out these tasks.

This enables the potential for innovation within the current power network, allowing for improvements in both the reliability of supply and the addressing of technical deficiencies in infrastructure and human resources. The integration of RE sources introduces variations that restrict their suitability for integration with the conventional power network, particularly at the transmission level when the network is already working at its maximum capacity [1], [2]. These hindrances pose a challenge to the incorporation of renewable energy sources into the power grid, thereby impeding the urgent worldwide need for a sustainable energy industry. The power transfer capacity of electrical wires at the transmission and distribution level is limited by the temperature restrictions they face. The act of overloading, which refers to the transmission of current above the thermal capacity of the electrical line, may lead to hastened degradation and reduced lifespan. This, in turn, increases the likelihood of experiencing faults. Hence, the need for dynamic ratings arises due to the fact that these thermal restrictions are contingent upon environmental factors [3]–[5].

Accordingly, the operating limitations of the power system are mostly determined by the voltage and frequency thresholds established for the power system. The network undergoes insulation damage, which subsequently leads to short-circuit problems and device malfunctions. In the case of over- and under-voltage circumstances, the system trips. Such issues were traditionally encountered in the context of broad-scale interconnection at both national and international levels. To address these challenges, voltage control devices, such as on-load tap changers, were used as a solution [6], [7]. Similarly, it is very important to take care of the system's frequency control needs, which demand real-time tracking that matches the profile of demand generation. Furthermore, a minor variation in frequency control leads to de-synchronization, which is mostly managed using automated generation control procedures.

Additionally, load shedding measures are used in emergency situations. Hence, given the unexpected output power variations shown by RE sources, there is a need for fast and responsive control techniques and equipment. It has been observed that many of the standard solutions produced thus far demonstrate inefficiency in addressing this requirement [8]–[10]. The proposed strategies for integrating RE sources mostly include the use of forecasting techniques, energy storage systems, spinning reserves, reliability measures, and system flexibility. These measures are often recommended to ensure the sustainable and economically feasible functioning of the entire energy system. However, most of these proposed solutions need the implementation of advanced communication systems to facilitate data collecting, data processing, and optimization.

The integration of electricity has become prevalent in several essential domains and areas of human well-being that are categorized as critical loads. It is essential to provide a dependable and secure power supply to these loads. In the past, the installation of redundant circuits was deemed vital, despite the associated high capital costs and environmental concerns. From this standpoint, the implementation of SG guarantees a better approach that enables algorithms to effectively identify faults after they occur while efficiently using each component in the power network. Consequently, the need for redundant circuits is eliminated. Hence, the implementation of SG supports the shift from a traditional power grid to a more advanced network that enables collaborative and adaptive interactions [11].

Furthermore, this approach facilitates and motivates the integration of prosumers' involvement in the energy industry, which is crucial for establishing a deregulated power grid that considers the constraints and demands of existing renewable energy systems. The activation of advanced bi-directional communication in smart grids enables the integration of sophisticated intelligent algorithms, hence improving the resilience and self-repairing capacity of the power network. Therefore, there is a global effort for current electrical sectors to transition into SG. The primary objective is to implement a systematic approach to attain a state of net-zero carbon emissions within the energy sector.

This is done to minimize the environmental effect while simultaneously developing technologically sophisticated power systems that are compatible with end-users. These considerations are not only driving the adoption of responsible business behavior, but several governments have also proactively implemented rules to restrict carbon emissions and provided legislative incentives to promote infrastructure modernization and sustainable growth transformation. Hence, the move from the conventional grid system to the SG has many possible advantages, including the following:

- Emission reduction of greenhouse gases.
- Enhanced reliability, self-healing capability, and resilience.
- Accommodation of all generation and energy storage types.
- Open market infrastructure in the electricity sector with prosumer participation.

- Improved system efficiency.
- Optimized operational accessibility of grid assets.
- Predictive system operation, performance, and maintenance.
- Improved real-time monitoring and security.

5.3.1 Defining Smart Grids

The overall concept of SGs can be visualized by a process of categorization, resulting in the identification of seven main categories, each of which may further include sub-categories [12]. These categories include applicants and actors. Actors inside an organization may be categorized as either internal or external. Actors include individuals, systems, or organizations that assume the role of initiating or engaging in actions throughout the seven domains. Actors include not just individuals in the performing arts but also gadgets, computer software and systems, and/or the organizations that own them. Actors engage in decision-making processes and engage in the exchange of information with their fellow actors. The applications are implemented with automation capabilities and are operated in alignment with the specific requirements of the stakeholders within the given area.

These applications have the potential to be executed by either a solitary individual or a collective group of individuals. The actors include many entities inside the system, such as control systems, stakeholders, communications centers, and hardware devices like smart meters and renewable energy systems. The applications include several duties performed by actors, such as energy management, energy arbitrage, and power quality maintenance, which are further classified into sub-categories. The attributes of a smart grid may be comprehensively and inclusively outlined as follows:

- Capability to integrate and sustain distributed network architecture of the power system.
- Able to coordinate new electric generation sources and their associated challenges, especially related to renewable energy sources both in technical and economic respect.
- Enable instantaneous balance of supply and demand at the device level.
- Digitalization of the communication network and ensuring reliability and security.
- Able to facilitate strategies targeted for enhancing demand-side response and demand-side generation.
- Protection and obviation capability towards vulnerabilities associated with cyber and infrastructural threats.
- Compatibility to integrate and coordinate the applicability of smart equipment and appliances.
- Provision to effectively integrate stationary and mobile energy storage systems.

The major difference between conventional grids and SGs are listed in Table 5.1, which highlights the capability and new facilitation a SG can introduce. Conclusively,

TABLE 5.1

Technical Differentiation between Conventional and Smart Grids [13]–[15]

Conventional Grids	Smart Grids
Mechanical	Digital
One-way communication	Two-way communication
Centralized bulk generation	Distributed generation
Reliance on coal and oil	Sustainable energy sources
Radial-dominated network	Deregulated and dispersed network
Lack of monitoring capability	High monitoring capability
Limited automation	High scope of autonomous control
Less security concern	High security vulnerability
Low response time	High response time

the framework of the SG enables a more rapid, efficient, and reliable version of the power system in comparison to the conventional contemporary network.

5.3.2 UTILIZATION OF SMART GRIDS IN POWER UTILITIES

Several governments and organizations have drawn up initiatives for the standardization and development of smart grid systems [16]. The development of a globally applicable standard is crucial for effectively designing all aspects of a SG. Currently, there is a need to upgrade the current standards to address the challenges encountered during the transition of implementing SG technology. These challenges include social, economic, and technological aspects. The IEEE P2030 proposes a comprehensive framework for characterizing the structural organization of smart grids using a system-level perspective.

This standard aims to depict the interoperable components of the smart grid, including energy systems, communication systems, and information systems. The method centered on interoperability offers a convenient means for service providers and users to discern the essential elements that enable efficient communication and data exchange across diverse information systems. Hence, it is also necessary to establish a standard that is founded upon the conceptual components of a smart grid, which has the capability to harmonize diverse smart grid infrastructures across various geographical and upcoming technologies. Moreover, the depiction of the SG as a sophisticated and extensive "system of systems" offers comprehensive guidance, encouragement, and design approaches across various academic and industrial research domains.

This facilitates the initiation of development and innovation within the electric network system and among end-users. The power utility network incorporates several contemporary technologies of smart grid, including geographical information systems, broad area management systems, meter data management, automated voltage regulation, automatic generation control, and advanced metering infrastructure.

The smart infrastructure system primarily comprises three subsystems, the energy, communication, and information systems. It enables bidirectional communication and the transmission of electrical current.

The use of two-way electric power flow eliminates the need for the traditional unidirectional flow. For example, traditional power systems operated on a centralized model of power production, where electricity was produced at a single location and then transmitted and distributed to end-users. The integration of SG enables users to contribute energy to the power system by installing solar panels in the smart panels or using electric cars. These devices are efficiently used to maintain power equilibrium and/or reduce peak loads in line with the demands of the network. Moreover, the phenomenon of energy flowing in the opposite way is relevant in the context of creating distributed generation systems. Broadly, the smart grid can be classified into:

- Smart energy system including the processes of electricity production, transmission, distribution, and consumption.
- Smart information systems are developed by integrating smart meters, innovative management techniques, and monitoring systems that are connected to the smart power grid.
- The establishment of a smart communication system is contingent upon the seamless integration and interconnectivity of wireless and wired networks and devices, that is the information system.

According to the guidelines set out by IEEE P2030, the effective management of the complex interoperability requirements of smart grid information and communication subsystems necessitates the segregation of technical considerations into separate and independent entities. The objectives for smart meters and smart grids are distinctly outlined, with specific goals to be attained within each of the primary facets, as shown in Table 5.2.

TABLE 5.2
Smart Grid and Smart Meter Solutions to Achieve Key Objectives

Objectives	Smart Grids	Smart Meters
Renewables	• Generation-demand mismatch • Flexible integration of diversified renewables	• Facilitate deregulated and distributed generation • Measurement and secure monitoring
Network	• Reduction of outages • Reduced system losses • Frequency and voltage regulation • Asset management and utilization	• Effective monitoring, processing, and prediction • Improve efficacy of the information and power flow
Customers	• Flexibility to incorporate future smart services associated with smart homes and buildings	• Support demand-side management • Facilitate seamless dynamic tariffs

5.4 ENERGY AND POWER SYSTEMS

The traditional electricity system is predicated upon a centralized and mostly radial network. Electricity is produced using electromechanical combustion generators, which are powered by either fuel-based engines or the kinetic energy of water. These generators are designed to be both physically separated and of significant size to ensure techno-economic gains. The electric power produced is conveyed to the substations across significant distances at elevated voltage levels via the use of the transmission network. Voltage reduction occurs at the distribution level, and afterwards, at the service level, additional voltage reduction is implemented to accommodate the specific needs of end-users. In contrast, the power flow of a smart grid enables more flexibility. For example, power generation can also be achieved at the distribution level through small- or medium-scale RE sources.

The use of SGs brings about significant changes to the production, transmission, and distribution phases of the power network. For visualization, the energy and power systems can be classified into generation, transmission, and distribution levels. The challenges at the generation level include integration of RE sources and energy storage solutions. Real-time grid monitoring and control are the challenges at the transmission level, along with dynamic ratings and power flow control. Finally, demand response, energy management, grid decentralization, automation, and self-healing are the aspects associated with the distribution level.

5.4.1 GENERATION LEVEL

The implementation of a smart power-generating system facilitates the establishment of distributed generation (DG). However, the implementation of DG poses significant challenges. The first concern about the widespread implementation of renewable energy sources is the introduction of unmanageable and uncertain power production, resulting in a mismatch between supply and demand. Furthermore, the present power generating and operating expenses associated with renewable energy sources are comparatively high when compared to those of existent traditional large-scale generation systems [17]. However, while renewable sources such as solar and wind energy may be more cost-effective in terms of generation when compared and optimally evaluated, there are additional expenses associated with auxiliary support systems such as advanced power electronics modules used for forecasting errors, computations of resource availability, and real-time tracking of power mismatch errors in grid. Hence, given the potential of DG, it is imperative to undertake systematic innovation aimed at mitigating variability and optimizing cost via the use of intelligent energy management systems. Currently, the integration of renewable energy sources into the power grid is limited, but it is anticipated that future smart grids will go through three distinct phases in order to accommodate a significant increase in the use of renewable energy [18]:

- The integration of renewable-based DGs and the establishment of several coordinated microgrids to facilitate the accommodation of diversified energy sources.

- Coexisting decentralized power network with the existing conventional infrastructure.
- Systematic mitigation of conventional generation systems and satisfying the load demand with increased DG integrations.

In addition, the advancement towards DG has given rise to the concept of virtual power plants (VPPs). The VPP includes a collection of DGs that are regulated by central controllers. The total capacity of the DGs is comparable to that of a traditional power generation system [19]. The operating mode exhibits improved power delivery capabilities during periods of high load demand and effectively adjusts to varying load levels. When comparing VPPs to traditional power plants, VPPs exhibit superior levels of efficiency and flexibility towards grid modernization.

During the phase of generation, the concept of flexibility encompasses the capacity to effectively integrate and accommodate the variability associated with RE technology. However, an efficient formulation of VPPs relies on intricate systems that need meticulous optimization and control mechanisms, along with state-of-the-art communication systems. Therefore, software development is needed for the incorporation of VPP technologies such as service-oriented architecture [20] and virtual energy buffers and provision [21] [22] and also for accommodating vehicle-to-grid functionalities [23].

5.4.2 Transmission Level

The driving forces behind smart transmission systems include both issues and opportunities. These issues include additional load demand, component aging, and restricted capacity. On the other hand, opportunities arise from improved technologies like enhanced power electronics, sensor devices, and information and communication technologies that will enable a higher degree of controllability. Deployment of dynamic line rating techniques, which operate with the incorporation of real-time weather conditions and grid parameters to optimize the transmission line's capacity, are among the objectives of modernization of the transmission level. Similarly, the integration of power flow control devices, such as flexible AC transmission systems, helps regulate power flow, mitigate transmission bottlenecks, and reduce line losses, resulting in improved transmission efficiency with better transmission level flexibility. Hence, the suggested future vision for smart grids entails the integration of advanced control centers equipped with enhanced analytical skills, efficient visualization tools, and comprehensive monitoring components. The transmission level associated with smart functionalities can be classified into three interactive components: smart transmission networks, substations, and smart control [24].

These components are derived from developing technologies, including ongoing development of materials (namely nanomaterials), power electronics, signal processing, and computers. Hence, the objective of incorporating smart substations and enhancing the transmission system mostly revolves around modernizing monitoring and measuring methodologies, given that the layout of high voltage infrastructure has remained relatively unchanged over the course of time [25]. The primary attributes include independent functioning, digitization, self-repairing capacity, predictive maintenance, and improved coordination. These characteristics provide operators and consumers with the potential to create a load-following power generating network that offers enhanced flexibility and dependability throughout the transmission phase of the electrical grid. In this step, flexibility entails the integration of intricate energy management algorithms.

5.4.3 DISTRIBUTION LEVEL

The integration of DG into the power grid at various voltage levels within the distributed smart grid requires the implementation of intelligent distribution networks (IDNs) and advanced delivery mechanisms for ensuring the power quality. Integrating DG, such as solar panels and wind turbines, into the distribution grid poses both technical and regulatory challenges. The variable nature of renewable energy generation creates fluctuations in power supply, which requires advanced grid management solutions. Managing these fluctuations and ensuring grid stability demand the implementation of reliable energy forecasting and real-time monitoring systems. Additionally, regulatory frameworks may need to be updated to enable seamless integration of DERs into the grid and address issues related to tariff structures, net metering, and grid access for prosumers.

However, this formulation involves the collection and communication of vast amounts of sensitive data, including real-time energy consumption patterns and behavioral data of consumers. Ensuring data privacy and cybersecurity is a critical challenge. Smart grids are susceptible to cyber threats and attacks, which could lead to unauthorized access, data breaches, and potential disruptions to grid operations. Building robust cybersecurity measures and privacy protocols is imperative to protect consumer data and instill confidence in smart grid technologies among both consumers and industry stakeholders. While this enhances grid flexibility and efficiency, it also poses challenges in maintaining grid resilience and reliability. The grids must be equipped with advanced monitoring and control systems to rapidly detect and isolate faults. Moreover, self-healing capabilities are pertinent for restoring power to affected areas during grid disturbances caused by natural disasters, faults, or cyber threats. Developing and deploying such resilient smart grid features require thorough testing and validation to ensure their effectiveness in real-world scenarios.

The proposition of using a smart distribution network (SDN) has been put out as a means to facilitate the implementation of intelligent residential environments, sometimes referred to as smart homes. Smart homes are integrated into a sophisticated communication and information network, enabling the implementation of intricate energy management algorithms that ease coordination between the utility and the grid. Smart houses consist of decentralized power generation, localized electrical loads, and a method for storing energy. The energy management between the utility and the end-users may be categorized as either localized or central. The research conducted in [26] examined two distinct situations involving smart home distribution networks, AC and DC. In both scenarios, the flow of power inside the network is regulated based on the information received. The DC power system operates by transmitting power via power packets that need the use of high-power switching devices. The authors highlight the potential use of silicon carbide junction gate field-effect transistors in establishing the framework for a direct current power smart home network [27]. The system described in [26], may be classified as an intelligent power router. The energy provided is partitioned into many payload units, each accompanied by a header and footer for the purpose of identifying its network flow route. This arrangement results in the formation of "energy packets". Hence, the viability of the energy packet idea is based on its capacity to regulate power via the manipulation of energy packet quantities. Consequently, most contemporary household gadgets operate on DC and are equipped with an integrated power conversion mechanism to transform AC voltage input. Therefore, the DC-system demonstrated has the potential capacity to provide simpler, more effective, and more efficient management of energy flow throughout the distribution phase.

5.5   COMMUNICATION INFRASTRUCTURE

The primary function of the intelligent communication system is to facilitate the interconnection and dissemination of data among the many devices, applications, and systems that make up the infrastructure of the intelligent grid. The comprehension of network integration, including utilities, wide area network, separate field, and local connectivity, remains ambiguous. [28]. However, in accordance with the envisioned future smart grid, the communication system should have the following characteristics:

- Maintain the quality of service of the data while ensuring the preservation of vital data integrity.
- Possess a high level of reliability and interoperability to facilitate the connectivity of diverse networks.
- Be widely accessible and possess a high level of coverage.
- Ensure security and privacy.

The design of communication solutions should prioritize the consolidation and gathering of data from various sources such as power lines, wireless sensor networks (WSNs), wireless technologies, public mobile networks, and both private and public optical fibers. This design should adhere to the IEC 62056 standard, which is widely adopted and enables seamless interoperability and compatibility among different manufacturers of smart meters. These technologies facilitate the implementation of volt-var control devices, supplementary services, automated fault detection, sophisticated distribution management systems (DMSs), and the eventual establishment of microgrids in both industrial and residential sectors.

Volt-var control using smart inverters with reactive power capacity is extensively implemented in the context of distribution systems. This is primarily due to the potential occurrence of voltage congestion resulting from the integration of distributed generators, which necessitates the provision of reactive power support. [29]. Hence, the measuring of voltage at the distribution level is relevant. Given the premise of the presence of a sophisticated communication system, the deployment of volt-var control will be implemented according to the voltage measurements of the MV buses, voltage measurements of the renewable integrated nodes, and information exchange between the control center and the MV-level generators.

5.5.1 COMMUNICATION TECHNOLOGIES

5.5.1.1 Wireless

Wireless technologies provide many technical advantages in relation to distant end applications, including reduced installation costs, enhanced mobility, and the ability to be rapidly deployed. [30]. Various wireless technologies, such as cellular communication, wireless mesh networks, wireless communication based on IEEE 802.15.4, satellite communication, and microwave communication, have emerged as prominent technologies. Satellite communication has comprehensive worldwide coverage, making it a viable option for the purposes of remote monitoring and control, particularly in rural settings [31].

The application of this system is shown by its implementation in distant wind energy systems [32]. Additionally, the communication system may be conveniently established to guarantee fail-safe operation, facilitate the transmission of crucial data, and serve as a backup communication system. However, it is important to note that satellite communication exhibits a significant level of communication latency, and the characteristics of the channel are greatly influenced by weather conditions and the occurrence of fading. The qualities mostly contribute to the deterioration of satellite communication performance within the context of the smart grid environment. A wireless mesh network (WMN) utilizes a mesh architecture to provide connectivity among radio nodes. Next-generation wireless communication and networking have been recognized as reliant on this technology [33].

Currently there are active working groups that are dedicated to the standardization and definition of wireless mesh networks, such as IEEE 802.11 and IEEE 802.16. The WMN has the capability to provide the fundamental framework for communication systems inside the smart grid infrastructure [34]. In order to fulfill the necessary criteria for the SG, a communication topology consisting of many gateways is proposed in [35]. The suggested framework relies on a multi-gate mesh network to provide a versatile design that can be expanded using hybrid tree routing, as specified by the IEEE 802.11s standard. The primary objective of this structure is to enhance the dependability and autonomous recovery capabilities of smart grids while also improving their overall throughput performance.

The primary advantages of WMN technology in the context of the SG include enhanced dependability, connection, and extensive coverage [31]. Wireless technologies that use the 802.15.4 standard are strongly recommended for implementation in smart grid infrastructure [34]. These technologies are developed upon ZigBee, ISA100.11a, and WirelessHART. Among the several technologies under consideration, ZigBee demonstrates a greater degree of applicability owing to its specific design for radio-frequency applications. Hence, it enables extended battery longevity and a reduced data transmission rate while maintaining a robust level of network security. The use of this technology is prevalent within the communication infrastructure pertaining to intelligent metering systems, as implemented by diverse electric utility companies. Moreover, it has been officially recognized as a communication standard by the U.S. National Institute of Standards and Technology [36], supporting advanced metering, demand response, and real-time pricing, among other smart grid activities [37].

5.5.1.2   Wired

It is anticipated that wired technologies using fiber-optics and powerlines would be included in the smart grid architecture. Fiber-optic communication has inherent resilience to electromagnetic and radio interference, making it a crucial component of forthcoming intelligent communication systems due to its high bandwidth capabilities [31]. The present communication technology has achieved extensive deployment, accompanied by a substantial surplus of unutilized space capacity. Nevertheless, the cost of installation for this system is significantly expensive. In contrast, power line communication (PLC) technology is widely used by various electric utility firms on a worldwide scale. However, more study is necessary to determine the suitable applicative breadth of this phenomenon. The presence of powerlines gives rise to a security risk [38]. Nevertheless, PLCs are the only wired technology that can rival wireless technology in terms of cost-effectiveness [39]. According to existing research, the use of PLC is highly suitable for facilitating communication in smart metering applications, particularly within the medium and low voltage levels. [40]. The reason for this is that the existing PLC communication network facilitates the smooth transfer of a significant volume of data between the utility and the grid.

5.5.2   MANAGEMENT OF END-TO-END COMMUNICATION

In the context of a comprehensive smart grid infrastructure, it is essential to provide a distinct and exclusive identification to each individual device. The management of end-to-end communication is a critical concern due to the integration of heterogeneous networks with many kinds of devices. Currently, there is a growing trend towards the use of TCP/IP technology as a reliable and standardized solution for addressing the end-to-end communication management challenges in smart grid systems. [41]–[43]. This technology offers a simple approach for integrating historical data from traditional communication systems, as well as other TCP/IP protocols that are incompatible, by using gateway functionality, semi-transparent tunneling, and encapsulation techniques. NIST also emphasizes and delineates the benefits of this technology in relation to its technical maturity, the availability of tools, its application in smart grid systems, and its widespread employment in both public and private networks [36].

5.6   INFORMATION FLOW

The development of a smart grid is dependent not only on modern technologies of power devices but also largely on highly developed computer analysis, optimization, control, and monitoring at the generation, transmission, and distribution levels. Concerns about the grid's interoperability of data interchange and process are addressed by the smart information, which focuses on components, applications, and systems that are already in use as well as those that will be developed in the future. [44]. Therefore, the intelligent information system enables the processes of information modeling, creation, optimization, integration, and analysis.

The intelligent information system consists of information metering and measuring capabilities that enable data creation from various end-users, such as sensors, phasor measurement units, and smart meters, inside a smart grid infrastructure. This knowledge is relevant to the advancement of the energy market and the optimization of energy resources. The primary objectives of the intelligent information solutions include facilitating remote meter reading, managing customer relationships, integrating tariff structures, and enabling software updates to support bidirectional communication. The primary advantages are the reduction of operational expenses, elimination of system losses, and enhancement of service continuity.

The indirect advantages include the decrease of peak power demand, the establishment of an optimum energy mix for generation, enhanced availability of fuel, and lower emissions of greenhouse gases. The customer-side technologies include the use of power metering for the purpose of establishing localized communication systems, often referred to as home area networks (HANs). HAN communication employs a combination of wireless and wired connections, as dictated by the underlying architecture. The ownership of the HAN communication infrastructure is intended to be held by the utility or a public/private network in which the utility has made investments, either independently or in collaboration with other businesses.

5.6.1 Metering, Monitoring, and Measurement

The domain of metering and measurement encompasses the areas of smart metering, measurement, and monitoring. Smart meters serve as the fundamental components of smart grid systems, enabling the collection of data from end-users and facilitating the control of device operations. Automatic meter infrastructures (AMIs) are extensively used for the establishment of smart metering networks inside smart grids. AMI technology enables the automated transmission and interchange of data related to data collection, diagnostics, and consumption from the smart metering device. This data is used for purposes such as invoicing, analysis, and optimization inside the central database system. The presence of two-way communication in AMI distinguishes it from AMR technology. Consequently, the whole set of information is readily accessible upon request and in real time [45].

A smart meter is capable of measuring and recording the amount of power used within a period of hours or less. This data is then sent to either the utility company or a central controller, where it is utilized for billing, monitoring, and optimization objectives. [46]. In addition, smart meters include the capability to be remotely removed and reconnected, enabling control of a user's electricity use. This feature underscores the potential relevance of smart meters in demand-side management. The smart meter provides special benefits to both the consumer and the utility. For example, individuals may make estimations and efficiently reduce their energy expenses, while utility companies can use data from smart meters to detect and enhance system efficiency during periods of high demand, therefore ensuring an optimum power distribution.

The monitoring and measurement of the grid status is a crucial function of the smart grid. The primary technical techniques that we emphasize on this topic are the use of sensors and phasor measuring units (PMUs). Sensors are being used in many applications for diverse monitoring and measuring purposes [47]. The utilization of sensor networks for the detection of tower collapse, conductor failure, and network congestion is formulated in [48]. This approach aims to obtain real-time information and evaluate the electrical and mechanical condition of the electrical network system at the generation, transmission, and distribution levels. This will allow us to develop a real-time diagnostic system to prevent impending and irreversible failures, as well as to identify appropriate control strategies and either apply them automatically or recommend them to the operators.

WSNs provide a viable and cost-effective solution for establishing a communication network in the realm of wireless technology [49]. The research in [50] explores the practical applications of WSNs in electric power systems, with a focus on the associated problems and possibilities. The research suggests that by implementing effective isolation measures, it is possible to identify a single contingency system and prevent it from leading to a catastrophic failure. The study in [51] introduces a closed-loop energy management method using WSNs. The effectiveness of this strategy is then shown via its application to an industrial energy system. The effectiveness of WSNs is emphasized due to their non-intrusive and cost-effective characteristics.

However, the sensor network system is subject to several requirements pertaining to its quality of service, maintenance, configuration, security, environmental conditions, and resource limits [31]. Previously, the primary emphasis of the phasor measurement unit was centered on the validation of system models and the study of events after they occurred. The acquisition of data in real-time on a large scale, specifically related to power quality characteristics of the power system, is crucial for implementing intelligent protection mechanisms and facilitating additional protective features inside the power grid [52]. The advancement and evolution of communication technologies have sparked a keen interest in the use of phasor measurement units for the evaluation and quantification of power quality inside smart grid infrastructure.

From a technical standpoint, it can be highlighted that PMU measurements are represented as complex values, including both magnitude and phase angle. These readings are used to determine the characteristics of sine waves in the context of electric power flow. Typically, the acquisition of the reading occurs from a decentralized source and then aligns with the temporal reference provided by the radio clock of the global positioning system (GPS). Consequently, a substantial quantity of placements will enable operators and power system planners to get a significant volume of precise grid data, thus facilitating the dynamic regulation of the system [53]. Hence, it is anticipated that PMUs can facilitate a cost-effective approach for the acquisition and analysis of accurate synchrophasor grid data, which can be utilized for enhancing grid performance, including power quality monitoring [54], monitoring of active distribution grid [55], detection of disturbance [56], fault event monitoring [57], and loss-of-mains protection [58].

5.6.2 PROCESSING AND MANAGEMENT

The smart grid infrastructure is expected to create a significant volume of data and information via various processes such as sensing, monitoring, and metering. The use of data modeling is crucial in order to develop a well-structured future energy system by effectively examining and establishing connections with past data. Hence, the optimization of smart information system performance necessitates the implementation of effective information management strategies including data modeling, processing, analysis, and integration. Data modeling plays a crucial role in preserving the integrity of information communication between two components within the smart grid system. The significance of data arises when it can be efficiently shared between two components of an application and afterwards fulfill their respective functions.

Therefore, developers may endeavor to preserve the semantic content and informational value of data inside a single application across different interfaces. However, the meaning and representation of the data might eventually be compromised throughout the process of exchange owing to compatibility issues. Data modeling plays a crucial role in ensuring the compatibility of a system with both the present conventional aspects and future advancements of the power grid. Hence, it is essential to use efficient data modeling techniques while addressing the intricate process of transitioning to smart grids on a wide scale. According to the IEEE P2030, ontology is a viable choice due to its growing popularity in data modeling and its ability to provide formal semantics via common understanding. Ontology, similar to a programming language, facilitates the establishment of a formal framework for data modeling, characterized by precise elucidation and standardization.

Furthermore, this data modeling technique offers the possibility to change data models with convenient exportability and translation to an extensible markup language, which facilitates a significant level of information interoperability. However, it is essential to establish universal standards to facilitate interoperability on a broader scope. This encompasses the integration of all components and devices inside the smart grid infrastructure, which serves as a comprehensive framework for generating compatible, modifiable, and transferable data for the evolving smart grid. Consequently, the examination of data is necessary to substantiate the amalgamation, manipulation, comprehension, association, and enhancement of a substantial volume of grid-based observational data sets.

Given the anticipated widespread installation of metering, monitoring, and measuring components, there is a substantial need for extensive data analysis. This analysis is crucial for effectively strategizing grid enhancements and developing tailored analytics for self-service modeling. Hence, the objective of information integration is to consolidate diverse conceptual, typographical, and contextual data representations originating from several sources. The concept of information integrability is significant in facilitating the interoperability of forthcoming smart grid systems. The integration of data provided by the new smart grid components with current applications is essential.

Additionally, it is important to blend the historical metadata of the conventional system, kept in the legacy system, with these new applications to facilitate new interpretations. Ensuring the preservation of data integrity necessitates the use of cross-verification mechanisms to ascertain its authenticity. Additionally, the existing utility grid infrastructure has a constrained capacity for the integration of information. This circumstance will inevitably give rise to a situation where there is a concentrated accumulation of information, resulting in the emergence of technical obstacles pertaining to the establishment of a smart grid. These obstacles are specifically concerned with customer operation, system planning, system integration, and power supply [59].

Moreover, the optimization of the information processes is of utmost importance in augmenting the efficacy of information flow. Once again, the anticipated magnitude of data collection and storage is substantial inside the future smart grid framework. This huge quantity of information may include redundancies and extraneous material. Hence, the use of sophisticated information technologies is necessary for reducing the strain on communication. Numerous potential proposals and investigations have been put forward in this regard. A data compression methodology that relies on the wavelet decomposition approach is presented in [60]. The presented methodology seeks to mitigate and minimize data congestion while improving the overall quality of data transmission. Accordingly, the study in [61] introduces a technique based on singular value decomposition analysis to investigate the possible decrease of network traffic in the electrical power grid. The formulated methodology facilitates the execution of suitable control measures required to uphold exclusive communication and transmission of pertinent data among the components of the intelligent power grid.

5.7 CONCLUSION

This chapter provides an overview of the smart grid concept and its associated infrastructure components. The implementation of smart grids requires careful organization and strategic planning due to their inherent complexity. Despite the promising nature of smart grid technology, it is essential to establish a comprehensive blueprint that encompasses advanced projection initiation, planning, development, and execution to achieve realistic and pragmatic deployment. The concept of a smart grid involves the integration of smart and intelligent management capabilities inside the system. However, majority of the implemented advancements in information and communication technology are focused on enhancing energy efficiency, achieving a balance between energy demand and production, optimizing the usage of energy assets, controlling emissions, and reducing operating costs. Concurrently, it is essential to initiate and facilitate the marketing of smart grid technologies to garner client acceptance and adoption of modern energy management frameworks while accelerating the development of secure and interoperable operation.

Security and privacy are crucial elements of smart grid technology that need regulatory supervision. Given the substantial expenses associated with renewable energy sources and the implementation of smart grid infrastructures, it is seen that utilities often prioritize cost reduction and profit maximization above the full use of the protective features offered by smart grids. Additionally, there is a potential danger of privacy breaches that arises when the utility company decides to delegate the management of the smart information system to other entities, such as a cloud storage provider. While such arrangements give improved information and communication flexibility, leading to improved interoperability, the trade-off is a diminution in the utility's control, presenting a threat to the privacy of its clients.

The quantification of smart grid success and the mitigation of additional investments in the future are both influenced by the element of interoperability. First, it is crucial to acknowledge the existence of multiple geographical, social, and diverse networks that are envisioned to be interconnected. Additionally, there are various technological possibilities in the domains of energy systems, information, and communication for the infrastructure of smart grids. Given these factors, it becomes imperative to develop a global standardized framework that can provide guidance for the establishment of interoperable smart grids. This facilitates the efficient gathering, handling, examination, enhancement, and transmission of data that is relevant for the integration and implementation of the primary goal of smart grids, the establishment of self-repairing, dependable, economical, adaptable, and environmentally friendly energy systems.

5.8 ACKNOWLEDGMENTS

The authors would like to express their profound gratitude to King Abdullah City for Atomic and Renewable Energy (K.A. CARE) for their financial support in accomplishing this work at King Fahd University of Petroleum & Minerals, Dhahran 31261, Saudi Arabia.

REFERENCES

[1] H. O. R. Howlader, H. Matayoshi, and T. Senjyu, "Distributed generation integrated with thermal unit commitment considering demand response for energy storage optimization of smart grid," *Renew. Energy*, vol. 99, pp. 107–117, 2016.

[2] K. Cheung, X. Wang, B.-C. Chiu, Y. Xiao, and R. Rios-Zalapa, "Generation dispatch in a smart grid environment," in *2010 Innovative Smart Grid Technologies (ISGT)*, Gothenburg, 2010, pp. 1–6, doi: 10.1109/ISGT.2010.5434781.

[3] I. Bilibin and F. Capitanescu, "Contributions to thermal constraints management in radial active distribution systems," *Electr. Power Syst. Res.*, vol. 111, pp. 169–176, 2014.

[4] F. De Angelis, M. Boaro, D. Fuselli, S. Squartini, F. Piazza, and Q. Wei, "Optimal home energy management under dynamic electrical and thermal constraints," *IEEE Trans. Ind. Inform.*, vol. 9, no. 3, pp. 1518–1527, 2013.

[5] S. Mohtashami, D. Pudjianto, and G. Strbac, "Strategic distribution network planning with smart grid technologies," *IEEE Trans. Smart Grid*, vol. 8, no. 6, pp. 2656–2664, 2017.

[6] A. Vaccaro, G. Velotto, and A. F. Zobaa, "A decentralized and cooperative architecture for optimal voltage regulation in smart grids," *IEEE Trans. Ind. Electron.*, vol. 58, no. 10, pp. 4593–4602, 2011.

[7] V. Loia and A. Vaccaro, "A decentralized architecture for voltage regulation in smart grids," *Proc.—ISIE 2011 IEEE Int. Symp. Ind. Electron.*, no. 1, pp. 1679–1684, 2011.

[8] J. Zhong et al., "Coordinated control for large-scale EV charging facilities and energy storage devices participating in frequency regulation," *Appl. Energy*, vol. 123, pp. 253–262, 2014.

[9] A. Y. S. Lam, K. C. Leung, and V. O. K. Li, "Capacity estimation for vehicle-to-grid frequency regulation services with smart charging mechanism," *IEEE Trans. Smart Grid*, vol. 7, no. 1, pp. 156–166, 2016.

[10] A. Keyhani and A. Chatterjee, "Automatic generation control structure for smart power grids," *IEEE Trans. Smart Grid*, vol. 3, no. 3, pp. 1310–1316, 2012.

[11] B. Mahdad and K. Srairi, "Blackout risk prevention in a smart grid based flexible optimal strategy using Grey Wolf-pattern search algorithms," *Energy Convers. Manag.*, vol. 98, pp. 411–429, 2015.

[12] I. Alotaibi, M. A. Abido, M. Khalid, and A. V Savkin, "A comprehensive review of recent advances in smart grids: A sustainable future with renewable energy resources," *Energies*, vol. 13, no. 23, p. 6269, 2020.

[13] L. Gelazanskas and K. A. A. Gamage, "Demand side management in smart grid: A review and proposals for future direction," *Sustain. Cities Soc.*, vol. 11, pp. 22–30, 2014.

[14] Q. Sun et al., "Review of smart grid comprehensive assessment systems," *Energy Procedia*, vol. 12, pp. 219–229, 2011.

[15] M. L. Tuballa and M. L. Abundo, "A review of the development of smart grid technologies," *Renew. Sustain. Energy Rev.*, vol. 59, pp. 710–725, 2016.

[16] F. Calise, N. Duic, A. Pfeifer, M. Vicidomini, and A. M. Orlando, "Moving the system boundaries in decarbonization of large islands," *Energy Convers. Manag.*, vol. 234, p. 113956, 2021.

[17] A. Al-Karaghouli and L. L. Kazmerski, "Energy consumption and water production cost of conventional and renewable-energy-powered desalination processes," *Renew. Sustain. Energy Rev.*, vol. 24, pp. 343–356, 2013.

[18] G. Pepermans, "European energy market liberalization: Experiences and challenges," *Int. J. Econ. Policy Stud.*, vol. 13, no. 1, pp. 3–26, 2019.

[19] A. Molderink, V. Bakker, M. G. C. Bosman, J. L. Hurink, and G. J. M. Smit, "Management and control of domestic smart grid technology," *IEEE Trans. Smart Grid*, vol. 1, no. 2, pp. 109–119, Sept. 2010, doi: 10.1109/TSG.2010.2055904.

[20] P. B. Andersen, B. Poulsen, M. Decker, C. Traeholt, and J. Ostergaard, "Evaluation of a generic virtual power plant framework using service oriented architecture," in *2008 IEEE 2nd International Power and Energy Conference*, Johor Bahru, Malaysia, 2008, pp. 1212–1217, doi: 10.1109/PECON.2008.4762651.

[21] L. H. Tsoukalas and R. Gao, "From smart grids to an energy internet: Assumptions, architectures and requirements," in *2008 Third International Conference on Electric Utility Deregulation and Restructuring and Power Technologies*, Nanjing, China, 2008, pp. 94–98, doi: 10.1109/DRPT.2008.4523385.

[22] T. Jin and M. Mecheloul, "Ordering electricity via internet and its potentials for smart grid systems," *IEEE Trans. Smart Grid*, vol. 1, no. 3, pp. 302–310, Dec. 2010, doi: 10.1109/TSG.2010.2072995.

[23] B. Jansen, C. Binding, O. Sundström, and D. Gantenbein, "Architecture and communication of an electric vehicle virtual power plant," in *2010 First IEEE International Conference on Smart Grid Communications*, Gaithersburg, MD, 2010, pp. 149–154, doi: 10.1109/SMARTGRID.2010.5622033.

[24] F. Li et al., "Smart transmission grid: Vision and framework," *IEEE Trans. Smart Grid*, vol. 1, no. 2, pp. 168–177, 2010.

[25] A. Bose, "Smart transmission grid applications and their supporting infrastructure," *IEEE Trans. Smart Grid*, vol. 1, no. 1, pp. 11–19, 2010.

[26] T. Takuno, M. Koyama, and T. Hikihara, "In-home power distribution systems by circuit switching and power packet dispatching," in *2010 First IEEE International Conference on Smart Grid Communications*, Gaithersburg, MD, 2010, pp. 427–430, doi: 10.1109/SMARTGRID.2010.5622079.

[27] T. Takuno, T. Hikihara, T. Tsuno, and S. Hatsukawa, "HF gate drive circuit for a normally-on SiC JFET with inherent safety," in *2009 13th European Conference on Power Electronics and Applications*, Barcelona, Spain, 2009, pp. 1–4.

[28] M. Souryal, C. Gentile, D. Griffith, D. Cypher, and N. Golmie, "A methodology to evaluate wireless technologies for the smart grid," in *2010 First IEEE International Conference on Smart Grid Communications*, Gaithersburg, MD, 2010, pp. 356–361, doi: 10.1109/SMARTGRID.2010.5622067.

[29] K. R. Babu and D. K. Khatod, "Improved volt/var control technique for over-voltage mitigation," in *2022 Second International Conference on Power, Control and Computing Technologies (ICPC2T)*, Raipur, India, 2022, pp. 1–6, doi: 10.1109/ICPC2T53885.2022.9776935.

[30] P. P. Parikh, M. G. Kanabar, and T. S. Sidhu, "Opportunities and challenges of wireless communication technologies for smart grid applications," in *IEEE PES General Meeting*, Minneapolis, MN, 2010, pp. 1–7, doi: 10.1109/PES.2010.5589988.

[31] V. C. Gungor and F. C. Lambert, "A survey on communication networks for electric system automation," *Comput. Networks*, vol. 50, no. 7, pp. 877–897, May 2006.

[32] U. D. Deep, B. R. Petersen, and J. Meng, "A smart microcontroller-based iridium satellite-communication architecture for a remote renewable energy source," *IEEE Trans. Power Deliv.*, vol. 24, no. 4, pp. 1869–1875, 2009.

[33] I. F. Akyildiz and X. Wang, "A survey on wireless mesh networks," *IEEE Commun. Mag.*, vol. 43, no. 9, Sep. 2005.

[34] B. A. Akyol, H. Kirkham, S. L. Clements, and M. D. Hadley, "A survey of wireless communications for the electric power system," Richland, WA (United States), Jan. 2010.

[35] H. Gharavi and B. Hu, "Multigate communication network for smart grid," *Proc. IEEE*, vol. 99, no. 6, pp. 1028–1045, 2011.

[36] "NIST framework and roadmap for smart grid interoperability standards, release 1.0," Gaithersburg, MD, Jan. 2010.

[37] P. Yi, A. Iwayemi, and C. Zhou, "Developing ZigBee deployment guideline under WiFi interference for smart grid applications," *IEEE Trans. Smart Grid*, vol. 2, no. 1, pp. 110–120, 2011.

[38] J. Li, C. C. Liu, and K. P. Schneider, "Controlled partitioning of a power network considering real and reactive power balance," *IEEE Trans. Smart Grid*, vol. 1, no. 3, pp. 261–269, Dec. 2010.

[39] S. Galli, A. Scaglione, and Z. Wang, "Power line communications and the smart grid," in *2010 First IEEE International Conference on Smart Grid Communications*, Gaithersburg, MD, 2010, pp. 303–308, doi: 10.1109/SMARTGRID.2010.5622060.

[40] D. W. Rieken and M. R. Walker, "Ultra low frequency power-line communications using a resonator circuit," *IEEE Trans. Smart Grid*, vol. 2, no. 1, pp. 41–50, 2011.

[41] C. Bennett and D. Highfill, "Networking AMI smart meters," in *2008 IEEE Energy 2030 Conference*, IEEE, Nov. 2008, pp. 1–8.

[42] F. Lobo et al., "How to design a communication network over distribution networks," in *CIRED 2009—The 20th International Conference and Exhibition on Electricity Distribution—Part 2*, Prague, 2009, p. 1, doi: 10.1049/cp.2009.0890.

[43] T. Sauter and M. Lobashov, "End-to-end communication architecture for smart grids," *IEEE Trans. Ind. Electron.*, vol. 58, no. 4, pp. 1218–1228, Apr. 2011.

[44] "IEEE guide for smart grid interoperability of energy technology and information technology operation with the electric power system (EPS), end-use applications, and loads," in *IEEE Std 2030-2011*, Sept. 2011, pp. 1–126, doi: 10.1109/IEEESTD.2011.6018239.

[45] S. Hatami and M. Pedram, "Minimizing the electricity bill of cooperative users under a quasi-dynamic pricing model," in *2010 First IEEE International Conference on*

Smart Grid Communications, Gaithersburg, MD, 2010, pp. 421–426, doi: 10.1109/SMARTGRID.2010.5622080.

[46] Y. Wang, Q. Chen, T. Hong, and C. Kang, "Review of smart meter data analytics: Applications, methodologies, and challenges," *IEEE Trans. Smart Grid*, vol. 10, no. 3, pp. 3125–3148, May 2019.

[47] I. F. Akyildiz, W. Su, Y. Sankarasubramaniam, and E. Cayirci, "A survey on sensor networks," *IEEE Commun. Mag.*, vol. 40, no. 8, pp. 102–105, 2002.

[48] R. A. León, V. Vittal, and G. Manimaran, "Application of sensor network for secure electric energy infrastructure," *IEEE Trans. Power Deliv.*, vol. 22, no. 2, pp. 1021–1028, Apr. 2007.

[49] B. Lu, T. G. Habetler, R. G. Harley, J. A. Gutierrez, and D. B. Durocher, "Energy evaluation goes wireless," *IEEE Ind. Appl. Mag.*, vol. 13, no. 2, pp. 17–23, 2007.

[50] V. C. Gungor, B. Lu, and G. P. Hancke, "Opportunities and challenges of wireless sensor networks in smart grid," *IEEE Trans. Ind. Electron.*, vol. 57, no. 10, pp. 3557–3564, Oct. 2010.

[51] B. Lu, T. G. Habetler, R. G. Harley, and J. A. Gutiérrez, "Applying wireless sensor networks in industrial plant energy management systems—Part I: A closed-loop scheme," *Proc. IEEE Sens.*, vol. 2005, pp. 145–150, 2005.

[52] A. Armenia and J. H. Chow, "A flexible phasor data concentrator design leveraging existing software technologies," *IEEE Trans. Smart Grid*, vol. 1, no. 1, pp. 73–81, 2010.

[53] J. De La Ree, V. Centeno, J. S. Thorp, and A. G. Phadke, "Synchronized phasor measurement applications in power systems," *IEEE Trans. Smart Grid*, vol. 1, no. 1, pp. 20–27, 2010.

[54] A. Carta, N. Locci, and C. Muscas, "GPS-based system for the measurement of synchronized harmonic phasors," *IEEE Trans. Instrum. Meas.*, vol. 58, no. 3, pp. 586–593, 2009.

[55] A. Borghetti, C. A. Nucci, M. Paolone, G. Ciappi, and A. Solari, "Synchronized phasors monitoring during the islanding maneuver of an active distribution network," *IEEE Trans. Smart Grid*, vol. 2, no. 1, pp. 82–91, 2011.

[56] J. Ma, P. Zhang, H. J. Fu, B. Bo, and Z. Y. Dong, "Application of phasor measurement unit on locating disturbance source for low-frequency oscillation," *IEEE Trans. Smart Grid*, vol. 1, no. 3, pp. 340–346, Dec. 2010.

[57] J. Zhu and A. Abur, "Improvements in network parameter error identification via synchronized phasors," *IEEE Trans. Power Syst.*, vol. 25, no. 1, pp. 44–50, Feb. 2010.

[58] D. M. Laverty, D. J. Morrow, R. J. Best, and P. A. Crossley, "Differential ROCOF relay for loss-of-mains protection of renewable generation using phasor measurement over internet protocol," in *2009 CIGRE/IEEE PES Joint Symposium Integration of Wide-Scale Renewable Resources Into the Power Delivery System*, Calgary, AB, Canada, 2009, p. 1.

[59] J. R. Roncero, "Integration is key to smart grid management," *IET Semin. Dig.*, vol. 2008, no. 12380, 2008.

[60] J. Ning, J. Wang, W. Gao, and C. Liu, "A wavelet-based data compression technique for smart grid," *IEEE Trans. Smart Grid*, vol. 2, no. 1, pp. 212–218, 2011.

[61] Z. Wang, A. Scaglione, and R. J. Thomas, "Compressing electrical power grids," in *2010 First IEEE International Conference on Smart Grid Communications*, Gaithersburg, MD, 2010, pp. 13–18, doi: 10.1109/SMARTGRID.2010.5622005.

6 Nuclear Power
Will It Be Part of the 21st-Century Sustainable Energy Transition?

Barry Solomon

6.1 INTRODUCTION

Several factors have historically made an energy source desirable. These include its local or regional supply, ownership and competing uses (e.g. is production restricted?), energy density, economic cost (e.g. is it cheap and easily accessible or costly and difficult to produce?), technological efficiency, environmental effects (e.g. air pollution, climate change, water pollution, waste, etc.) and more recently whether it is renewable or non-renewable. In the case of nuclear power, its proponents generally identify its lack of air pollution and greenhouse gas emissions as its main selling points, along with the domestic availability of uranium fuel resources in the case of major producers such as the U.S. and Canada. Also, from the 1950s until the early 1970s, nuclear power was promoted as a low-cost energy source. Upon close scrutiny, the first claim holds up better than the other two. For example, nuclear power generation does not emit conventional air pollution, though it does release a small amount of radioactive air pollution (Solomon, 2020). The lifecycle greenhouse gas emissions of nuclear generation are indeed small, similar to the emissions from a range of renewable energy technologies used for electricity generation (Sovacool, 2008; Warner & Heath, 2012; Turconi et al., 2013). Regarding the domestic production of uranium resources for domestic nuclear plants, however, this has depended upon market conditions, as some countries with plentiful uranium resources have shifted to greater reliance on lower-cost imports. For instance, the U.S. has since the early 1990s imported most of its uranium fuel from Kazakhstan, Russia, Canada, Australia and elsewhere (EIA, 2021a). As for the cost of nuclear power, its construction costs have rapidly escalated since the 1960s in the U.S. and Germany and to a lesser extent in France, Canada and Japan[1] (Cohn, 1997; Lovering et al., 2016).

Energy transitions have normally been extremely slow (Kramer & Haigh, 2009; Smil, 2010; Solomon & Krishna, 2011). The first major global energy transition, from wood to fossil fuels, occurred at different speeds in different economic sectors over 80 to 400 years (Fouquet, 2010). However, transitions can occur more quickly on national and regional scales, especially for individual sectors, greater use of energy

DOI: 10.1201/9781032715438-6

efficiency technologies and in smaller and less complex societies (Verbong & Geels, 2007; Solomon & Krishna, 2011; Bromley, 2016).

Nuclear power was first conceived as a possible commercial energy source during the efforts to win World War II during the 1940s. Separate research programs were underway in the U.S., UK and USSR, which not so coincidentally were the first three countries to demonstrate the peaceful use of nuclear energy. The U.S. opened a very small experimental breeder reactor in Idaho in December 1951, and two years later President Dwight Eisenhower delivered his famous "Atoms for Peace" lecture before the United Nations General Assembly in December 1953 (Weiss, 2003; WNA, 2020a). But it was the former Soviet Union that was first to successfully build and connect a nuclear power plant to an electricity grid in June 1954 in Obninsk, with a 5 MWt plutonium production reactor (IAEA, 2017). This was followed by commercial nuclear plants in the UK (1956) and U.S. (1957) (Blowers et al., 1991, p. 4). As will be noted in the next section of this chapter, global nuclear power production appeared to peak around 50 years later, in 2006, when it provided less than 15% of the global electricity supply and a little over 5% of the total primary energy consumption. This raises the distinct possibility that further expansion of nuclear power has been stunted, most likely due to its high costs and its various controversies, such as risk of accidents and nuclear weapons proliferation and inadequate waste disposal, among others (Hsu, 2021; Blowers et al., 1991; Whitfield et al., 2009).

This chapter will explore the potential for nuclear power to play a role in a 21st-century energy transition to help mitigate climate change. In the next section I will provide some historical background. While first commercialized in the late 1950s, nuclear power developed slowly, as only 14 nations had operating nuclear power plants by 1970 (many of them small), which eventually peaked at 32. Following this, I will provide a short history of nuclear power development by country, as well as the overall trends in nuclear power usage in the electricity mix through 2021. These developments were disrupted by three major nuclear accidents: Three Mile Island in the U.S. in 1979, Chernobyl in Ukraine in 1986 and Fukushima-Daiichi in Japan in 2011. The next section will discuss a series of bottom-up case studies of nuclear power's planned role in the few countries with significant nuclear expansion plans, especially China, India and South Korea, along with all other countries or regions that seek to add or expand their nuclear capacities. The penultimate section will review the findings of two recent comprehensive global energy scenario projections and models with respect to the role that nuclear power is likely to play in a global energy future over the mid to long term. The chapter will close with a summary and my conclusions.

6.2 A SHORT HISTORY OF NUCLEAR POWER DEVELOPMENT

6.2.1 1950s–1960s

This section will provide a short history of global nuclear power development, starting with the first reactors commissioned until the present day. The first four developers of commercial nuclear power were also the first four countries to build and test nuclear weapons: the U.S. (1945), USSR (1949), UK (1952) and France (1960). Thus, there was initially a close nexus between the military and civilian aspects of nuclear power, which U.S. President Eisenhower sought to change with his "Atoms for Peace" speech to the United Nations on December 8, 1953 (Weiss, 2003; Cirincione,

2007; Jasanoff & Kim, 2009). The power plant development order was slightly different, as the first non-experimental reactors opened in the USSR (1954), UK (1956), US (1957) and France (1962), respectively. One important factor that held up the development of nuclear power in the U.S. was the industry's concern about liability for potential accidents, which was addressed by passage of the Price-Anderson Act in 1957, a law written to partially compensate the nuclear industry and general public for liability claims following any accidents (Meek, 1978). Nevertheless, by 1960 there were 17 commercial nuclear reactors in operation in these countries with a combined generating capacity of 1,200 MWe, equivalent to one large nuclear power station today. Development accelerated in the 1960s, and by the end of the decade a total of 90 nuclear reactor units in 14 countries were operating with a combined capacity of 16,500 MWe (Char & Csik, 1987). New countries to add nuclear power plants in the 1960s after the initial four included Belgium, Italy, Sweden, Spain, Germany, Switzerland, the Netherlands, Canada, Japan and India (Fischer, 1997). By 1970, however, nuclear power accounted for only 2% of global electricity capacity.

6.2.2 1970s

While rapid nuclear power development continued in the early 1970s, with an average of 25–30 new nuclear reactor units being sited each year, two significant events on the world stage as well as a growing social development would begin to slow the pace. First, the oil embargo of the Organization of Petroleum Exporting Countries (OPEC) was imposed from October 19, 1973, until March 18, 1974, and the resultant price hikes and inflation led to much higher energy and electricity prices, declining economic growth, and declining electricity demand growth (Toth & Rogner, 2006). After initially boosting the growth of nuclear power as an alternative to oil (especially in France), this eventually led to the delay and cancellation of many coal-fired and nuclear power plant orders, especially in the U.S. (Ellis & Zimmerman, 1983; De Bondt & Makhija, 1988). Second, the accident at the U.S.'s Three Mile Island nuclear power plant in Pennsylvania occurred on March 28, 1979, the first such major accident in the world. In addition, in the mid to late 1970s there already was growing public opposition to the further development of nuclear power, due to the perceived accident risk and lack of radioactive waste disposal plans (Hohenemser et al., 1977; Kasperson et al., 1980). By 1980 there was a total of 253 nuclear reactor units operating with 135,000 MWe generating capacity in 22 countries, for an eightfold growth over the decade (Char & Csik, 1987). New countries to add nuclear power plants in the 1970s included Bulgaria, Slovakia, Ukraine (Chernobyl, which has since been decommissioned), Kazakhstan, Armenia, Finland, South Korea, Pakistan, Taiwan and Argentina (Fischer, 1997). However, this robust growth concealed the forces that were beginning to slow the advancement of nuclear power.

6.2.3 1980s

The 1980s was the last decade of rapid growth in nuclear power development, albeit much slower growth than in the 1970s as capacity and generation grew less than threefold (Char & Csik, 1987; Häfele, 1990). As the decade advanced, there were many fewer construction starts each year as construction costs rose in several key countries, though numerous past nuclear builds were completed (Char & Csik, 1987). The real price of electricity that had experienced rapid price growth following the OPEC oil embargo peaked in 1982 and then declined after the 1980–82 recessions for the rest of the 1980s (and 1990s). However, lower energy prices did not help nuclear power growth because the disastrous accident at the Chernobyl nuclear power plant in Ukraine, which occurred on April 26, 1986, had a chilling effect on the industry. France commissioned the world's largest breeder reactor, the Superphénix in 1985, but it was shut down only 13 years later after very high costs, poor performance and a less than 10% capacity factor.

New countries to add nuclear power plants in the 1980s were limited to Hungary, the Czech Republic, Lithuania, Slovenia, Brazil and South Africa (Fischer, 1997). Public opposition to nuclear power continued and grew in intensity in many nations, nuclear costs continued to escalate and the radioactive waste disposal problem remained unsolved (Renn, 1990; Van der Pligt, 1992; Blowers et al., 1991; Rosa & Dunlap, 1994). By the end of 1989, there were 426 nuclear reactor units operating with 318,271 MWe generating capacity in 27 countries. The U.S. and France alone accounted for almost half of this capacity (Häfele, 1990).

6.2.4   1990s

By the late 1980s and throughout the 1990s (and ever since), global warming and climate change became a prominent scientific issue and concern, which led to many calls to greatly increase the use of nuclear power in order to mitigate the most adverse effects (e.g., Niehaus & Mueller, 1990; Brunner, 1991; Peters & Slovic, 1996; Sailor, 2000). However, while nuclear power growth continued during the 1990s the pace greatly slowed, with its contribution to the overall global electricity mix peaking in 1996 at 17.7%. The global peak generation of nuclear power occurred 23 years later, at 2,669 terawatt-hours in 2019 (Table 6.1). Thus, it does not appear that concerns with global climate change have led to any noticeable effect on nuclear power's contribution to the global electricity profile. Only three countries added nuclear power capacity in the 1990s: Mexico (1990), Romania (1996) and most importantly China (1991).

6.2.5   21st Century

Very few new countries have added nuclear power generating capacity thus far in the 21st century. Those that have done so are Iran (2011) and most recently Belarus and United Arab Emirates in 2020 (WNA, 2021). While some analysts have continued to predict a renaissance in nuclear power development (e.g., Stulberg & Fuhrmann, 2013), it has yet to occur. Moreover, in addition to the high cost of nuclear power generation, another significant accident occurred, further setting back the industry. This accident was the disaster that occurred following the magnitude 9.0–9.1 Tohoku undersea megathrust earthquake and tsunami on March 11, 2011, at Fukushima-Daiichi in Japan, though the Chernobyl accident was actually more serious (Steinhauser et al., 2014). Even so, as noted subsequently, the Fukushima accident has led some major countries deciding to phase out the use of nuclear power.

TABLE 6.1
International Nuclear Energy Use over Time (Terawatt-Hours), Selected Years

Year	Generation	% of Total Electricity
1971	111	2.1%
1975	381	5.9%
1980	684	8.6%
1985	1,426	15.2%
1990	1,910	17.0%
1995	2,211	17.6%
1996	2,293	**17.7%**
2000	2,451	16.8%
2001	2,518	17.0%
2002	2,547	16.5%
2003	2,519	15.7%
2004	2,620	15.6%
2005	2,627	15.1%
2006	2,661	14.7%
2007	2,610	13.7%
2008	2,599	13.5%
2009	2,562	13.4%
2010	2,629	12.8%
2011	2,518	11.6%
2012	2,347	10.8%
2013	2,364	10.6%
2014	2,420	10.6%
2015	2,448	10.6%
2016	2,489	10.5%
2017	2,517	10.3%
2018	2,570	10.2%
2019	2,669	10.4%
2020	2,596	10.5%

Source: International Energy Agency, World Electricity Generation Mix by Fuel, 1971–, Paris; BP
    Statistical Review of World Energy, London, var. years.

6.2.6 NUCLEAR POWER PHASEOUTS AND MORATORIUMS

Numerous countries and states have created policies of varying degrees to place
moratoriums on nuclear power development or to phase it out. A complete phaseout
of nuclear power use is obviously a significant confounding factor that would hinder
the prospects for nuclear power to play a major role in an energy transition, though
such policies can change. This section will provide an overview of the current policies on nuclear phaseouts and moratoriums worldwide. Unless otherwise stated,
most of the information in the next two paragraphs is based on the recent review of
nuclear phase-out policies by Carrara (2020).

Sweden was the first country that decided to phase out nuclear power. Sweden's policy was based on the results of a non-binding national public referendum on the future of nuclear power. The referendum was held in March 1980, a year after the Three Mile Island accident in the U.S. (Granberg & Holmberg, 1986). This phaseout decision was later repealed by the Swedish Parliament in 2009–10. However, seven large nuclear reactors have been shut down or decommissioned in Sweden with no new ones built to replace them (Johansson, 2021), though two of the shutdown reactors will be reopened. Besides Sweden, several other European countries have adopted nuclear moratoriums and phaseouts. These include Italy, Germany, Spain, Belgium and Switzerland (Hsu, 2021). Spain initially placed a moratorium on building new nuclear plants in 1984. Italy passed a referendum in 1987 after the Chernobyl accident of 1986 and completed its phaseout in 1990. Belgium approved a nuclear phaseout in 2003. The German and Swiss phaseouts were approved after the Fukushima accident in 2011. The German phaseout was completed on April 15, 2023, though no end date has yet been set in Switzerland. Belgium and Spain will complete their phaseouts by the end of 2035. Austria also built a nuclear power plant in the late 1970s, but it was never opened.

There have been very few nuclear phaseout decisions taken outside of Europe. More than a dozen U.S. states have moratoriums on the construction of new nuclear power plants, though the country allows new builds elsewhere (Jasanoff & Kim, 2009; Maury, 2019). In Asia, South Korea reversed its nuclear expansion plans in 2017 and initially planned to phase out nuclear power over 40 years but then reversed course again in 2021 by deciding to continue with nuclear power (Hosokawa, 2021). Taiwan also decided in 2017 to phase out nuclear power, in its case by 2025 (Wilcox, 2021). In addition, the Philippines built a nuclear power plant in the early 1980s, but the facility never opened.

6.3   NATIONAL/REGIONAL CASE STUDIES

6.3.1   INTRODUCTION

Almost all of the existing nuclear power generating capacity is fairly evenly divided between Europe, Asia and North America (Table 6.2). Decisions to construct, as well as decommission, nuclear power plants have long-lasting implications. Since the 1970s it had historically taken 7 to 12 years to build most nuclear plants, but in a few countries such as the U.S. construction times have often been longer, lasting two decades or more in some cases.[2] Once built, nuclear power plants are normally expected to remain in service for 40 years, though some plants have closed early. In recent years, however, these estimates have changed. For example, new nuclear plants can occasionally be built in 5 to 6 years (or even less), and a growing number of plants are having their operating licenses renewed and extended so they can run for up to 80 years (WNA, 2020b; Lovering et al., 2016; DOE, 2021). At the same time, however, there is a growing number of nuclear reactors being decommissioned, as many of the first generation of nuclear facilities that have reached the end of their operating licenses cannot be cost-effectively upgraded to remain in service. Thus,

TABLE 6.2

Nuclear Power Capacity by Continent, 2021 (GWe)

Europe	151.197
Asia	127.186
North America (incl. Mexico)	110.699
South America	3.525
Africa	1.860
Total	394.467

Source: IAEA (International Atomic Energy Agency). 2021; Operational & long-term shutdown reactors. Available at: https://pris.iaea.org/PRIS/WorldStatistics/OperationalReactorsByCountry.aspx.

several factors complicate an assessment of the future of nuclear power in any given country.

In this section, I will provide a series of national and regional (multi-national) case studies of expansion plans for nuclear power in the world' nuclear-powered countries and regions and where decisions made today will determine the future of nuclear power over the next 40 years and beyond. Unless otherwise indicated, most of the material in this section will be drawn from IAEA (2021) and WNA (2021). We begin with China.

6.3.2 CHINA

China was the last major country to develop nuclear power and, as noted earlier, commissioned its first nuclear power plant only in 1991. Since then, as its economy rapidly expanded, it has built many new reactors and power plants and today ranks second in total nuclear capacity, only trailing the U.S. (Table 6.3). China's nuclear plants are located along the eastern seaboard of the country, from the northeast to the southeast. Given its large and growing demand for electricity, however, China meets only around 5% of its total electricity demand with nuclear, with most of the rest coming from coal (though with a rapidly expanding contribution from wind and solar power as well). In the last decade, China has increased its commitment to reducing greenhouse gas emissions, which has stimulated faster development of nuclear and renewable sources of energy and a gradually decreasing reliance on coal.

As of this writing, China had by far the most nuclear power plants and reactors under construction, although its plans were scaled back somewhat in light of the Fukushima Daiichi nuclear disaster in 2011. A moratorium on approving new nuclear projects in China was announced on March 17, 2011, but after a safety review it was lifted in October 2012 (Hutchinson, 2018). Twenty-three reactors are currently under construction, with a rated capacity of 23,724 MWe, and many more units are planned. Perhaps complicating the nuclear situation in China was an incident that occurred at Unit 1 of its Taishan nuclear plant in Guangdong in June 2021, which

TABLE 6.3

Nuclear Power Generation Leaders in 2021 (GW-h)

U.S.	771,638	19.0%
China	383,205	4.9%
France	363,394	69.0%
Russia	208,443	20.0%
South Korea	150,456	28.0%
Canada	86,780	14.0%
Ukraine	81,126	55.0%
Germany	65,444	12.0%
Japan	61,304	4.0%
Spain	54,218	20.8%
Sweden	51,426	30.8%
Belgium	47,962	50.8%
UK	41,789	15.0%
India	39,758	3.0%
Czech Republic	29,044	36.6%

Source: Nuclear Energy Institute, 2021. Top 15 nuclear generating countries. Available at: www.nei.org/
resources/statistics/top-15-nuclear-generating-countries; IAEA (International Atomic Energy
Agency). 2021; Nuclear share of electricity generation in 2020. Available at: https://pris.iaea.org/
PRIS/WorldStatistics/NuclearShareofElectricityGeneration.aspx.

ultimately led to the plant being shut down for over a year because of damaged fuel rods and the need to vent radioactive gases (van der Made, 2021).

6.3.3 EAST ASIA

Besides China, not surprisingly, the two countries in East Asia most affected by the Fukushima Daiichi accident are Japan and South Korea. While Japan once had the third-highest nuclear power capacity in the world, with 54 reactors and 48.8 GWe of generating capacity, it closed all of its nuclear power plants one year after the accident (Suzuki, 2015). Following significant internal debate about the future of nuclear power, as of today Japan has only 12 of its reactors operational. These nuclear facilities account for 11 GWe of generating capacity and generate only 6.1% of Japan's electricity, well down from 29% before the accident. Only two reactors are currently under construction, rated at 2,653 MWe.

The nuclear situation in South Korea is quite different from Japan. Initially after the Fukushima accident, South Korea also seriously debated its nuclear future, and President Moon Jae-in in 2017 cancelled several projects that were in the planning states as well as life extensions on existing power plants. Moreover, the Moon government expressed a preference for an eventual phaseout of nuclear power in favor of natural gas and renewable energy (Solomon & Li, 2020, p. 156). However, as noted earlier, South Korea reversed course in 2021 by deciding to continue with nuclear power. There are currently two nuclear reactors under construction in South Korea, rated at 2,680 MWe.

6.3.4 SOUTHERN AND WESTERN ASIA

South Asia is also the location of substantial nuclear power development. India, as has been the case with China, experienced significant economic growth in the decade preceding the coronavirus pandemic. This has resulted in a rapidly increasing demand for electric power (Bandyopadhyay & Rej, 2021). While India's nuclear capacity is a small fraction of China's (Table 6.3), it has the second-largest new capacity under construction: eight reactors, with a rated capacity of 6,028 GWe. Nuclear power plants are also under construction in Bangladesh. While Bangladesh does not have any operating nuclear plants, two large reactors are under construction in Pabna. Iran also has a nuclear reactor unit under construction, as it is building Unit 2 at the Bushehr Nuclear Power Plant. Finally, Turkey (in Asia Minor) is also an aspiring new entrant to the nuclear power club, with the third-largest capacity under construction (4,456 GWe). When its four nuclear reactors being built at Akkuyu Bay are completed, Turkey will become yet another nuclear-powered country in Asia.

6.3.5 FORMER SOVIET UNION

Perhaps surprisingly, given the Chernobyl accident and its legacy, the world region outside of Asia with the largest nuclear power development plans is the former Soviet Union. Russia has two large breeder reactors operating with capacity factors over 70%. Russia, Ukraine and Belarus all have nuclear power plants under construction. Russia's reliance on nuclear power for electricity generation is similar to that of the U.S. at around 20% (Table 6.3). In addition, based on legislation passed in 2001 all nuclear facilities are state-run by Energoatom. Most of the existing nuclear plants in Russia are in Europe, with a few located in Asia. While the Chernobyl accident did not immediately slow down the nuclear development plans in Russia, only two new reactors were commissioned in the 1990s. The construction activity and reactor openings picked up again the 21st century, though the pace has stagnated again more recently because of high prices and lower demand growth for electricity. There are currently just three nuclear reactors under construction in Russia, with a rated capacity of 2,700 GWe.

Ukraine has the largest nuclear power plant in Europe, the Zaporizhzhia station, with six reactors totaling 6,000 MWe. Although Ukraine shut down all of the RBMK reactors at the Chernobyl power station, it is much more reliant on nuclear power than Russia (Table 6.3). However, given the very large number of nuclear power plants that came online in the 1980s, very few reactors have been built there more recently (Kasperski, 2015). There are currently two reactor units under construction in Ukraine, with a rated capacity of 2,070 GWe. Finally, while Belarus, along with Ukraine, was also heavily impacted by the Chernobyl accident (Jacob et al., 2006), it was able to commission its first nuclear power plant in November 2020, at Ostrovets. A second reactor unit opened at the same site in May 2023.

6.3.6   MIDDLE EAST

Interest in nuclear power has existed in the Middle East for many years, though the first nuclear reactors in the region were only commissioned in 2020–2023, when the first three units at the Barakah nuclear power plant entered service in the United Arab Emirates (UAE). The UAE has developed its nuclear power program in close consultation with the International Atomic Energy Agency (IAEA), and the power plant is being built by an international consortium led by the Korea Electric Power Company. A fourth reactor is under construction at the Barakah site, which when opened will result in a combined rated capacity of 5,668 MWe. Saudi Arabia and Egypt also have expressed strong interest in nuclear power, possibly as part of a cold war against Iran and its nuclear program (Miller & Volpe, 2018). However, no nuclear power plants are currently being built in Saudi Arabia, other than a very small research reactor, and its nuclear plans have been delayed. Critics have argued that it would not be cost effective for Saudi Arabia to develop nuclear power (Ahmad & Ramana, 2014), though the country currently plans to install an ambitious 17–19 GWe of nuclear power capacity by 2040. Egypt, for its part, plans to build 3,600 MWe of nuclear power capacity with the assistance of Russia, but construction has yet to begin.

6.3.7   UNITED KINGDOM AND EUROPE

Other than the reactors being added to existing nuclear power plants in France, Finland and Slovakia, the UK is the only other nation in Europe (outside of the former Soviet Union) that is currently expanding nuclear power. Once highly reliant on nuclear power in its electricity mix, the UK now generates only 14.2% of its electricity from nuclear sources, a little over half the peak of 26% in 1997. Indeed, the UK has shut down 36 of its nuclear power reactors, and just nine remain in operation (primarily advanced gas-cooled reactors, as opposed to the more common pressurized water reactors). Moreover, more than a third of the existing capacity is expected to be retired by 2025, and all but one of the remaining reactors closed by 2030. Two pressurized water reactors are being built in England with French and Chinese financing and will be called the Hinkley Point C power plant in Somerset, which will add 3,440 MWe of capacity. When Hinkley Point C is expected to open in 2026, it will have been over 30 years since the last nuclear power plant was commissioned in the UK. Plans for additional nuclear plants have not advanced due to a lack of financing (Colombo, 2020).

6.3.8   UNITED STATES AND SOUTH AMERICA

While not as dramatic as in the UK, the U.S. has also experienced a decline in its reliance on nuclear power. While still the largest user of nuclear-generated electricity in the world, the U.S. has seen only minimal new plant openings since the large growth period of the 1970s and 1980s. Moreover, the U.S. nuclear reactor fleet is the oldest in the world, at over 40 years (Carrara, 2020). Indeed, following the Three Mile Island accident in 1979, the U.S. Nuclear Regulatory Commission did not issue any new construction permits until 2012, a period of over 30 years. Thus, over this period the U.S. had a de facto nuclear power moratorium, when only power plants that were already under construction before the Three Mile Island accident were completed, with many projects being scrapped. The U.S. had a peak of 112 nuclear power plant units operating in 1990–1991 (out of 177 that had received construction permits), although nuclear energy generation peaked more recently in 2019 with only 96 to 98 nuclear units in service due to higher capacity factors (EIA, 2021b). Yet only four nuclear reactors were commissioned in the 1990s, and only one thus far in the 21st century. There is currently one additional reactor being built at the Vogtle nuclear power plant in Georgia, which would add 1,250 MWe of generating capacity. In addition, plans are under way to build small modular nuclear reactors (SMRs) in Wyoming and Idaho, of 350 MWe each or less (Caponiti, 2021; Gheorghiu, 2021). SMRs have also been proposed in Canada, the UK, Poland and China (Mignacca & Locatelli, 2020).

Finally, there are some nuclear expansion plans in South America as well. Brazil is building a third unit of 1,405 MWe capacity at its existing Angra Nuclear Power Plant, while Argentina is constructing a small (29 MWe) pressurized water reactor unit adjacent to the two-unit Atucha Nuclear Power Plant in Lima, along with a dry cask spent fuel storage facility.

6.3.9 Plant Retirements and Conclusions

Despite the increasing trend in life extensions at nuclear power plants, reactor retirements also have been increasing. According to the IAEA, two out of every three nuclear power reactors worldwide have been in operation for over 30 years and are scheduled for retirement in the foreseeable future, and new construction seems unlikely to replace them all (IAEA, 2020). These estimates are in flux, however, as last-minute life extension financing can prolong the lifetime of some nuclear facilities, as recently happened in the case of four large nuclear reactors at two power plants in Illinois (Gardner, 2021). Nonetheless, closure of two thirds or even half of existing nuclear power capacity worldwide would amount to losing ~197 to 262 GWe as of October 2021, and according to the IAEA, only ~53.9 GWe of new capacity is currently under construction at 51 reactor units (around 13.7% of existing capacity), well short of what is required to replace the likely plant closures, if not expand nuclear power capacity. Thus, based on our review of new nuclear power capacity being built in various countries around the world, it is difficult to see how installed capacity will be able to keep up with expected plant retirements in the next few decades, unless a construction boom occurs in the near future and/or additional government subsidies forestall closure of many other nuclear power plants.

6.4 NUCLEAR PROJECTION SCENARIOS

6.4.1 Introduction

The future role of nuclear power in the electricity mix will be based on many factors: the price of nuclear power and competing fuels such as natural gas, coal and renewable sources of energy; the price elasticity of demand for electricity (both short-term and long-term); the retirements and life extensions of existing power plants; the degree of electrification of the energy system; resolution of the radio-active waste problem; government policies on climate change, among others; and future nuclear accidents (Blowers et al., 1991; IAEA, 2020; Hsu, 2021). In the last 45 years there has been a major accident at a commercial nuclear power plant around once every 15 years, though it is impossible to forecast the future accident rate and severity. Nevertheless, such accidents have major effects on the nuclear power development.

Past projections or forecasts of the future use of nuclear power have generally been poor and are fraught with a high degree of uncertainty. Muellner et al. (2021) provide an extensive review of projections of nuclear power growth rates that were made over the past 50 years or so. They found that all of the past projections of the nuclear power plant build rate overestimated the actual rate of capacity additions, often by a very large amount (e.g. a 1974 projection by the IAEA found that the maximum global capacity in 2000 would have been 5,300 GWe in its most opti-mistic case and 3,600 GWe in its most likely case, while the actual nuclear capac-ity in 2000 ended up at 350 GWe). The authors also found that the most accurate projections were those made only 10–15 years beforehand and usually the lower growth scenarios. Indeed, most long-term projections beyond 20 years, nuclear or otherwise, should be considered with a large degree of caution (Carvallo et al., 2018). With this in mind, several key recent projections of nuclear power growth will now be reviewed.

6.4.2   U.S. ENERGY INFORMATION ADMINISTRATION PROJECTIONS

The U.S. Energy Information Administration (EIA), the statistical and analytical arm of the U.S. Department of Energy, issues a comprehensive Annual Energy Outlook (AEO) for the U.S. energy system and a parallel International Energy Outlook (IEO) for the world (EIA, 2021c, 2021d). The AEO is based on the EIA's economic and energy model, the National Energy Modeling System (NEMS), which contains 13 modules (EIA, 2019). The IEO is based on the EIA's World Energy Projection System Plus (WEPS+), an integrated international economic and energy model, which also contains 13 modules (EIA, 2020).

In the Reference (base) case for the AEO, nuclear power's share of the U.S. elec-tricity mix falls from 19.7% in 2020 to 11% in 2050. Total nuclear power generation falls as well, from 785 billion kilowatt-hours in 2020 to 594 billion kilowatt-hours in 2050. This occurs largely for reasons already discussed, especially the increas-ingly attractive economics of natural gas and renewable sources of electricity vs. nuclear power. According to the IEO, nuclear power generation will increase by 15% worldwide from 2020 to 2050, with growth in Asia slightly exceeding the decline in capacity in Organization for Economic Co-operation and Development (OECD) countries (where nuclear generation is projected to fall by almost one third). However, the Reference case projection shows nuclear power generation and use peaking in 2040 and the nuclear energy portion of the global electricity mix falling to 8.6% in 2040 and 7.2% in 2050, compared to 10.5% in 2020. In the international

Reference case scenario, nuclear power generation is projected to total 3,084 billion kilowatt-hours in 2040 and 3,025 billion kilowatt-hours in 2050. By way of comparison, actual global nuclear power generation was 2,630 billion kilowatt-hours in 2020 (Table 6.1). The EIA also modeled high and low oil price scenarios as well as high and low economic growth. While the two oil price scenarios had little effect on the projected nuclear power generation, the low and high economic growth scenarios resulted in nuclear power generation varying more, from 3,045 to 3,125 billion kilowatt-hours in 2040 and 2,984 to 3,075 billion kilowatt-hours in 2050. In these two scenarios for nuclear power, its share of the electricity mix falls to 9.5% and 7.8%, respectively, in 2040.

6.4.3 MUELLNER ET AL. (2021) AND IAEA (2020) PROJECTIONS

Muellner et al. (2021) project global nuclear power use in 2040 based on their own study, and compare their results to those of the International Atomic Energy Agency (IAEA). These findings can be readily compared to the global IEO projections of the EIA discussed in the last section, as well as to each other, for 2040. The Muellner et al. methodology is simple and transparent. The authors made a detailed case-by-case evaluation of nuclear power plans for all countries. Unless more information was available, they assumed a 60-year reactor operating lifetime going forward. All announced nuclear power projects, expected construction periods, life extensions of existing plants and reactor retirements were determined. They found that for the next 20 years ~70% of nuclear power generating capacity will be from existing plants and plant lifetime extension projects, and only ~30% from new builds. The IAEA also publishes its own nuclear power projections based on currently known industry and country plans and identifies low- and high-growth scenarios (IAEA, 2020). The IAEA high scenario assumes 400–500 new nuclear reactor units by 2040, compared to just 51 reactors currently being built worldwide. Notably, the IAEA lowered its projections following the Fukushima accident.

Muellner et al. project 3,013 billion kilowatt-hours of nuclear power generation in 2040. This result is between the IAEA low and high projections of 2,804 and 4,977 billion kilowatt-hours, though much closer to the IAEA's low projection. The Muellner et al. projection is slightly below the EIA's IEO global projection for 2040, whose most favorable scenario for nuclear power, as noted earlier, was 3,125 billion kilowatt-hours, well below the IAEA high nuclear scenario. Thus, there is close consensus between the EIA, Muellner et al. and IAEA low scenarios, with the IAEA high scenario appearing to be the outlier. Muellner et al. also considered an extreme nuclear expansion case scenario to determine the maximum hypothetical level of greenhouse gas emissions that could conceivably be offset by nuclear power. This scenario assumed that all of the world's fossil-fueled power plants would be replaced by nuclear power plants. In this highly unlikely (and perhaps impossible) scenario, the authors found that uranium mining for nuclear fuel would not be able to keep up with the massive growth and that even if such a scenario were feasible, it would still result in 70+% of projected global greenhouse gas emissions from other sectors in 2040 (Muellner et al., 2021).

6.4.4 CARRARA (2020)

Carrara (2020) models nuclear power developments differently than the previously discussed studies, though he starts at the same point as Muellner and colleagues by assessing countries' current plans. A dynamic optimization, integrated assessment model called WITCH (world induced technical change hybrid) is used. WITCH combines a top-down intertemporal optimal economic growth model (with a simplistic description of the macroeconomy based on capital and labor production factors) with a bottom-up, detailed description of the energy sector. While WITCH does not provide an accurate portrayal of the dynamics of the world economy, such models are common and can be instructive. Six long-term nuclear scenarios were modeled, from 2015–2100: business as usual, with no climate mitigation policies or other technological constraints, and five mitigation policy scenarios designed to meet the 2015 Paris (climate change) Agreement target of limiting global temperature increases to 2° C above pre-industrial levels by 2100. The policy scenarios assume, unrealistically, that a global carbon tax is imposed worldwide starting in 2020, and four of the five cases consider various levels of nuclear phaseout and abandonment (e.g. only OECD countries vs. the whole world, phaseout vs. the immediate abandonment of nuclear power).

The long-term scenario results of the WITCH model should be considered in the context of the Muellner et al. (2021) finding that past nuclear scenarios or projections which have been most accurate were for 10–15 years or less, with the lowest growth scenarios the most accurate. As a result, we will only consider the model results through 2040. Only two of the six scenarios are worth considering—business as usual and OECD nuclear phaseout with a global carbon tax. The other four scenarios, involving an immediate nuclear abandonment or global phaseout of nuclear power or unconstrained global growth of nuclear power, seem implausible. Moreover, even an OECD-wide nuclear phaseout along with a global carbon tax seems implausible but will be discussed here for comparison purposes.

WITCH begins by estimating global nuclear power generation as 2,778 billion kilowatt-hours in 2020, when actual nuclear generation was only 2,630 billion kilowatt-hours. This may be due in part to depressed demand for electricity during the first year of the COVID-19 pandemic, with the model projections being made largely pre-pandemic. The results of the business as usual and OECD phaseout cases are similar. Nuclear generation projections for 2040 are 6,667 and 6,250 billion kilowatt-hours for the business as usual and OECD phaseout cases. The nuclear generation share for the two scenarios in 2040 diverge more, at 12% and 17.5%, respectively, since there is a large difference in total electricity generation in the two scenarios. Since both scenarios result in total nuclear power generation well above the IAEA high case, these findings stretch credulity and would require a very large increase in nuclear plant construction in the 2020s beyond current plans. The results are presumably an artifact of the integrated assessment modeling approach, which has been subject to a variety of criticisms. In particular, the model may underestimate the cost of nuclear power vs. renewable sources of electricity, which would compound even more over the 85-year time frame of WITCH (Gambhir et al., 2019).

6.5   SUMMARY AND CONCLUSION

Historical experience has indicated that global energy transitions have normally been extremely slow, although exceptions have occurred in the case of some countries and world regions, especially for individual sectors, greater use of energy efficiency and smaller and less complex societies. There is no evidence that indicates that future energy transitions will be any different, although supportive public policy and plans can help. Ever since the post–World War II period, nuclear power was assumed by many observers to be a likely facilitator of a modern clean energy transition, though as of this writing it has yet to happen. Moreover, the growth of nuclear power in the 21st century has greatly slowed from the peak expansion period of the 1970s and 1980s, with the modest exception of Asia. According to the country-by-country review of nuclear power expansion plans, based on the small level of nuclear power plant construction activity that is currently underway (less than 15% of existing capacity) and the expected growth in plant retirements, this slow growth is not likely to change anytime soon.

Based on the nuclear power plant construction activity, planned plant retirements, government policies and market mechanisms, the consensus among the major projections of nuclear power use indicates that global nuclear power generation will most likely be between 2,804 and 3,125 billion kilowatt-hours in 2040, compared to 2,630 billion kilowatt-hours in 2020. However, because of the faster overall growth in electricity consumption over this period, the percentage share of nuclear power in the global electricity mix falls from 10.5% in 2020 to between 7.8% and 9.5% in 2040. Nuclear projections beyond 2040 are considered unreliable.

I conclude that nuclear power will not be part of a sustainable energy transition, at least through the middle of the 21st century. This finding is robust, based on the consistency of the country-by-country review of nuclear power expansion plans with the results of the most reliable projections of nuclear power generation through 2040. While longer-term projections of nuclear power generation are considered unreliable, there is a small chance that this finding could prove erroneous. This could occur if dramatic increases in nuclear power reactor orders and construction activity occurs in the very near future in several major countries (U.S., France, Russia, China, India, Japan, South Korea, etc.) and there is much faster deployment of new or improved nuclear technology such as small modular reactors, better and lower-cost breeder reactors and so on. While such a scenario is possible, historical experience with nuclear power suggests that this is highly unlikely. Consequently, other sustainable electric power technologies based on renewable sources of energy are more likely to meet the demand that would have been met by nuclear power.

NOTES

1 Construction costs of new nuclear builds have actually stabilized or even slightly declined since the 1980s in Japan, as well as South Korea and India (Lovering et al., 2016).

2 Iran's first nuclear power plant, Bushehr Unit 1, took 36 years to build. Unit 2 has been under construction off and on for 45 years. See WNISR (2019). In the case of Unit 2 of the Watts Bar Nuclear Power Plant of the Tennessee Valley Authority in the U.S., commissioned on October 19, 2016, it took 43 years to build the reactor. See Groskopf (2016).

REFERENCES

Ahmad, A. and M.V. Ramana. 2014. Too costly to matter: Economics of nuclear power for Saudi Arabia. *Energy* **69**: 682–694.

Bandyopadhyay, A. and S. Rej. 2021. Can nuclear energy fuel an environmentally sustainable economic growth? Revisiting the EKC hypothesis. *Environmental Science and Pollution Research*. https://doi.org/10.1007/s11356-021-15220-7.

Blowers, A., D. Lowry and B.D. Solomon. 1991. *The International Politics of Nuclear Waste.* London: Macmillan.

Bromley, P.S. 2016. Extraordinary interventions: Toward a framework for rapid transition and deep emission reductions in the energy space. *Energy Research & Social Science* **22**: 165–171.

Brunner, R.D. 1991. Global climate change: Defining the policy problem. *Policy Sciences* **24**(3): 291–311.

Caponiti, A. 2021. Next-Gen nuclear plant and jobs are coming to Wyoming. *U.S. Department of Energy*, 7 June. Available at: www.energy.gov/ne/articles/next-gen-nuclear-plant-and-jobs-are-coming-wyoming

Carrara, S. 2020. Reactor ageing and phase-out policies: Global and regional prospects for nuclear power generation. *Energy Policy* **147**: 111834.

Carvallo, J.P., P.H. Larsen, A.H. Sanstad and C.A. Goldman. 2018. Long term load forecasting accuracy in electric utility integrated resource planning. *Energy Policy* **119**: 410–422.

Char, N.L. and B.J. Csik. 1987. Nuclear power development: History and outlook. *IAEA Bulletin* **29**(3): 19–25.

Cirincione, J. 2007. *Bomb Scare: The History and Future of Nuclear Weapons.* New York: Columbia University Press.

Cohn, S.M. 1997. *Too Cheap to Meter: An Economic and Philosophical Analysis of the Nuclear Dream.* Albany, NY: SUNY Press.

Colombo, C. 2020. United Kingdom nuclear energy. *Market Intelligence*, 15 July. Available at: www.trade.gov/market-intelligence/united-kingdom-nuclear-energy

De Bondt, W.F.M. and A.K. Makhija. 1988. Throwing good money after bad: Nuclear power plant investment decisions and the relevance of sunk costs. *Journal of Economic Behavior* **10**(2): 173–199.

DOE (U.S. Department of Energy). 2021. What's the lifespan for a nuclear reactor? Much longer than you might think, 16 April. Available at: www.energy.gov/ne/articles/whats-lifespan-nuclear-reactor-much-longer-you-might-think

EIA (Energy Information Administration). 2019. *The National Energy Modeling System: An Overview 2018.* Washington, DC: U.S. Department of Energy.

EIA (Energy Information Administration). 2020. *World Energy Projection System (WEPS): Overview.* Washington, DC: U.S. Department of Energy.

EIA (Energy Information Administration). 2021a. Nuclear explained: Where our uranium comes from. Available at: www.eia.gov/energyexplained/nuclear/where-our-uranium-comes-from.php

EIA (Energy Information Administration). 2021b. Nuclear energy. *Monthly Energy Review*, September 2021. Available at: www.eia.gov/totalenergy/data/monthly/pdf/mer.pdf

EIA (Energy Information Administration). 2021c. Annual energy outlook 2021: With projections to 2050. Available at: www.eia.gov/outlooks/aeo/pdf/AEO_Narrative_2021.pdf

EIA (Energy Information Administration). 2021d. International energy outlook 2021: With projections to 2050. Available at: www.eia.gov/outlooks/ieo/pdf/IEO2021_Narrative.pdf

Ellis, R.P. and M.B. Zimmerman, 1983. What happened to nuclear power: A discrete choice model of technology adoption. *The Review of Economics and Statistics* **65**(2): 234–249.

Fischer, D. 1997. *History of the International Atomic Energy Agency: The First Forty Years.* Vienna: International Atomic Energy Agency.

Fouquet, R. 2010. The slow search for solutions: Lessons from historical energy transitions by sector and service. *Energy Policy* **38**(11): 6586–6596.

Gambhir, A., I. Butnarm, P.H. Li, P. Smith and N. Strachan. 2019. A review of criticisms of integrated assessment models and proposed approaches to address these, through the lens of BECCS. *Energies* **12**(9): 1747.

Gardner, T. 2021. Illinois approves $700 million in subsidies to Exelon, prevents nuclear plant closures. *Reuters*, 13 September. Available at: www.reuters.com/world/us/illinois-senate-close-providing-lifeline-3-nuclear-power-plants-2021-09-13/

Gheorghiu, J. 2021. Nuscale to consider Xcel to operate its advanced nuclear reactors. *Utility Dive*, 17 August. Available at: www.utilitydive.com/news/nuscale-to-consider-xcel-for-nuclear-plant-operator-of-its-small-modular-re/605068/

Granberg, D. and S. Holmberg. 1986. Preference, expectations, and voting in Sweden's referendum on nuclear power. *Social Science Quarterly* **67**(2): 379–392.

Groskopf, C. 2016. Slow build: The United States' newest nuclear power plant has taken 43 years to build. *Quartz*, 11 May. Available at: https://qz.com/681753/the-united-states-newest-nuclear-power-plant-has-taken-43-years-to-build/

Häfele, W. 1990. Energy from nuclear power. *Scientific American* **263**(3): 136–145.

Hohenemser, C., R. Kasperson and R. Kates. 1977. The distrust of nuclear power: Nuclear power is assessed hypercritically because of its unique history, complexity, and safety management. *Science* **196**(4285): 25–34.

Hosokawa, K. 2021. Small is beautiful in South Korea's pivot back to nuclear power. *Nikkei Asia*, 2 September. Available at: https://asia.nikkei.com/Business/Energy/Small-is-beautiful-in-South-Korea-s-pivot-back-to-nuclear-power

Hsu, J. 2021. Nuclear power looks to regain its footing 10 years after Fukushima. *Scientific American*, 9 March. Available at: www.scientificamerican.com/article/nuclear-power-looks-to-regain-its-footing-10-years-after-fukushima/

Hutchinson, I. 2018. Ghosts of Fukushima: The evolution and future of China's nuclear power. *China Business Review*, 23 October. Available at: www.chinabusinessreview.com/ghosts-of-fukushima-the-evolution-and-future-of-chinas-nuclear-power/

IAEA (International Atomic Energy Agency). 2017. From Obninsk beyond: Nuclear power conference looks to future. Available at: www.iaea.org/newscenter/news/obninsk-beyond-nuclear-power-conference-looks-future

IAEA (International Atomic Energy Agency). 2020. *Energy, Electricity, and Nuclear Power Estimates for the Period up to 2050, 2020 Edition.* Vienna: IAEA.

IAEA (International Atomic Energy Agency). 2021. Power reactor information system: Under construction reactors. Available at: https://pris.iaea.org/PRIS/WorldStatistics/UnderConstructionReactorsByCountry.aspx

Jacob, P., I. Bogdanova, E. Buglova et al. 2006. Thyroid cancer risk in areas of Ukraine and Belarus affected by the Chernobyl accident. *Radiation Research* **165**(1): 1–8.

Jasanoff, S. and S.H. Kim. 2009. Containing the atom: Sociotechnical imaginaries and nuclear power in the United States and South Korea. *Minerva* **47**: 119–146.

Johansson, B. 2021. Energy governance in Sweden. In: M. Knodt and J. Kemmerzell, eds. *Handbook of Energy Governance in Europe.* Cham, Switzerland: Springer.

Kasperski, T. 2015. Nuclear power in Ukraine: Crisis or path to energy independence? *Bulletin of the Atomic Scientists* **71**(4): 43–50.

Kasperson, R.E., G. Berk, D. Pijawaka, A.B. Sharaf and J. Wood. 1980. Public opposition to nuclear energy: Retrospect and prospect. *Science, Technology, & Human Values* **5**(2): 11–23.

Kramer, G.J. and M. Haigh. 2009. No quick switch to low-carbon energy. *Nature* **462**(7273): 568–569.

Lovering, J.R., A. Yip and T. Nordhaus. 2016. Historical construction costs of global nuclear power reactors. *Energy Policy* 91: 371–382.

Maury, L.A. 2019. The law of nuclear power in the warmth of the Anthropocene. *Rutgers Computer & Technology Law Journal* **45**(2): 98–133.

Meek, D.W. 1978. Nuclear power and the Price-Anderson Act: Promotion over public protection. *Stanford Law Review* **30**(2): 393–468.

Mignacca, B. and G. Locatelli. 2020. Economics and finance of Small Modular Reactors: A systematic review and research agenda. *Renewable and Sustainable Energy Reviews* **118**: 109519.

Miller, N.L. and T.A. Volpe. 2018. Abstinence or tolerance: Managing nuclear ambitions in Saudi Arabia. *The Washington Quarterly* **41**(2): 27–46.

Muellner, N., N. Arnold, K. Gufler, W. Kromp and W. Renneberg. 2021. Nuclear energy—the solution to climate change? *Energy Policy* **155**: 112363.

Niehaus, F. and T. Mueller. 1990. Nuclear power programs and strategies worldwide: The issue of climate change. In: *VDI/VDE Technical Meeting on Perspectives of Nuclear Power and Carbon Dioxide Abatement.* Dusseldorf: VDI-Berichte, pp. 175–192.

Peters, E. and P. Slovic. 1996. The role of affect and worldviews as orienting dispositions in the perception and acceptance of nuclear power. *Journal of Applied Social Psychology* **26**(16): 1427–1453.

Renn, O. 1990. Public responses to the Chernobyl accident. *Journal of Environmental Psychology* **19**(2): 151–167.

Rosa, E.A. and R.E. Dunlap. 1994. Nuclear power: Three decades of public opinion. *Public Opinion Quarterly* **58**(2): 295–324.

Sailor, W.C. 2000. Nuclear power: A solution to climate change? *Science* **288**(5469): 1177–1178.

Smil, V. 2010. *Energy Transitions: History, Requirements, Prospects.* Westport, CT: Praeger.

Solomon, B.D. 2020. Is nuclear power green? A review. *International Journal of Green Technology* **6**: 64–73.

Solomon, B.D. and K. Krishna. 2011. The coming sustainable energy transition: History, strategies, and outlook. *Energy Policy* **39**: 7422–7431.

Solomon, B.D. and F. Li. 2020. Environmental equity and nuclear waste repository siting in East Asia. In: Z. Chen, W. Bowen and D. Whittington, eds. *Development Studies in Regional Science: Essays in Honor of Kingsley E. Haynes.* New York: Springer, pp. 147–166.

Sovacool, B. 2008. Valuing the greenhouse gas emissions from nuclear power: A critical survey. *Energy Policy* **36**(8): 2950–2963.

Steinhauser, G., A. Brandl and T.E. Johnson. 2014. Comparison of the Chernobyl and Fukushima nuclear accidents: A review of the environmental impacts. *Science of the Total Environment* **470–471**: 800–817.

Stulberg, A.N. and M. Fuhrmann, eds. 2013. *The Nuclear Renaissance and International Security.* Stanford: Stanford University Press.

Suzuki, T. 2015. Nuclear energy policy issues in Japan after the Fukushima nuclear accident. *Asian Prospect* **39**(4): 591–605.

Toth, F.L. and H.-H. Rogner. 2006. Oil and nuclear power: Past, present, and future. *Energy Economics* **28**(1): 1–25.

Turconi, R., A. Boldrin and T. Astrup. 2013. Life cycle assessment (LCA) of electricity generation technologies: Overview, comparability and limitations. *Renewable and Sustainable Energy Reviews* **28**: 555–565.

Van der Made, J. 2021. China nuclear reactor closes for maintenance after talks with French engineers. *RFI (Radio France Internationale)*, 30 July. Available at: www.rfi.fr/en/

science-and-technology/20210730-china-nuclear-reactor-closes-for-maintenance-after-talks-with-french-engineers-taishan-radioactivity-leak

Van der Pligt, J. 1992. *Nuclear Energy and the Public*. Hoboken, NJ: Blackwell Publishing.

Verbong, G. and F. Geels. 2007. The ongoing energy transition: Lessons from a socio-technical, multi-level analysis of the Dutch electricity system (1960–2004). *Energy Policy* **35**(2): 1025–1037.

Warner, E.S. and G.A. Heath. 2012. Life cycle greenhouse gas emissions of nuclear electricity generation: Systemic review and harmonization. *Journal of Industrial Ecology* **16**(1): S73–S92.

Weiss, L. 2003. Atoms for peace. *Bulletin of the Atomic Scientists* **59**(6): 34–44.

Whitfield, S.C., E.A. Rosa, A. Dan and T. Dietz. 2009. The future of nuclear power: Value orientations and risk perception. *Risk Analysis* **29**(3): 425–437.

Wilcox, J. 2021. Phasing out nuclear power in Taiwan. *Nuclear Engineering International*, 18 August. Available at: www.neimagazine.com/features/featurephasing-out-nuclear-power-in-taiwan-9010839/

WNA (World Nuclear Association). 2020a. Outline history of nuclear energy. Available at: https://world-nuclear.org/information-library/current-and-future-generation/outline-history-of-nuclear-energy.aspx

WNA (World Nuclear Association). 2020b. World nuclear performance report. Available at: www.world-nuclear.org/getmedia/3418bf4a-5891-4ba1-b6c2-d83d8907264d/performance-report-2020-v1.pdf.aspx

WNA (World Nuclear Association). 2021. Country profiles. Available at: www.world-nuclear.org/information-library/country-profiles.aspx

WNISR (World Nuclear Industry Status Report). 2019. Iran: Construction restart of Bushehr-2, 14 November. Available at: www.worldnuclearreport.org/Iran-Construction-Restart-of-Busheer-2.html

7 Distributed Generation Systems
Types and Applications

Nikita Gupta, Seethalekshmi K.

7.1 INTRODUCTION

Distributed generation (DG) systems are basically small-scale power generation systems that use different technologies such as renewable and non-renewable for producing electrical energy near the load or consumer site. The following definition is generally agreed upon by everyone in the literature for DG: "A generation plant connected directly into the grid at distribution voltage level or on the customer side of the meter" [1]. Distributed generation is also called dispersed or decentralized generation. Generally, DG systems are categorized based on size and location, but in some countries, they are also categorized based on other parameters, such as usage of renewable sources, cogeneration, and dispatchability.

The Department of Energy (DOE) considers a generation capacity up to 100 megawatts (MW) in size of DG units distributed generation [2]. In New Zealand, the capacity of generating units less than 5 MW is categorized as DG [3]. According to the Gas Research Institute, the range of 25 kW to 25 MW is considered DG [4]. DG size is defined as "ranging from a few kilowatts (kW) to over 100 MW" in [5]. The Electric Power Research Institute (EPRI) defines distributed energy resources as small generation units from a few kW up to 50 MW and/or energy storage devices typically situated near load sites or distribution and sub-transmission substations [6].

DG systems can help and support the delivery of a clean, reliable power supply. DG systems provide numerous benefits such as low transmission losses and are modular in nature. The integration of a DG in a grid is shown in Figure 7.1.

DG systems operate either in grid-interconnected mode or islanded mode. In grid-interconnected mode, the frequency and voltage of the system are controlled by the grid. In islanded mode, the main grid is disconnected from the DG side and DG systems meet the local power demand. Most DG systems are powered by renewable energy resources.

7.2 DISTRIBUTED GENERATION TECHNOLOGIES

There are many DG technologies that are used in power systems depending on resources such as non-renewable and renewable, shown in Figure 7.2. These are as follows.

128

DOI: 10.1201/9781032715438-7

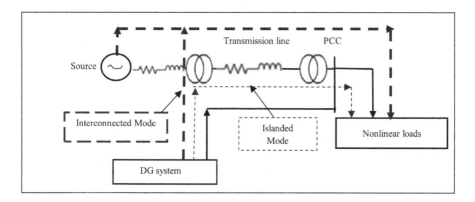

FIGURE 7.1 DG-integrated grid.

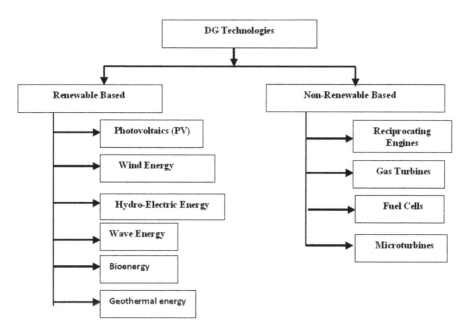

FIGURE 7.2 Classification of DG sources.

7.2.1 PHOTOVOLTAICS

Photovoltaic (PV) systems are also called solar panels. These panels consist of tiny cells and convert the radiation of light into electric energy [7]. These cells can produce a direct-current (DC) supply that is converted into an alternating-current (AC) supply with the help of inverters. Photovoltaic systems are reliable technology and require minimal maintenance for operation. Photovoltaic systems have numerous applications such as households, small businesses, community lighting, agriculture, and healthcare.

7.2.2 Wind Energy or Power

Wind is a popular, sustainable source of power that is produced from harnessing the wind with windmills or wind turbines. Wind rotates the blades of the turbine around a rotor, which spins a generator to produce electricity [7].

7.2.3 Hydroelectric Energy

This form of energy is produced by harnessing the power of water in motion such, as water flowing over a waterfall, for producing electricity [7]. Hydroelectric power plants exploit the energy of falling water: a hydraulic turbine converts the energy of flowing water into mechanical energy, and then it is converted into electrical energy.

7.2.4 Wave Energy

This form of energy is produced by exploiting the variations in air pressure occurring in the waves of the sea or ocean. The periodic ups and downs of ocean waves are converted into mechanical energy and then into electricity.

7.2.5 Bioenergy

This form of energy is derived from organic materials (biomass) such as sugarcane, grasses, straw, corn, and soyabean. There are three ways to extract energy from biomass: burning, bacterial decay, and conversion to gas or liquid fuel. Then the extracted energy is converted into heat and electricity [7].

7.2.6 Geothermal Energy

This energy comes from heat produced during formation of the planet and radioactive decay of materials. This energy is basically stored in rocks and fluids in the center of the earth. The steam comes from hot water reservoirs situated few miles or more below the surface of earth, and it rotates a turbine and spins a generator to produce electricity [7].

7.2.7 Reciprocating Engines

These can be fueled either by diesel or natural gas, with varying emission outputs [7]. Reciprocating engines are available in variety of sizes. These engines provide a continuous power supply or backup emergency power. These engines convert heat pressure released during combustion of fuel mixed with air into mechanical energy.

7.2.8 Combustion Gas Turbines

Combustion gas turbines provides numerous advantages such as low emissions, low installation costs, and less maintenance. But they have less electric efficiency, which limits turbines to merely peaking units and applications in combined heat and power (CHP).

7.2.9 FUEL CELLS

These are electrochemical cells that converts fuel's chemical energy and oxidizing agents into electricity. There are various types of fuel cells in the development stage with a range from 5–1000 kW. These cells include proton exchange membranes, phosphoric acid, molten carbonate, alkaline, solid oxide, and direct methanol [7].

7.2.10 MICROTURBINES

These are new distributed generation technologies that consist of a combustor, compressor, turbine, and generator. These are basically combustion turbines that give heat and electricity on a small scale. Most microturbine units are designed in such a way to provide continuous operation and high electric efficiency.

7.3 ADVANTAGES OF DG SYSTEMS

A DG system has many benefits compared to centralized generation [8]. The major benefits are:

- DG integration increases the reliability of power supplied to the customers, using localized sources.
- DG integration reduces the transmission and distribution losses in systems.
- DG integration improves the voltage profile and gives good power quality.
- DG integration also supports voltage stability and can withstand high loading situations.
- The installation of DG is less time consuming and has a short payback period.
- DG technologies such as micro-turbines and combined heat and power have low pollution and good overall efficiency.
- DG technologies based on renewable energy such as photovoltaics and wind turbines reduce greenhouse gases.
- DG provides high security, social welfare, and profit maximization.
- DG improves energy efficiency, reduces environmental impact, and lessens carbon emissions.
- DG systems are modular and flexible technologies.

7.4 CHALLENGES OF DG SYSTEM INTEGRATION IN AN EXISTING GRID

DG integration provides many advantages but poses different technical as well as economic challenges that affect the whole operation of the integrated system [8]. Some challenges of DG integration are as follows:

a. Voltage level
b. Protection system
c. Stability
d. Power quality

7.4.1 VOLTAGE LEVEL

For proper function of the power supply system and consumer equipment, the distribution grid voltage must be kept within a specific range. The voltage profile is not much influenced when the power injected by the DG is equal to the load or is lower than that of the feeder. If the power supplied by the grid and the current flowing through the feeder decrease, a reduction of voltage occurs. However, when the generated power exceeds the load of the feeder, a rise in voltage will occur. This rise in voltage is produced due to the reversed power flow in the circuit. The effect of the reversed power flow gets stronger when the DG injects reactive power as well.

However, flexible AC transmission system (FACTS) controllers can be used for controlling voltage levels [9].

7.4.2 PROTECTION SYSTEMS

The protection system in power distribution in the case of centralized generation consists of only a simple overcurrent protection scheme, as there is only one supply source and the current flow direction is defined [10]. The incorporation of DG leads to multiple sources of fault current due to which the overcurrent protection philosophy fails. The contribution of DG to the fault current strongly depends on the way the DG unit is connected to the distribution grid and the type of DG. There are many problems produced in protection systems, such as unsynchronized reclosing, coordination issues, islanding problems, fault level issues, blinding of protection, and false tripping.

To protect the system, an automatic recloser is applied in distribution grids. Integration of DGs can disturb the automatic reclosing process because during the opening time of recloser, a small island is created and the connected DG units either accelerate or decelerate resulting in unsynchronized automatic reclosing. Another problem, islanding, is produced when a distributed generator continues to supply power even though electrical grid power from the electric utility is down [11].

When a short circuit occurs, both the grid and the DG unit contribute to the fault current. Due to this, the total fault current increases as DG provides additional input power to the system. However, this decreases the grid contribution and can cause poor fault current detection. It is also possible for the short-circuit to remain undetected because the grid contribution to the current never reaches the required pickup value of the current of the feeder protection relay. This is known as blinding of protection. Due to this, sometimes false tripping will occur as the DG unit contributes to the fault in an adjacent feeder connected to the same substation.

7.4.3 STABILITY

Due to the integration of DG, power oscillations are produced, which increase the instability of the system. Most DG technologies, such as solar, wind, geothermal, and non-renewable (e.g., internal combustion engines, microturbines, and fuel cells) are of considerably smaller scale and are connected to distribution grids, so the impact of DG systems on a power system's transient stability is negligible [12]. However, the dynamic behavior of the power system is affected if the integration is on a large scale. The use of technology also affects the dynamics of a power system [13–16].

7.4.4 Power Quality

Power quality is defined as "the concept of powering and grounding sensitive electronic equipment in a manner suitable for the equipment" by the Institute of Electrical and Electronic Engineers Standard (IEEE Std 1100–2005) [17]. The power quality depends on factors like type of DG sources, DG interfacing technologies, location of DG in distribution system, and environmental conditions. There are most common power quality issues are flicker [18], harmonics [19], voltage sags, voltage swells, and voltage interruptions [20]. According to IEEE 1159–1995 [19], harmonics in a power system are defined as "sinusoidal voltages or currents having frequencies that are integer multiples of the frequency at which the supply system is designed to operate". Flicker is a visible change in brightness of a light due to rapid fluctuations in the voltage of the power supply. Harmonics are sinusoidal components of a periodic wave with a frequency that is an integral multiple of the fundamental frequency. Voltage fluctuations are repetitive or random variations in the magnitude of the supply voltage. Supply may be interrupted due to different faults occurring either in lines or in power equipment like transformers.

7.5 ANALYSIS AND CONTROL METHODS OF POWER QUALITY ISSUES USED IN DG SYSTEM INTEGRATION WITH AN EXISTING GRID

With the advancement in power electronics and digital control technology, the power quality of a DG integrated system can easily be analyzed and controlled [21]. To improve power quality (PQ) aspects related to DG, first we need to analyze PQ issues and then study the control methods that have been used by different researchers in their literature surveys.

7.5.1 Analysis of Power Quality Problems in DG Systems

In a competitive electricity market, electricity utilities are required to guarantee consumer interest by providing a high-quality power supply. Digital signal processing (DSP)–based schemes are popular in analyzing different PQ issues. There are parametric and non-parametric methods for analysis. The methods of analysis of PQ issues in DG systems are shown in Figure 7.3.

The major parametric methods employed are of four types: estimation of signal parameters via rotational invariance technique (ESPRIT) [22], multiple signal classification (MUSIC) [23], Kalman filtering (KF) [24], and the Prony method [25].

The major non-parametric methods are of six types: discrete Fourier Transform (DFT) [26], Stockwell transform (S-transform) [27], Short Time Fourier Transform (STFT) [28], Wavelet Transform (WT) [29], Wavelet Packet Transform (WPT) [30], and Hilbert-Huang Transform (HHT) [31].

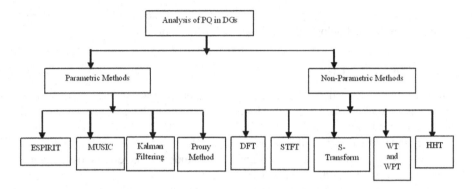

FIGURE 7.3   PQ analysis methods in DG systems.

7.5.1.1   Estimation of Signal Parameters via Rotational Invariance Technique

This technique uses a sinusoidal or complex exponential model that splits a covariance matrix into noise as well as a signal subspace [22]. The frequencies depending on the signal-subspace are estimated. Initially the frequency components are estimated; then the amplitude and phase angle of each frequency component can be evaluated from the eigen relationship of the signal. The ESPRIT algorithm involves following steps:

a. Build the covariance matrix R_x from samples of the signal x_n.
b. R_x is decomposed into two matrices, the signal space matrix R_{xs} and the noise space matrix R_{xn}.
c. The selection matrices C_1 and C_2 are used for estimating sub-matrices R_2 and R_2 from R_{xs} by using:

$$C_1 = \begin{bmatrix} I_{N_S} & 0I_{d_s} \end{bmatrix} \qquad (7.1)$$

$$C_2 = \begin{bmatrix} 0_{d_s} & I_{N_S} \end{bmatrix} \qquad (7.2)$$

where I_{N_S} is the identity matrix, the size of the matrix is $N_s \times N_s$, and the distance between the two submatrices d_s is almost equal to 1.

4. To find the matrix φ, least squares estimation is utilized.
5. The components of frequency are extracted from the diagonal elements of the matrix $\varnothing$, which are the eigen values of the matrix φ.

- **Merits of ESPIRIT:** High resolution.
- **Demerits of ESPIRIT:** High computational burden on controllers.

7.5.1.2   Multiple Signal Classification

Multiple signal classification is one of the most effective parametric methods of noise subspace and is shown in [23]. This algorithm estimates the directions of arrival (DOA) and frequencies of different signals. The frequencies of the various incident signals are calculated using the spectral peak searching method. MUSIC can be termed a super-resolution approach. The orthogonality of signal and noise subspace makes this method noise resistant. According to the harmonic frequencies, the steering vector is:

$$a(\omega) = \begin{bmatrix} 1 & e^{j\omega} & \text{........} & e^{j(p-1)\omega} \end{bmatrix} \tag{7.3}$$

where p is a scalar integer. According to the orthogonal characteristics and steering vector of noise subspace, we can get:

$$f(\omega) = a^H(\omega).V_n.V_n^H a(\omega) = 0 \tag{7.4}$$

P_{MUSIC} is as follows:

$$P_{MUSIC} = \frac{1}{f(\omega)}, \omega = 1,2,....M \tag{7.5}$$

Then the harmonic frequencies are calculated from minima of $f(\omega)$ by searching over ω using a fine grid.

- **Merits of MUSIC:** Accuracy is better with short data length.
- **Demerits of MUSIC:** Prior knowledge of frequency search is required for operation.

7.5.1.3   Kalman Filtering

A Kalman filter adopts a number of mathematical equations, reported in [24], for estimating the harmonic content in an input signal. KF is also known as an optimal estimator and is recursive in nature. The performance of KF is good in the presence of parametric errors and has better convergence characteristics. The mathematical model is as follows:

$$y_{n+1} = \phi_n y_n + \omega_n \tag{7.6}$$
$$x_n = h_n y_n + \eta_n \tag{7.7}$$

where y_n is the state vector, x_n is the measurement at time t_n, ω_n is the model error, η_n is the measurement error, φ_n is the state transition matrix, and h_n is the output matrix.

The Kalman filter algorithm begins with the initial estimation at y_0. It calculates the previous error and then improves the initial estimate. The recursive equation for updating the state variables is:

$$y_n = y_n^0 + K_n(x_n - h_n y_n^0) \tag{7.8}$$

where the matrix K_n' is the Kalman gain.

- **Merits of KF:** This filter can adjust the Kalman gain according to the actual measurement to obtain the optimal solution.
- **Demerits of KF:** This filter has the filter dropping off problem and is insensitive to an abrupt change in the conditions of state variables when the parameters of estimation remain invariable for a long period. For accurate modelling, Kalman filtering needs prior information about the data.

7.5.1.4 Prony Method

Prony is a parametric method used in [25] for extracting sinusoids by linear equations for finding the recurrence equation's coefficients that are satisfied by the signal. Let $x(n)$ be the data samples that are estimated by a function $g(x)$:

$$g(x) = \mathrm{Re}^{rx} + Se^{sx} + \ldots \ldots Ve^{vx} \tag{7.9}$$

where $R, S \ldots V$ and $r, s, \ldots v$ are the closest match signal constants.

- **Merits of Prony method:** This method can detect all three attributes of a signal: amplitude, frequency, and phase.
- **Demerits of Prony method:** This method is computationally unstable.

Parametric methods require prior information about the system for analysis, which is a difficult task, as the system is time varying and non-linear. The most common issues with these techniques are inaccuracy due to model mismatch and order selection of an appropriate model. Parametric methods like ESPRIT, Prony, and KF give good resolution and can detect inter-harmonics. However, they are more prone to noise.

7.5.1.5 Discrete Fourier Transform

Discrete Fourier transform is reported in [26] for estimating the frequency spectrum of signals. It converts a discrete time signal to a frequency signal with a finite range of data samples. It can be used to represent a discrete sequence as the equivalent frequency domain representation and linear time-invariant discrete time system and for developing various computational algorithms. DFT is used only for periodic signals. The DFT of a signal is:

$$X_{DFT}(k) = \sum_{n=0}^{N-1} x(n).e^{-j(\frac{2\pi}{N})kn}, k = 0,1,......(N-1) \quad (7.10)$$

where signal $x(n)$ has N samples, and k is the frequency index. This technique is generally employed in frequency estimation in phasor measurement units (PMUs).

- **Merits of DFT:** Suitable for stationary and non-stationary signals with added windowing.
- **Demerits of DFT:** Gives only frequency information.

7.5.1.6 Short Time Fourier Transform

The short time Fourier transform is a Fourier-related transform, used to estimate the sinusoidal frequencies and phase components of a signal's local sections with respect to time, and is presented in [28] to calculate the phase component and also frequency of local sections of signal with respect to time. The signal is split into small fragments that are assumed to be stationary in nature. A sonogram or spectrogram is used to visualize the representation of spectrum of the frequencies. The STFT for signal $x(n)$ is:

$$X_{STF}(m, f_k) = \sum_{n=0}^{N-1} h(n-m)x(n)e^{-j(\frac{2\pi}{N})kn}, k = 0,1,....(N-1) \quad (7.11)$$

where f_k stands for the kth harmonic frequency, and m is an integer for window position on the time scale.

- **Merits of STFT:** Calculation of both phase components and frequency of local sections of signal.
- **Demerits of STFT:** Predetermined resolution with respect to time.

7.5.1.7 Stockwell Transform

The Stockwell transform (S-Transform) is a generalization of STFT, extending the continuous wavelet transform and mitigating some of its disadvantages. S-transform using spectrogram analysis is reported in [27] and provides a good visual analyzation of signal. A key feature of the S-transform is that it uniquely combines absolutely referenced local phase information with frequency-dependent resolution of the time-frequency space, allowing one to define the meaning of the phase in a local spectrum setting. The S-transform is derived as the phase correction of CWT, with the window being the Gaussian function.

$$S_x(t,f) = \int_{-\infty}^{\infty} x(\tau)|f|e^{-\pi(t-\tau)^2 f^2} e^{-j2\pi f\tau} d\tau \tag{7.12}$$

- **Merits of S-Transform:** Better resolution.
- **Demerits of S-Transform:** At high-frequency events, resolution is not satisfactory.

7.5.1.8 Wavelet Transform and Wavelet Packet Transform

Wavelet transform uses wavelets to decompose any stationary and non-stationary signals for detailed analysis with various frequency-time resolution [29]. It becomes more suitable than the Fourier technique when the exact frequency components of the signal are not clear. WT is a multilevel representation used to bifurcate a given function or continuous-time signal into different scale components. Usually, a frequency range is assigned to represent each scale component. This resolution enables one to study each scale component accordingly. These scaled or translated copies (known as "daughter wavelets"), formed of a finite-length or fast-decaying oscillating waveform (known as the "mother wavelet"), enable their study. The wavelet transform technique is advantageous over the traditionally used Fourier technique for representing those functions that contain discontinuities and sharp peaks. It also accurately deconstructs and then reconstructs finite non-periodic and/or non-stationary signals. DWT decomposes signal into multiresolution levels, known as multi resolution analysis (MRA). This technique consists of various filter banks for performing wavelet analysis and MRA. The decomposition of the input wavelet is done in such a way that the original signal can be reconstructed. The various frequency bands of the signal are represented by the wavelet coefficients. The continuous wavelet transform is:

$$X_{WT}(\tau,S) = \frac{1}{\sqrt{|S|}} \int_{-\infty}^{+\infty} x(t)\psi(\frac{t-\tau}{S})dt \tag{7.13}$$

where the signal $x(t)$ has mother wavelet $\psi(t)$.

For dyadic grid, scale,

$$S = 2^m,$$

and translation,

$$\tau = n2^m$$

where n and m are integral values. This method is especially useful for real-time identification of PQ events in DG-incorporated grids.

Wavelet packet transform is reported in [30] and splits a signal into uniform frequency bands. A method of distortion estimation of frequency ranging from 2 to 9 kHz using WPT is proposed in [30]. This scheme has an instrument for time-frequency analysis of the distortion in voltage waveforms using wavelets that introduces a new performance index to assess the distortion in this frequency range.

- **Merits of WT:** Provides multi resolution analysis (MRA) and is applicable for real-time applications.
- **Demerits of WT:** High computational burden.

7.5.1.9 Hilbert-Huang Transform

The Hilbert-Huang transform is applied on non-stationary signals. It has two steps: In the first step, empirical mode decomposition (EMD) is done in which the input signal is decomposed into a finite number of intrinsic mode functions (IMFs) in a manner such that the highest frequency in the signal gets captured at the first IMF. In the second step, Hilbert transform (HT) is applied on the IMF, which provides orthogonal IMFs displaced by 90 degrees in phase. Each set thus found is used to estimate the instantaneous frequency and amplitude with respect to time. HHT, reported in [31], uses adaptive IMF for the basis function for disintegration. The spline fitting, in creating both upper and lower envelopes, is of the utmost importance for the accuracy and effectiveness of this technique. For precise information on instantaneous frequency, over-sampled data is required. The signal $x(t)$ of the jth IMF C_j is:

$$x(t) = \sum_{j=1}^{n} C_j(t) + r_n \tag{7.14}$$

IMFs can be represented as:

$$HC_j(t) = d_j(t) = \frac{1}{\pi} P \int_{-\infty}^{+\infty} \frac{C_j(t')}{(t-t')} dt' \tag{7.15}$$

where $HC_j(t)$ stands for the Hilbert transform of a function with any real value, D_j is the Hilbert transform, and P is the Cauchy principal value. The equation in terms of HT and IMF is:

$$Z(t) = C_j(t) + jD_j(t) = A(t)e^{j\theta(t)} \tag{7.16}$$

where amplitude is $A(t)$ and phase is $\theta(t)$.

- **Merits of HHT:** This method is applicable to non-stationary signals and has good adaptability.
- **Demerits of HHT:** Oversampled data is required.

Non-parametric techniques have good frequency resolution and do not require prior information about the system. The accuracy of these methods are higher compared to parametric methods. These methods are easily applicable for both non-linear and time-varying systems.

7.5.2 CONTROL OF POWER QUALITY PROBLEMS IN DG SYSTEMS

The improve system performance and utilize DG in existing grids, the quality of power must improve. If the quality of power is poor, that adversely affects many factors related to the system such as power factor and system capacity. Therefore, the system needs control techniques that are used to improve power quality and protect sensitive loads from any adverse effects. The main methods used to improve power quality issues are the usage of passive harmonic filters, active harmonic filters, and FACTS controllers [32]. Control techniques for PQ used in DG systems are shown in Figure 7.4.

7.5.2.1 Control of Harmonics

The majority of DG systems use power electronic inverters at the interface, which causes harmonics in the power system. Total harmonic distortion (THD) at point of common coupling (PCC) must be maintained per the IEEE standards [19]. Harmonic filters such as passive filters [33], active filters [34] and FACTS controllers [35] can be installed in order to minimize the harmonics generated in switching operations.

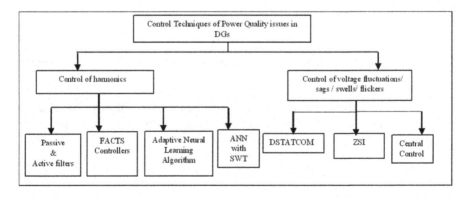

FIGURE 7.4   Control techniques for power quality issues in DG systems.

7.5.2.1.1 Passive and Active Filters

Passive filters employ passive elements such as inductance (L), capacitance (C), and resistance (R). The values of capacitors and inductors are selected to give low impedance paths at selected frequencies. These filters are tuned at harmonic frequencies that are to be attenuated.

A high-order LCL filter is used to interconnect an inverter to the utility grid to filter out higher-order harmonics produced by the inverter [33].

- **Merits of Passive Filters:** More reliable and low running losses.
- **Demerits of Passive Filters:** These filters are very bulky in size and can cause resonance at selected frequencies.

Active filters are a very efficient method of eliminating more than one harmonic at a time. These filters consist of active elements such as op-amps, along with passive elements. A method of flexible control of DG units is preferable over installation of filters. Active harmonic filtering using a current controlled scheme with two parallel control branches is used in [34]. The first control branch is responsible for DG unit fundamental current control, and the second one is employed to compensate local load harmonic current or feeder resonance voltage.

- **Merits of Active Filters:** These filters are adaptive and eliminate more than one harmonic at a time.
- **Demerits of Active Filters:** The reference signals generated in these methods may be affected by the variations in magnitude of voltage at PCC. But these filters are more expensive than passive filters.

7.5.2.1.2 FACTS Controllers

According to the IEEE [35], FACTS are controllers based on power electronic devices and other static equipment that carry out control of one or multiple parameters of an AC transmission system to increase the controllability, thereby increasing the power transfer capacity of the system.

FACTS controllers such as unified power-quality conditioners (UPQCs) are used as a power quality conditioner that consists of a series inverter, a shunt inverter, and a distributed generator connected in the DC link through a rectifier. However, the performance of the UPQC depends on a proper control algorithm that helps in creating reference signals of both voltage and current. Many control algorithms, such as the instantaneous active-reactive (IAR) power method [36], synchronous reference frame (SRF) method [37], unit vector template (UVT) method [38], and exponential composition algorithm (ECA) [39] are used.

7.5.2.1.2.1   Instantaneous Active-Reactive Power Method   Instantaneous active-reactive power theory is used to produce voltage as well as current reference signals, and it maintains the DC offset value at zero. This control strategy has both series and shunt inverter controls as well as a positive sequence detector. The positive sequence detector and voltage reference generator use the source voltage and current to estimate the instantaneous active and reactive power by αβ0 transformation. To find the constant components of power and positive-sequence fundamental components of the source voltage, the values of instantaneous power (active and reactive) are sent through the low-pass filter (LPF). Then, an inverse αβ0 transformation is applied on the nominal instantaneous powers to derive the reference voltages, and further, these reference signals are utilized in the pulse width modulation (PWM) generator for controlling the gate pulses of series and shunt inverters [36].

7.5.2.1.2.2   Synchronous Reference Frame Theory   In this method, current signals at the load side are transformed to the d-q frame. The obtained fundamental current components have ripples as a value of the DC offset. By removing these offset DC values, the harmonics present in the d-q transformed signal can be removed. Further, reference signals are generated that are utilized by the propotional integrator (PI) controller and further in the PWM generator for limiting the gate pulses of series and shunt controllers [37].

7.5.2.1.2.3   Unit Vector Template Generation Algorithm   In this method, distorted supply voltages are used to extract unit vector templates. A phase lock loop (PLL) is used for generating unit vector templates, and then the supply voltage is multiplied by these unit vector templates, and the reference load voltage is generated. Error is computed between measured and reference voltage and is sent to a hysteresis controller to generate the triggering pulses for the series controller. For generation of triggering pulses for the shunt controller, the actual DC link voltage is compared with the reference DC link voltage, and the error is calculated and then sent to the PI controller that creates reference current signals. Error is computed between actual and reference currents and is sent to the hysteresis controller to generate triggering pulses for the shunt controller [38].

7.5.2.1.2.4   Exponential Composition Algorithm–Based Approach   This method utilizes a second-order biquad filter that includes a low-pass filter, high-pass filter (HPF), and band-pass filter (BPF) for discriminating the real and reactive parts of the load current. The BPF is used to extract the reactive part of the load current that has cut-off frequencies as 49.9 and 50.1 Hz, whereas LPF is used to extract the fundamental part of the load current that has a cut-off frequency of 50 Hz [39].

The merits and demerits of UPQC algorithms are presented in Table 7.1.

7.5.2.1.3   Artificial Neural Network–Based Controllers

It is a mathematical model of an interconnected group of artificial neurons that has a connectionist approach for computation. It is a self-adaptive, multilayer complex

TABLE 7.1

Types of UPQC Algorithms

UPQC Algorithm	Merits	Demerits
Instantaneous active-reactive	Maintains DC offset to zero	Limitations in cut-off frequencies
Synchronous reference frame	Synchronizing circuits not required	Not applicable for unbalanced supply
Unit vector template	By removing DC offset, easy removal of harmonics is possible	Needs complex PLL
Exponential composition algorithm	Simple process	Needs filters

structure that consist of several artificial neurons and layers [40]. A neural network converts input into a meaningful output. An artificial neural network (ANN) consists of three layers:

a. Input layer stores input data.
b. Hidden layer processes the inputs from the input layer and depends on weights.
c. Output layer gathers computation results.

7.5.2.1.3.1 UPQC with Artificial Neural Network and Synchrosqueezing Wavelet Transform–Based Controller This method is used for minimizing voltage sag, voltage swell, and THD in a distributed generation system. Synchrosqueezing wavelet transform (SWT) is a time-frequency analysis method that analyses multi-component signals with oscillating modes. A SWT-based feature extraction technique is used in shunt controllers and gives a time-frequency relation of signals to the ANN model. The potential and performance of UPQC is carried out with an ANN-based controller and ANN with a SWT-based controller under various working conditions in the time and frequency domains [41]. This control scheme helps in decreasing load-side THD in DG integrated grids.

- **Merits of ANN and SWT-based UPQC:** It provides immediate control of PQ issues, high accuracy, good dynamic response, and fast processing of reference signals.
- **Demerits of ANN and SWT-based UPQC:** Burden on controllers.

7.5.2.2 Voltage Fluctuations/Sags/Swells/Flickers

To mitigate voltage fluctuations/sags/swells, a distributed static compensator (DSTATCOM) [42] and Z-source inverter (ZSI) [43] are used.

7.5.2.2.1   D-STATCOM with Positive-Sequence Admittance and Negative-Sequence Conductance

A DSTATCOM with positive-sequence admittance is used in recovering the positive-sequence voltage, whereas negative sequence conductance is used to balance the voltage profile [42]. The DSTATCOM has a tuning control circuit that dynamically adjusts the admittance and conductance commands to maintain both positive- and negative-sequence voltages at permissible levels with respect to the power variation of DG systems or loads. The control is achieved by adjusting both admittance and conductance separately, and a balance between D-STATCOM rating and required improvement on power quality can be achieved. It provides reactive power compensation, low switching losses, and improved reliability.

- **Merits of D-STATCOM:** Controls voltage fluctuations and maintains high penetration of PVs.
- **Demerits of D-STATCOM:** Needs tuning circuit for successful operation.

7.5.2.2.2   Z-Source-Inverter-Based Flexible Distributed Generation Systems

It consists of an X-shaped impedance network with inductors that creates safe shoot-through of inverters and boosts voltage at the Z-source capacitor. When the inverter operates at full capacity, the controller reduces the harmonics of the injecting current, and during partial capacity working, the differential capacity improves the THD of the voltage. It provides a unique buck-boost feature to the inverter and optimizes power at the output in a better way than conventionally used voltage source inverter (VSI) [43].

- **Merits of ZSI Inverter:** More flexible in interfacing DG with main grid.
- **Demerits of ZSI Inverter:** Complexity is high.

7.6 COMPARISON OF ANALYSIS AND CONTROL METHODS FOR POWER QUALITY ISSUES USED IN DG SYSTEM INTEGRATION WITH EXISTING GRIDS

The comparison of analysis and control methods for power quality is summarized in Tables 7.2 and 7.3.

This comparison enables researchers to choose the required analysis and control technique that can be employed in DG systems for maximum utilization of energy.

TABLE 7.2
Comparison of Analysis Methods

Type of Issue	Analysis Technique	Merits	Demerits
Parametric methods	Modified exact model order estimation of signal parameters by use of rotational invariance technique	Advantageous from the fault diagnosis and analysis point of view No storage requirements High resolution Capability of inter-harmonic detection	High computational burden and takes more time
	Multiple signal classification	Good accuracy with shorter length of data High resolution	High storage requirements and takes more time
	Kalman filtering	This filter can adjust Kalman gain according to the actual measurement for obtaining the optimal solution	For accurate modelling, Kalman filtering needs prior information on the data
	Prony method	This method can detect all three attributes of a signal: amplitude, frequency, and phase	This method is computationally unstable
Non-parametric methods	Fast Fourier transform and discrete Fourier transform	Suitable for stationary signal Used for periodic signal to find its frequency content Applied to non-stationary signals but with added windowing to focus on certain periods	Does not provide time information for the signal Does not provide exact amplitude and phase values for harmonics whose frequencies different from the window function frequency
	Short time Fourier transform	Used to find the phase components and sinusoidal frequency of local sections of signal with change in time	Fixed resolution all the time
	Stockwell transform	Works in both normal and noisy conditions Better performance	Fixed width of window
	Wavelet transform and WPT	Insensitive to regular signal behavior but sensitive to disturbances or irregularities	It has a batch processing step It fails to detect notches under noisy conditions (20 dB)
	Hilbert-Huang transform	Provides information about the phase and amplitude in both time and frequency scales Good adaptive ability	Accuracy is dependent on spline fitting, and it requires oversampled data for exact definition of instantaneous frequency

TABLE 7.3
Comparison of Control Methods

Type of Issue	Control Technique	Merits	Demerits
Power quality (harmonics)	Passive filters	More reliable and low running losses	Removes only one harmonic at a time and is bulky in size
	Active filters	Adaptive	Expensive and high running losses
	Hybrid filters	Suppresses parallel resonances	More complex circuit
	FACTS controllers (UPQC)	Operates in islanding and integrated modes and used as power conditioning device	High conduction losses
	ANN with SWT	High accuracy, good dynamic response, and fast processing of reference signals	ANN is only used for feature extraction
Power quality (voltage fluctuations, sags, swells, and flickers)	DSTATCOM	Controls only voltage fluctuations	Needs tuning control circuit for operation
	ZSI	Reduces switching losses and provides high reliability	Output range is limited
	Centralized controller	Controls flow of power and voltage fluctuations, sags, and swells	Proper selection of communication link is quite difficult

7.7 CONCLUSION

This chapter provides the basic types and challenges of DG systems that are introduced when integrating with an existing grid. The need for DG systems is increasing day by day as electricity consumption increases rapidly. A grid-interconnected system will reduce the cost of power generation per unit for consumers as well as generation centers and increase the revenue of transmission and distribution organizations. The power crisis in India per the data available in March 2022 is a shortage of around 623 million units [44]. Implementation of DG systems for existing grid electricity will help with maximum power generation, which could solve the electricity scarcity in India, and electricity would be available 24/7 for rural and urban areas in countries like India with enhanced power quality.

To maximize the utilization of DG systems in existing grid adaptive control techniques in systems used for control of protection, stability and power quality are necessary. This chapter also shed light on power quality issues related to the integration of distributed generation sources and mitigation techniques. These issues are mitigated by basic inverter control strategies as well as various sophisticated control schemes based on FACTS devices, also discussed in the chapter.

REFERENCES

[1] Mohit Bajaj and A. K. Singh, 2019. Grid integrated renewable DG systems: A review of power quality challenges and state-of-the-art mitigation techniques. *International Journal of Energy Research*, Vol. 44, Issue 1, pp. 26–69.

[2] The US Department of Energy, 2003. *Office of Distributed Energy Resources*, online publications available at: www.eere.energy.gov/der/.

[3] Distributed Generation in Liberalised Electricity Markets, 2002. *International Energy Agency*, online publications available at: https://www.iea.org/reports/.

[4] Gas Research Institute, 1998. *Distributed Power Generation: A Strategy for a Competitive Energy Industry*, Chicago: Gas Research Institute.

[5] D. Sharma and R. Bartels, 1998. Distributed electricity generation in competitive energy markets: A case study in Australia. *The Energy Journal* (Special issue: Distributed Resources: Toward a New Paradigm of the Electricity Business, The International Association for Energy Economics, Clevland, Ohio, USA), Vol. 18, pp. 17–40.

[6] The Electric Power Research Institute, 2002, online publications available at: www.epri.com/.

[7] T. Adefarati and R.C. Bansal, 2016. Integration of renewable distributed generators into the distribution system: A review. *Renewable Power Generation, IET*, Vol. 10, Issue 7, pp. 873–884.

[8] Edward J. Coster, Johanna M. A. Myrzik and Bas Kruimer, 2011. Integration issues of distributed generation in distribution grids. *Proceedings of the IEEE*, Vol. 99, Issue 1, pp. 29–39.

[9] Narain G. Hingorani and Laszlo Gyugyi, 2011. *Understanding Facts: Concepts and Technology of Flexible AC Transmission System*. Wiley-IEEE Press, online publications available at: https://onlinelibrary.wiley.com/.

[10] Manohar Singh and S. G. Srivani, 2016. Adaptive protection coordination scheme for power networks under penetration of distributed generation resources, *IET Generation, Transmission & Distribution*, Vol.10, Issue.15, pp 3919–3929.

[11] Basanta Kumar, Ria Nandi and Bibhusmita Mahanta, 2016. Islanding detection in distributed generation. *International Conference on Circuit, Power and Computing Technologies (ICCPCT)*, pp. 1–5. doi: 10.1109/ICCPCT.2016.7530295

[12] Konstantinos F. Krommydas and Antonio T. Alex, 2015. Modular control design and stability analysis of isolated PV source/battery-storage distributed generation systems. *IEEE Journal Emerging and Selected Topics in Circuits and Systems*, Vol. 5, Issue 3, pp. 372–382.

[13] Muhammad Babar Rasheed and Muhammad Awaish, 2016. Transient stability analysis of an islanded microgrid under variable load. *IEEE 19th Conference on Network Based Information System (NBIS)*. doi: 10.1109/NBiS.2016.79

[14] Iraklis P. Nikolakakos and Hatem H. Zeineldin, 2016. Stability evaluation of interconnected multi-inverter microgrids through critical clusters. *IEEE Transactions on Power Systems*, pp. 3060–3072.

[15] W. Al-Saedi, S. W. Lachowicz, D. Habibi and O. Bass, 2012. Stability analysis of an autonomous microgrid operation based on particle swarm optimization. *IEEE International Conference on Power System Technology (POWERCON)*, pp. 1–6. doi: 10.1109/PowerCon.2012.6401291

[16] Diptargha Chakravorty and Jinrui Guo, 2017. Small signal stability analysis of distribution networks with electric springs. *IEEE Transactions on Smart Grid*, Vol. 10, Issue 2, pp. 1543–1552. doi: 10.1109/TSG.2017.2772224

[17] IEEE Std 1100–2005 (Revision of IEEE Std 1100–1999), 2006. *IEEE Recommended Practice*. doi: 10.1109/IEEESTD.2006.216391

[18] Shailendra Kumar Sharma and Ambrish Chandra, 2017. Voltage flicker mitigation employing smart loads with high penetration of renewable energy in distribution systems. *IEEE Transactions Smart Grid*, Vol. 8, pp. 414–424.

[19] IEEE 519–1992, 1993. *IEEE Recommended Practices and Requirements for Harmonic Control in Electric Power Systems*. doi: 10.1109/IEEESTD.1993.114370

[20] Po-Chen Chen and Reynaldo Salcedo, 2012. Analysis of voltage Profile problems due to the penetration of distributed generation in low-voltage secondary distribution. *IEEE Transactions on Power Delivery*, Vol. 27, Issue 4, pp. 2020–2028.

[21] Nikita Gupta and Seethalekshmi K., 2019. Review of analyzing techniques in technical challenges related to distributed generation. *International Journal of Applied Engineering Research*, Vol. 14, Issue 2, pp. 311–316.

[22] Antonio Bracale and Guido Carpinelli, 2009. An ESPRIT and DFT-based new method for the waveform distortion assessment in power systems. *20th Int. Conf. and Exhibition on Electricity Distribution, CIRED*, pp. 1–4. doi: 10.1049/cp.2009.0746

[23] T. Cai, S. Duan and C. Chen, 2010. Real-value MUSIC algorithm for power harmonics and interharmonics estimation. *International Journal of Circuit Theory*, Vol. 39, pp. 1023–1035.

[24] C.I. Chen, G.W. Chang, R.C. Hong and H.M. Li, 2010. Extended real model of Kalman filter for time-varying harmonics estimation. *IEEE Transactions on Power Delivery*, Vol. 25, pp. 17–26.

[25] Zhijian Hu, Jianquang Guo, Mei Yu, Zhiwei Du and Chao Wang, 2006. The studies on power system harmonic analysis based on extended prony method. *IEEE International Conference on Power System Technology*, pp. 1–8. doi: 10.1109/ICPST.2006.321738

[26] R.I. Diego and J. Barros, 2010. Subharmonic measurement using DFT and Wavelet-Packet Transform in an IEC extended framework. *Measurement*, Vol. 43, pp. 1603–1608.

[27] M. Jaya Bharata Reddya, Rama Krishnan Raghupathy, K. P. Venkatesha and D.K.M Mohanta, 2013. Power quality analysis using Discrete Orthogonal S-transform (DOST). *Digital Signal Processing*, Vol. 23, Issue 2, pp. 616–626.

[28] Francisco Jurado, 2002. Comparison between discrete STFT and wavelets for the analysis of power quality events. *Electric Power System Research*, pp. 183–190.

[29] Nikita Gupta, Seethalekshmi K. and Stuti Shukla Datta, 2020. Wavelet based real-time monitoring of electrical signals in distributed generation (DG) integrated system, *Engineering Science and Technology, an International Journal, Elsevier*, Vol. 24, Issue 1, pp. 212–228.

[30] Julio Barros and Ramon I. Diego, 2018. Analysis of harmonics in power systems using the wavelet-packet transform. *IEEE Transactions on Instrumentation and Measurement*, Vol. 57, Issue 1, pp. 63–69.

[31] Ashika Gururani, 2016. Microgrid protection using Hilbert-Huang Transform based-differential scheme. *IET Generation Transmission and Distribution*, pp. 3707–3716.

[32] Nikita Gupta and Seethalekshmi K., 2018. A review on key issues and challenges in integration of distributed generation system, *5th International Conference on Electrical, Electronics and Computer Engineering (UPCON)*, 2–4 November. doi: 10.1109/UPCON.2018.8597014

[33] Aleksandr Reznik, Marcelo Godoy Simoes, Ahmed Al Durra and S. M. Muyeen, 2014. LCL filter design and performance analysis for grid-interconnected systems. *IEEE Transactions Industry Applications*, Vol. 50, Issue 2, pp. 1225–1232.

[34] Jinwei He, Yun Wei Li, Frede Blaabjerg and Xiongfei Wang, 2014. Active harmonic filtering using current controlled, grid-connected DG units with closed-loop power control. *IEEE Transactions Power Electronics*, Vol. 29, Issue 2, pp. 642–653.

[35] IEEE, 1997. Proposed terms and definitions for flexible AC transmission system (FACTS). *IEEE Transactions on Power Delivery*, Vol. 12, Issue 4, pp. 1848–1853.

[36] B. Han, B. Bae, H. Kim and S. Baek, 2006. Combined operation of unified power-quality conditioner with distributed generation. *IEEE Transactions Power Delivery*, Vol. 21, Issue 1, pp. 330–338. doi: 10.1109/TPWRD.2005.852843

[37] M. Kesler and E. Ozdemir, 2011. Synchronous-reference-frame-based control method for UPQC under unbalanced and distorted load conditions. *IEEE Transactions on Industrial Electronics*, Vol 58, Issue 9, pp. 3967–3975.

[38] Y. Lu, G. Xiao, X. Wang, F. Blaabjerg and D. Lu, 2016. Control strategy for single—phase transformer less three—leg unified power quality conditioner based on space vector modulation. *IEEE Transactions on Power Electronics*, Vol. 31, Issue 4, pp. 2840–2849.

[39] S. Sindhu, M. R. Sindhu and T.N.P. Nambiar, 2015. An exponential composition algorithm based UPQC for power quality enhancement. *In Procedia Technology, Elsevier,* Vol. 21, pp. 415–422.

[40] M. B. I. Reaz, F. Choong, M. S. Sulaiman and F. Mohd-Yasin, 2007. Prototyping of wavelet transform, artificial neural network and fuzzy logic for power quality disturbance classifier. *Electric Power Components and Systems,* Vol 35, Issue 1, pp. 1–17.

[41] Nikita Gupta and Seethalekshmi K., 2021. Artificial neural network & synchrosqueezing wavelet transform based control of power quality events in distribution system integrated with distributed generation sources. *International Transactions on Electrical Energy Systems (Willey),* Vol. 31, Issue 1, pp. 1–20.

[42] Tzung-Lin Lee, Shang-Hung HuYu and Hung Chan, 2013. D-STATCOM with positive-sequence admittance and negative-sequence conductance to mitigate voltage fluctuations in high-level penetration of distributed-generation systems. *IEEE Transactions Industrial Electronics,* Vol. 60, pp. 1417–1428.

[43] Chandana Jayampathi Gajanayake, Mahinda Vilathgamuwa D and Frede Blaabjerg, 2019. Z-source-inverter-based Flexible distributed generation system solution for grid power quality improvement. *IEEE Transactions Energy Conversion,* Vol. 24, pp. 695–704.

[44] Power Crisis in India, *Business Today Desk,* May 2022, online available at www.businesstoday.in.

8 Role of Hydrogen and Fuel Cells in Energy Transition

Muhammad Haseeb Hassan, Syeda Youmnah Batool, Hafiz Ahmad Ishfaq, Saeed-ur Rehman

8.1 INTRODUCTION

A remarkable increase in energy reliance has been observed in the recent past. The rise in the significance of energy is due to the larger contribution of energy to performing routine activities in everyday life. According to the prediction of various statistics, the world will face about a 25% rise in energy demand in the next 20 years (Pudasainee, Kurian, and Gupta 2020). At this moment, coal, petroleum oil, and gas are responsible to generate around 84% of the global energy (Iordache, Gheorghe, and Iordache 2013; Nejat Veziroglu 2012; Veziroğlu and Şahin 2008; Rusman and Dahari 2016; Sun et al. 2018). Among fossil fuels, coal is the most abundant fossil fuel present on earth and is a nonrenewable resource (Figure 8.1). It is anticipated that the supply of coal reserves will run out in the next 150 years, natural gas in 60 years, and petroleum reserves in the upcoming 40 years (Midilli et al. 2005). Alarmingly, a decrease in conventional sources of energy such as fossil fuels has given rise to global warming, which is today classified as a major emerging challenge worldwide.

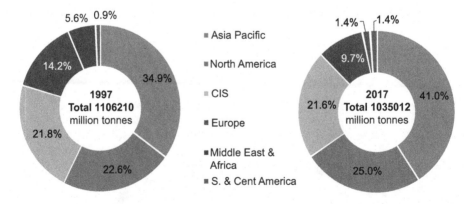

FIGURE 8.1 Distribution of proven reserves in 1997 and 2017 (British Petroleum Report 2018).

DOI: 10.1201/9781032715438-8

Ecologists believe that if this paradigm continues, it will be a cause of severe deterioration in quality of life standards.

Subsequently, the incredible growth in the demand for energy with the potential to cause damage in terms of human life encouraged the theory of an alternative energy source that would retain all the characteristics of conventional fuel but with no or less environmental impact. The concept of incorporating hydrogen as fuel was proposed by a British scientist, J.B.S. Haldane, in the year 1923. In the early 1970s, J. O'M. Bockris, a well-known scientist, was the first to call this idea "hydrogen economy". Hydrogen can be an environmentally friendly source of energy with no release of pollutants such as NO_x, SO_x, CO_x, or carbon dioxide. In the existing discussion of energy strategy, hydrogen is positioned as a unique element in terms of energy supply, particularly in mobile applications. Hydrogen is the most commonly available element in space and makes up about 75% of the total mass of the universe as well as more than 90% of all atoms. Hydrogen was found by the Englishman Henry Cavendish in 1766, and the name hydrogen was given by the Frenchman Antoine Lavoisier in 1787.

Hydrogen is the first and the lightest element present in the periodic table, with the highest known calorific value for any fuel, as shown in Table 8.1 (Abe et al. 2019). It has prominently higher calorific values when compared with fossil fuels such as petroleum and natural gas. Moreover, hydrogen is the cleanest fuel because it produces only water upon burning. At 143.0 MJ/kg, hydrogen has a high energy density by weight (three times larger than gasoline), but at 0.0108 MJ/L, it has the lowest energy density by volume, which is over 3000 times smaller than gasoline (Figure 8.2). The energy density by volume is only 1/4th of that of gasoline, even when it is highly compressed and stored in a solid or liquified state. Hydrogen has the potential to transport renewable energy over long distances and store it for the long term, for example, from wind power or solar electricity.

This chapter examines the application of hydrogen as a fuel in combination with fuel cells and their foreseeable role in future energy transition. A fuel cell is a promising electrochemical device that uses a process of electrochemical oxidation to convert hydrogen directly into electricity, generating pure water as a byproduct. In contrast, electricity from an external source such as a renewable energy source is utilized by a reversible fuel cell, which splits water into hydrogen and oxygen (fuel cell electrolysis). The chapter is divided into six segments. The first section highlights the importance of hydrogen in nature, followed by a description of different viable methods for producing hydrogen. Key challenges associated with hydrogen storage and transportation are discussed in the third part. The fourth part underlines the significance of fuel cell technology, considering it one of the primary components of a hydrogen economy. The fifth section features several applications of fuel cell technology, followed by the sixth part, which includes the current challenges and future policies to deal with the barriers in the way of clean and efficient energy transition via hydrogen and fuel cells.

8.1.1 PRIMITIVE MODEL OF HYDROGEN ECONOMY

In its simplest form, a hydrogen economy is about the role of hydrogen to transport and store energy from renewables over long distances and in larger amounts,

TABLE 8.1
Various Hydrogen Production Methods and Their Advantages, Disadvantages, and Efficiency

Hydrogen Production Method	Advantages	Disadvantages	Efficiency
Steam reforming	Mature technology	Generates CO_2, CO, unstable H_2 supply	74–85
Partial oxidation	Developed technology	Produces petroleum coke and other heavy oils with H_2 production	60–75
Autothermal reforming	Existing infrastructure and already established	Generates CO_2, utilization of fossil fuels	60–75
Biophotolysis	Requires mild working conditions, CO_2 consumption with the production of O_2	Use of expensive materials, lower production of H_2, requires sunlight	10–11
Dark fermentation	No need of sunlight, simple method, no O_2 limitation, CO_2-neutral	Low H_2 yields, large volume of reactor, elimination of fatty acids	60–80
Photo fermentation	CO_2 neutral, recycling of wastewater	Low efficiency, requires sunlight, large volume of reactor, low yield of H_2, O_2-sensitivity	0.1
Gasification	Inexpensive feedstock, widely used, neutral CO_2	Fluctuating H_2 yield, seasonal availability, and formation of tar	30–40
Pyrolysis	Inexpensive feedstock, widely used, neutral CO_2	Fluctuating H_2 yield, seasonal availability, and formation of tar	35–50
Thermolysis	O_2 byproduct, clean and sustainable	Corrosion problems, high capital costs, use of toxic elements	20–45
Photolysis	Zero emissions, O_2 byproduct, availability of feedstock	Needs sunlight, low efficiency, non-effective photocatalytic material	0.06
Electrolysis	Green and developed technology, O_2 byproduct, existing infrastructure	H_2 storage and transportation issues	60–80

respectively (Figure 8.3). There are two ways in which hydrogen can be used. Primarily, it can be incorporated to produce electricity with the help of fuel cells, and furthermore, due to its distinctive property of clean combustion, it can replace natural gas in industry and can run ships, aircraft, and trains.

8.1.2 SUSTAINABLE HYDROGEN ECONOMY

The notion of sustainability relates to consistency of prosperity and growth. The philosophy behind a sustainable hydrogen economy relates to growth in four different aspects, technological development, economic growth, environmental compatibility, and social issues. The technological aspect covers a number of indicators;

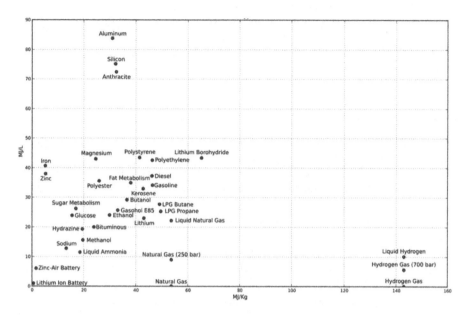

FIGURE 8.2 Comparison of energy densities by weight (MJ/kg) versus volume (MJ/L) for many common fuels and other useful materials (Eberhardt 2002).

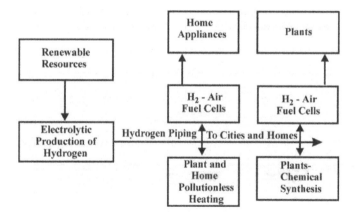

FIGURE 8.3 A hydrogen economy. A solar–hydrogen energy system (J.O.M. Bockris 2002).

however, the most common are validity, reliability, flexibility, and quality. Likewise, the economic aspect mainly includes capital cost, cost related to operations and maintenance activities, and cost of transportation and storage. The environmental element largely consists of energy consumption efficiency, utilization of fossil fuel, ecological footprints, and comparative reduction of harmful gases. The social part is quite significant, as it deals with the principles of occupational health and safety and overall security of the energy supply. The sustainability of the hydrogen supply chain is essential for the growth of the hydrogen industry. For this, the economic element is the most important of all, followed by technological and social factors. Paying coherent attention to these aspects can potentially enhance the sustainability of the hydrogen supply chain.

8.2 HYDROGEN PRODUCTION

Since hydrogen is present abundantly on our planet, it is almost always found only in a bonded form with other elements such as water, carbohydrates, and hydrocarbons. Thus, using hydrogen as fuel is not a simple task, as it first must be liberated from the bonded state, which demands a certain amount of energy. There are multiple standardized and popular routes available to generate hydrogen. Various methods for hydrogen production, along with their advantages, disadvantages, and efficiency, are summarized in Table 8.1.

Hydrogen can be produced directly from primary as well as from secondary energy sources. Hydrogen can be produced from renewable sources, fossil fuels, and nuclear energy. Apart from this, there are other methods still at the research and development stage, particularly those based on biomass, but also biological hydrogen production. Decomposition of fossil fuels or biomass into hydrogen with the help of thermochemical, chemical, and biological means is the key approach. Moreover, dissociation of water by employing electrical or thermal energy obtained from nuclear or renewable energy is also an important alternative to extract hydrogen (Figure 8.4). Separation from hydrocarbons is achieved via steam reforming. Partial oxidation can also serve this purpose.

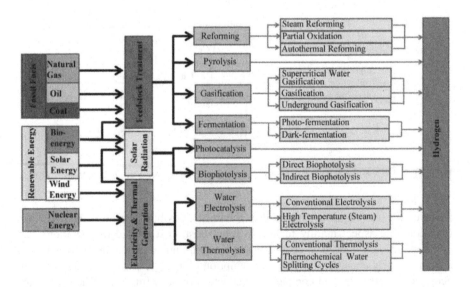

FIGURE 8.4 Routes for hydrogen production from fossil fuels, nuclear energy, and renewable energy (Scipioni, Manzardo, and Ren 2017).

8.2.1 Steam Reforming of Natural Gas

Hydrogen can be separated from natural gas via reaction with steam, oxygen, or both. Steam reforming is considered a popular way to produce hydrogen (Figure 8.5). As a starting step, desulfurization of natural gas takes place, which is then followed by mixing of gas with steam and preheating before moving into the reformer. The reformer is composed of pipes containing a nickel catalyst. In this manner, the transformation of a natural gas–steam mix into a synthesis gas of hydrogen, carbon dioxide, carbon monoxide, and water takes place. Afterward, the synthesized gas enters a shift converter. The heat required during this process is in a temperature range of 800–900° C, with a pressure of approximately 20 to 40 bar. The following reaction takes place:

$$CH_4 + H_2O \rightarrow CO + 3H_2 \quad \Delta H = 206 kj/mol,$$

With the help of steam in a ferric-oxide catalytic converter, the carbon monoxide from the syngas is converted into hydrogen and carbon dioxide. Later, the gas is cooled down to ambient temperature and purified or treated. The removal of carbon dioxide and any remaining residues, such as carbon monoxide, occurs in a pressure-swing adsorption (PSA) column. The hydrogen produced with this method has a purity of at least 99.9 vol.%.

8.2.2 Coal Gasification

To produce hydrogen from solid fuels, a combination of pyrolysis and gasification is employed. Various gasification tools and devices are currently available. A mixture slurry of coal and water is provided as a feed. The gasification occurs in the presence of air or oxygen. In a typical gasification process, first, the fuel is converted into a combustible gas in the presence of air, oxygen, or steam. This phenomenon is called pyrolysis. Next, ash particles and sulfur contents are removed from the raw synthesized gas before it is delivered to the shift section. Hydrogen is produced by treating the obtained syngas in a similar method as that for the natural gas steam reforming technique.

8.2.3 Biomass Gasification

A clean and sustainable resource for hydrogen production can be achieved through biomass. The net carbon emission of the whole cycle is neutral because the biomass

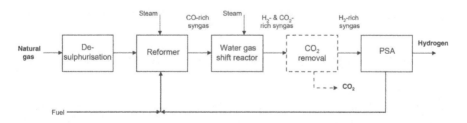

FIGURE 8.5 Schematic of hydrogen production via natural gas steam reforming (Wietschel and Ball 2009).

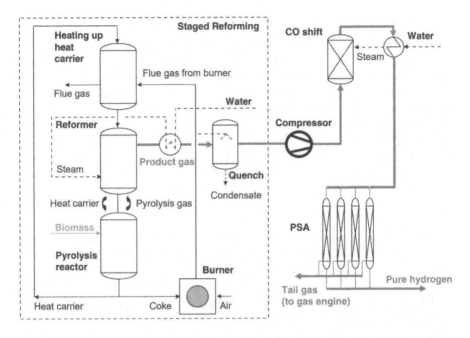

FIGURE 8.6 Layout of a plant to produce high-purity hydrogen based on the staged-reforming process from D.M.2

acts as a carbon sink during the growing phase. An attractive method for biomass gasification is the so-called staged reforming method developed by the German company D.M.2 (Figure 8.6) and the indirectly heated circulating-bed gasifier developed by the Battelle Columbus Laboratory (BCL) in the US (Wietschel and Ball 2009).

The gasification process can use a variety of biomass resources, such as agricultural residues and wastes, or specifically grown energy crops. The produced gas is largely composed of CO and hydrogen. The goal is to separate pure hydrogen through CO shift and PSA. The tail gas is attained from the PSA plant, which is used for the generation of electricity and heat (e.g., in a gas engine).

8.2.4 Electrolysis

One of the most efficient and capable processes for producing hydrogen is electrolysis, and the relevant equipment is termed an electrolyzer. In electrolysis, renewable water is used as a reactant, and direct current electricity is supplied to the process. Consequently, hydrogen is generated along with pure industrial grade oxygen as a by-product. Electrolysis is currently being used for producing pure hydrogen for electronics, pharmaceuticals, food, and other industries, and it is regarded as a potential method to produce hydrogen fuel. Electrolysis follows the fact that a water molecule consists of two atoms of hydrogen, bonded with an atom of oxygen. Hence, when external energy in the form of electrical current is provided, decomposition of water takes place. It involves two reactions that occur at the two electrodes: hydrogen is formed at the cathode and oxygen at the anode. Ionic conduction occurs through an electrolyte. An electrolyte is used as a separator between the reactions going on at both electrodes, and it confirms the isolation of hydrogen and oxygen gases. The primary advantage of using electrolysis is that it is a clean and green process that produces hydrogen without the evolution of any of the harmful gases that may hinder the balance of the environment.

$$\textbf{Anode: } 2OH^- \rightarrow \frac{1}{2}O_2 + H_2O + 2e^-$$

$$\textbf{Cathode: } 2H_2O + 2e^- \rightarrow H_2 + 2OH^-$$

Various electrolyzers are distinguished based on their operating conditions or the ionic agents. Common examples are alkaline water electrolyzers, polymer electrolyte membrane electrolysers (PEM electrolyser), and high-temperature electrolyzers using solid oxide electrolysis cell (SOEC).

8.3   HYDROGEN STORAGE AND DISTRIBUTION

Hydrogen storage and distribution are undeniably the most challenging tasks for the establishment of an economical hydrogen economy. Particularly, hydrogen storage is considered an extraordinarily crucial issue to be resolved. Effective storage systems for mobile applications are essential for an efficient and feasible and sustainable hydrogen economy. Following are some of the major challenges in storing H_2 (Mori and Hirose 2009; Schüth 2009).

- H_2 requires a tank with a large volume and weight due to low density, which correspondingly limits the amount of storage.
- Upon encountering oxygen, H_2 can explode.
- The cost of hydrogen is comparatively higher than that of other fuels such as petroleum.
- The refueling of H_2 fuels is slow.
- The lifetime and durability of H_2 storage systems are not up to the mark.

In mobile applications, special characteristics are needed for storage design because a mobile vehicle can manage a limited amount of weight and volume. Therefore, balance is required when designing storage systems for H_2 for mobile applications.

8.3.1   Possibilities of Hydrogen Storage

Hydrogen has certain unique properties that contribute to the need to store H_2 in special conditions. Pressurized hydrogen, storage in solids, liquid hydrogen, and hybrid storage systems are some of the prominent proposals for storing hydrogen on board vehicles.

8.3.1.1   Pressurized Hydrogen Storage

The most conventional method used for the storage of hydrogen is pressurized hydrogen (CGH$_2$) storage. In this mode of storage, challenges are faced due to the low density of hydrogen gas. Keeping in view the storage requirements, various types of cylinders are employed during operations. The usage of metallic cylinders is quite common in industrial applications where a pressure range of 20–30 MPa is required. In contrast, hydrogen tanks for mobile applications can store gas from 35 to 70 MPa. In correspondence to the storage conditions, the metallic part of gas bottles can be replaced by lightweight and highly stable fiber resin composite with a polymer lining inside the vessel. The purpose of the polymer liner is to act as a permeation barrier for H_2 gas, as there is always a risk of gas leakage when stored at high pressures. Therefore, the cylinder material must have a high tensile strength, low density, and no permeability to hydrogen. Mechanical strength and robustness are considered key parameters during manufacturing of these cylinders. An additional protective layer is applied to cover the outer shell in order to avert any potential mechanical failure. Although the cylinders and tanks are manufactured in compliance with sophisticated safety standards, apprehensions related to the likelihood of tank rupture cannot be overlooked.

8.3.1.2   Liquified Hydrogen Storage

To achieve higher storage densities, liquification of hydrogen (LH_2) with a density of 70.8 kg/m^3 is advantageous. For liquification to be attained, the gas must be cooled down to 21 K (approx. –252° C). It is considered a standard method of storage. This increases the density and the hydrogen storage efficiency. Hydrogen liquification is an energy-demanding process that consumes about 30% more energy compared to using compressed hydrogen. It is advisable to store the cryogenic liquid at a temperature of 20 K. However, normally after a certain time period, evaporation starts in liquified hydrogen due to inevitable heat entrance in the container/tank. Even the incorporation of highly effective vacuum insulation and heat radiation barriers cannot restrict it. Overpressure should be released from the tank in order to avoid buildup of pressure. This can be done via a catalytic converter. Boiling off hydrogen gas is a critical concern and can impose special effects on safety, overall cost, and energy efficiency as well.

A system developed by Linde minimizes evaporation losses. The system draws in the surrounding air, which is dried and then liquefied by the energy released as the hydrogen increases its temperature. The cryogenically liquefied air (–191° C) flows through a water-cooling jacket surrounding the inner tank and, thus, acts as a refrigerant and as a result causes a substantial delay in the temperature increase of the hydrogen (Bossel, Eliasson, and Taylor 2003)

8.3.1.3   Hydrogen Storage in Solids

Among various fuel storage mechanisms, a common idea is to store hydrogen in a solid state. However, there must be a synchronization of the thermal properties of a system with the operating conditions. Generally, there are two fundamental bonding methods that are endorsed for achieving high-density hydrogen storage.

1. Physisorption (adsorption of hydrogen) of molecular hydrogen to the inner surface of a highly porous material with a larger surface area through weak van der Waals forces. The weak van der Waal forces can be easily broken

down by slight heat. Adsorption has been studied on various nanomaterials, for example, nanocarbons, metal organic frameworks, and polymers. Insulated cryovessels are required to store hydrogen by physisorption. During this technique, hydrogen keeps its molecular form, and the interaction between the hydrogen molecules and the surface is in the lower range of 5–8 kJ/mol H_2. Consequently, less heat management is necessary to perform the required function.

2. Chemisorption (absorption of hydrogen) involves the chemical bonding of hydrogen atoms with a host matrix. Hydrogen molecules are first dissociated into hydrogen atoms and then bonded chemically with solids. Hence, hydrogen is unified in the lattice of a metal, an alloy, or a chemical compound. The achievement of high volumetric storage density is a principal benefit of storing hydrogen in chemisorbed form. However, there is also a disadvantage in storing hydrogen in atomic form with a metal matrix. There is a compulsion to split and recombine the hydrogen molecules that requires significant heat control. During the refueling of the hydride, heat is released from the material that is to be removed.

8.3.2 POSSIBILITIES OF HYDROGEN DISTRIBUTION

The principal choices for appropriate transportation and distribution of hydrogen are either through pipelines or trailers (gaseous and liquid). Criteria for the selection of each option depend on the distances during delivery and volume efficiency offered by the mode of transportation.

8.3.2.1 Transport of Gaseous Hydrogen

To be transported in gaseous form either by pipelines or by trailers, hydrogen needs to be compressed. The following options are considered to transport hydrogen through pipelines.

1. Development of a purpose-built distribution network.
2. Usage of exiting natural gas pipelines for hydrogen distribution.
3. Transportation by mixing natural gas and hydrogen during transportation and separation at delivery points.

There are several dangers when it comes to adapting natural gas pipelines to transport hydrogen or the blend of hydrogen and natural gas. Due to the unique chemical and physical properties possessed by hydrogen, it can diffuse rapidly through seals and most materials as well. This may cause deterioration of steel, which can further result in the formation of internal cracks and material fractures. The phenomenon is called "hydrogen embrittlement" (Castello, Tzimas, and Moretto 2005). Hence, pipelines must be developed with (non-porous) high property stainless steel. This can also be a cause of high investment for hydrogen pipelines as compared to those for natural gas with the same diameter. Moreover, owing to lower molecular weight and viscosity, hydrogen flows faster but with relatively less energy due to its lower heating value. Therefore, to supply the same amount of energy, hydrogen pipelines must be operated at high pressure or should have a larger diameter. During the incident of leakage, hydrogen can be more vulnerable in terms of ignition, with a high range of flammability, which gives rise to serious safety concerns when used in a confined space.

As for the blend of hydrogen and natural gas, evaluation suggests that 30 vol.% of hydrogen might be added to natural gas pipelines while making no alteration in the existing transmitting system. Subsequently, separation can be done by using appropriate membranes. However, worries about high hydrogen loss and high energy retribution during the separation process make the option of membrane separation unconventional.

8.3.2.2 Transport of Liquid Hydrogen

Liquid hydrogen is transported via semitrailer in highly insulated cryogenic cylindrical vessels. There is another concept for the transportation of liquid hydrogen known as maritime hydrogen transport. Among various concepts of maritime transport of hydrogen, the small waterplane area twin hull (SWATH) carrier concept is quite viable. According to this concept, 8150 t of liquified hydrogen can be transported. Transportation in such high quantities can potentially make a significant reduction in the overall transportation cost.

8.4 ROLE OF FUEL CELLS IN HYDROGEN ECONOMY

Hydrogen and fuel cells create a valuable combination simultaneously. The commercialization of fuel cells has a vital part in the hydrogen economy and energy revolution. Hydrogen is the leading justification behind the high electricity to heat ratio with superior conversion efficiency. As compared to the emissions produced by bio-based methane or renewables while generating electricity, the exhaust of a fuel cell produces zero emissions when hydrogen is used as a fuel.

A fuel cell differs from conventional energy conversion devices because it converts chemical energy directly into electrical energy rather than following a three-phase conversion process (Figure 8.7).

8.4.1 Principles of Fuel Cells

The potential energy conversion efficiency of fuel cells is the most attractive point of all. The theoretical efficiency can be extracted by the famous Carnot cycle. A Ragone plot is displayed in Figure 8.8 that shows that fuel cells have the highest energy density of various types of energy devices.

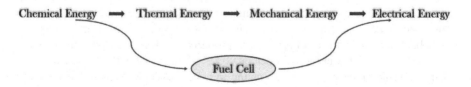

FIGURE 8.7 Comparison of conventional and electrochemical energy conversion devices.

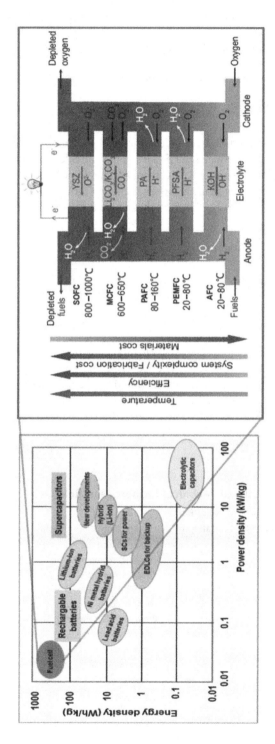

FIGURE 8.8 (a) Ragone plot. (b) Working principles of various fuel cell technologies and their general trend in the relationship between materials cost, fabrication cost, system complexity, efficiency, and operating temperature.

Permitted by S. Wang and Jiang (2017); Muthukumar et al. (2020).

Principally, a fuel cell works like a battery that continues to supply power until the fuel is provided. The fundamental concept of a fuel cell is an electrolyte surrounded by two electrodes, an anode (fuel electrode) and a cathode (air electrode) on each side and an interconnect.

The electrochemical reaction of the fuel cell is shown in Eq. 8.1.

$$2H_2(g) + O_2(g) \rightarrow 2H_2O + energy \tag{8.1}$$

Preferably, fuel is supplied at the anode side, which is converted to H^+ ions and electrons, whereas oxygen/air is reduced to O^{2-} ions at the cathode side. These O^{2-} ions are passed through a dense electrolyte, which only allows ionic conduction while the electrons move across the interconnect (external circuit) for power generation. Eqs. 8.2 and 8.3 reflect the reactions occurring at that anode and cathode, respectively.

$$\textbf{Anode: } H_2(g) \rightarrow 2H^+ + 2e^- \tag{8.2}$$

$$\textbf{Cathode: } \frac{1}{2}O_2(g) + 2H^+ + 2e^- \rightarrow H_2O \tag{8.3}$$

8.4.2 Types of Fuel Cells

There are a variety of fuel cells, which differ primarily with respect to electrolytes, the applicable chemical reaction, and the concerned operating temperatures. The names of these fuel cell categories originate from the type of electrolyte employed in each type. Various types of fuel cells with corresponding properties are given in Table 8.2.

TABLE 8.2
Summary of Operational and Technical Properties of Relevant Fuel Cells

Fuel Cell Type	Operating Temperature ° C	Material/ Catalyst	Electrolyte	Mobile Ion	Efficiency %
PEMFC	30–100	Carbon/platinum	Proton-conducting polymer membrane (e.g., Nafion)	H^+	40–50
AFC	50–200	Carbon/nickel	30–50% Potassium hydroxide	OH^-	50–60
DAFC	20–90	Platinum/ ruthenium	Proton-conducting polymer membrane	H^+	20–50
PAFC	~ 200	H^+	Concentrated Phosphoric acid	H^+	40–50
MCFC	~ 650	Stainless steel/ nickel	Carbonated metal (Li_2CO_3, K_2CO_3)	CO_3^{2-}	>60
SOFC	500–1000	Ceramic/ perovskites	Ion-conducting ceramic (e.g., YSZ)	O^{2-}	50–60

8.5 APPLICATIONS OF HYDROGEN AND FUEL CELLS

Legal requirements to reduce pollutant emissions involving greenhouse gas emissions are continuously increasing and turning out to be non-negotiable. Some countries, including Germany, are making significant efforts to abandon nuclear energy. Political and economic consequences can be anticipated with this promising shift of energy systems.

To achieve these statistics, hydrogen in combination with fuel cells is likely to present a tremendous range of applications in energy evolution. However, the need is to specify the relevant type of fuel cell technology in harmony with the application. Fuel cells are able to generate energy for low-range power devices to a broader range of electrical energy. The main role of fuel cells, along with the related benefits and drawbacks, is shown in Table 8.3. Since the 1990s, the transportation sector has contributed a lot to pollutant emissions, specifically hydrocarbons, nitrogen oxide, and different sulfur-based substances. This fact has drawn worldwide attention to the reduction of CO_2 in the first place. Though automotive transportation is not the only and not the biggest CO_2 emitter, the automobile is the focal point of CO_2 regulations. With the passage of time, regulations regarding CO_2 emissions have become more sophisticated and rigorous.

8.5.1 AUTOMOBILES

In the recent past, a disturbing escalation in the overall concentration of particulates (e.g. SO_x and NO_x emissions) in the environment has given rise to statistics of

---

TABLE 8.3
Benefits, Limitations, and Applications of Some Fuel Cells

Fuel Cell Type	Benefits	Drawbacks	Applications
PEMFC	Wide power range, quick start-up, high power density	Slow ORR, water management issues, CO poisoning	Vehicles, mobile equipment, low-power CHP systems
AFC	Inexpensive, rapid start-up, simple heat management, high activity	Low tolerance to CO_2, requires pure oxygen	Military, submarines, space vehicles
DAFC	No CO_2 emissions, availability of fuel, use of cheap fuel, quick start-up	Crossover of fuel, use of expensive Pt and Ru catalysts, cathode poisoning, toxic fuel	Portable electronic systems with low power, long operating times
PAFC	High tolerance to CO_2, inexpensive due to lower Pt usage	Slow start-up, intolerant to CO, low power density, limited materials	Large numbers of 200-kW CHP systems
MCFC SOFC	High efficiency, supports internal reforming, fuel diversification, can be used with gas turbines	Poor long-term stability, long start-up time, use of selected materials, cathode poisoning	All sizes of distributed generation, auxiliary power, electric utilities

premature deaths in European countries (Ferreira et al. 2017). Global well-being is being affected because of the increase in harmful emissions. Consequently, diesel-operated vehicles are facing a severe prohibition, especially in France, the UK, and some states of Germany (Staffell et al. 2019). Over time, it has been established as a convincing argument that the CO_2 emissions targets can only be achieved by having a certain number of zero-emission-vehicles, such as battery electric vehicles (BEVs) and fuel cell electric vehicles (FCEVs). Battery electric vehicles, however, have some limitations when it comes to vehicle size, weight of the vehicle, charging time, and range. Therefore, hydrogen fuel cell electric vehicles are the most suitable technology to fill this void to reach the target of minimum possible emissions. When used jointly, hydrogen and fuel cells have many viable applications in the field of transportation, including FCEVs, hydrogen powertrains, fuel cell buses, trucks, and motorbikes.

The incorporation of zero-emission vehicles is the only promising approach to reach the prescribed targets of CO_2 emissions. Fuel cell vehicles and powertrains exhibit a complete capacity to deliver required mobility services while doing no harm to the planet. However, for the desired results to be achieved, the most essential thing is implementation of the whole concept as intended. Maintaining compliance with road safety regulations, this technology needs to be developed in accordance with the characteristics and features, no less than the conventional vehicles present in the market.

Considering the general perceptions and concerns of a customer, some typical requirements can be stated as follows:

- The location, fitting, and response of all relevant hydrogen components during collision or crash.
- Sensitivity and promptness of hydrogen sensors.
- Reliability of tests to check the strength and robustness of fuel cylinders/ tanks.
- Addition of technical services and solutions according to the vehicle design and category.
- Suitability of vehicle in comparison with environmental conditions.
- Economical influences and ease of purchase.

In addition to these requirements, it is extremely important to be concerned about the following technical and design aspects.

- The fuel cell vehicle takes the lead when compared to a typical battery car in terms of driving distance. It can cover a distance in the range of 400 to 500 km.
- Another attribute of a fuel cell car is that it allows quick refueling activity, that is, around 3 mins. BEVs, however, consume several hours during recharging, and even with fast charging, the time is still unmatched with the FCEV.
- The lifetime of a battery is strongly influenced by overcharging and the local climate. In contrast to this behavior, fuel cells cars demonstrate cold start aptitude and high discharging without compromising the lifetime.
- In the customer's view, fuel efficiency and depreciation cost of a vehicle are generally noticed as the most important operating costs. The economic practicability of a vehicle is the total running cost, depending on the variation

in the vehicle's value. These days, FCEVs normally have high operating and capital costs as compared to BEVs. But FCEVs are expected to be a cost-effective technology in the upcoming days with the increase in their demand.

• Finally, there are some key attributes that need to be considered for provision of unique performance experience to customers: power to weight ratio, maximum speed, power of system, and overall dynamics. In the general perception, FCEVs are much quieter, with fewer vibrations, and offer effortless gear shifting.

8.5.2 CITY BUSES

Fuel cell buses are earning enormous attention around the globe. As a result, technology is becoming stable and mature. The fuel cell technology is very suitable for application in city buses. Hydrogen fuel cell buses represent the best alternative solution to the internal combustion engine for urban traffic in European bus cities. In terms of performance, flexibility, and infrastructure cost per km, the fuel cell is a considerably better solution when compared to conventional engines. It is estimated that the total cost of ownership (TCO) of fuel cell buses will experience an increase of 10–20% compared to that of diesel by the year 2030. Regarding timely deployment of fuel-cell hydrogen buses, a prominent lead has been taken by European countries by crossing a milestone of around 7 million kilometers of operating experience. China has the biggest bus market around the globe and had already planned to launch around 300 fuel cell buses in Foshan City. These buses are quite reliable and consistent, as they have been operated for more than 18,000 hours in London. Furthermore, the Department of Energy (DOE), United States, has established a target of 25,000 operating hours for fuel cell buses. Out of several buses in California, ten fuel cell buses have passed 12,000 hours of operation, and one of them has successfully reached 22,400 hours of operating time, which is close to the defined goal.

8.5.3 HEAVY-DUTY VEHICLES

It is possible to achieve full electrification of a heavy-duty vehicle. However, it is important to consider the specific usage profiles and hence the corresponding requirements of the powertrain. Trucks have the substantial capacity to take on fuel cell technology, though it requires enhanced durability because of long traveling distance, with a target of a 50,000-hour lifetime of the stack. For instance, in the case of a waste disposal vehicle, low ranges are required with a maximum power need of 250 kW. Besides this, low fuel and high efficiency costs are also critical. The heavy goods vehicle market is particularly cost sensitive with very little or no backing from the country's administration. Hence, it is pertinent to contemplate that fuel cell trucks have lower acceptance than fuel cell buses. No effective designs and arrangements are witnessed with fuel cell powertrains when compared to the diesel engine. Yet the ban on diesel trucks in some major cities has diverted interest in fuel cell trucks.

8.5.4 TRAINS AND SHIPS

The Jet Propulsion Laboratory (JPL), USA, published the initial study and application of fuel cell technology as a train power unit in 1995. The European

Railway Research Institute (ERRI) also conducted a study regarding the feasibility of using fuel cell technology in trains under EU conditions. Fuel cell technology is a wonderful substitute for the diesel engine, especially in the regions where smoke, dust, and dirt should be avoided, such as in mining activities or underground railway stations, including London, Munich, or Paris. Hydrogen trains can be used on the roads, which are comparatively hard to electrify economically. Germany has begun the testing of fuel cell trains that have roof-mounted tanks for hydrogen. Likewise, in order to get rid of the constraint of line electrification, Alstom has declared a plan to convert electric and diesel trains to hydrogen by 2040 (Gerlici et al. 2018) in the UK. Hydrogen power trains are, however, 50% more costly than diesel-powered trains, but the economic feasibility depends on the lower fuel cost.

For vessels and ships, the application of fuel cells depends on the size and usage. Incorporation of a fuel cell engine might be a decent drive train in a small vessel or yacht, but very high energy is required in case of a big ship, and the more capable and efficient power train is still diesel. *Alsterwasser* is a hydrogen ship that was launched in Hamburg in November 2008 with 12 integrated hydrogen tanks.

8.5.5 AEROSPACE

For an airplane, a power generation system is required for different situations that can emerge. To keep the aircraft controls operative, the power generation system is independent from the main engines and the auxiliary power units (APUs). In certain circumstances, an appropriate ram air turbine (RAT), a hydraulic pump or an electric generator, is deployed and provides power for the flight controls until landing. These specific situations can potentially arise due to contaminated fuel or kerosene lava dust in the atmosphere, which can result in the failure of all internal combustion engines, main engines, and APU. Fuel cell modules can serve as a suitable alternative for this. Reliability and functionality tests can also easily be performed at any time. In 2016, the International Civil Aviation Organization approved limiting emission in air applications at 2020 levels but mainly by offsetting carbon instead of low-emission fuels (Lyle 2018). For this purpose, some ideas of hybrid electrics are being examined, though a reduction in emissions has been seen. Biofuels can be fine alternatives because of their superior power density to batteries or hydrogen, but they are not emission free and may remain expensive, with their low accessibility.

There are some serious concerns associated with the usage of hydrogen for aircraft technology. Liquified hydrogen is to be used as a propulsion fuel for a sustained fuel supply according to the desired range. As fuel cells do not have the ability to take off, combustion turbines are therefore needed. Moreover, the use of hydrogen in jet applications produces water vapor, causing radiation forcing and contributing to global warming. This poses a serious question on the climate benefits of using hydrogen for aviation. The deployment of hydrogen is therefore not considered feasible before 2050 with the exception of low-flying jet applications (Staffell et al. 2019). Hence, there is more room for R&D to open the doors for the usage of hydrogen in low-emission aviation applications.

8.5.6 Uninterruptible Power Supply

In our highly technological environment, the steady supply of energy is essential for many applications. IT and communication technology have become indispensable for handling processes, workflows, and tedious business processes. Moreover, constant provision of energy is also imperative with the growing demand for public safety results and efficient security surveillance. As a solution to this problem, intricate and sophisticated infrastructures are often equipped with an uninterruptable power supply (UPS). In general, DC voltages are used. However, this means that with the rise in voltage, the number of batteries increases. The job of the rectifier is to provide the connected load of the application and to recharge the battery bank, so the rectifier must be large enough to fulfill the tasks. It's a benefit that power supply unit (PSU) losses occur only once, and even small "network wiper" sags or frequency changes in the primary network do not reach sensitive electronics system.

However, the solution has three disadvantages as well. First, batteries are electrochemical devices subject to intense degradation. Second, the maximum possible battery time can be determined after a real test. Last, the weight of batteries (especially for lead-acid batteries) increases proportionally with the increase in backup time. Hence, an exchange of cost and expense takes place after a specified time, according to the type of battery featured. Therefore, it is considered a temporary solution for short durations of minutes to a few hours and at low power.

For such applications, fuel cells are undeniably a potential candidate, with the following advantages.

- The electrochemical process is not subjected to aging during standby.
- Considerably reduced maintenance and renewal costs as compared to a battery UPS.
- Emission-free and quiet operation.

A comparison of a modular fuel cell with conventional battery UPS and diesel engine generators is shown in Figure 8.9, with a required bridging time on the x-axis and required power along the y-axis.

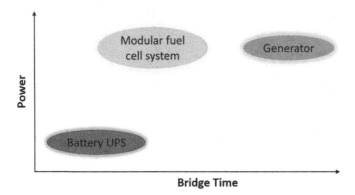

FIGURE 8.9   Comparison of some backup technologies.

8.6 OPPORTUNITIES AND CHALLENGES FOR SUSTAINABLE DEVELOPMENT

Hydrogen is an widely existing element in the environment. For decades, it was also a vital and commonly used input for industries such as in the synthesis of ammonia. Today, hydrogen has been emphasized as a strategically indispensable composition in the energy planning of the main countries of the world. It can also serve as an ideal form of energy which is expected to replace fossil fuel combustion to realize sustainable development in the energy sector. Undoubtedly, hydrogen has emerged as a very popular source of energy, and since the 1990s, the technologies have been rapidly developed to play an important role in the sustainability of the hydrogen economy. However, it will still require several years to manage the challenges associated with the use of hydrogen. The unique properties of hydrogen, such as high energy density and combustion without emissions, are the reasons for its involvement in sustainable development.

The world's energy consumption grew rapidly during the 20th century and is expected to maintain growth in the 21st century as well. This energy intake trend has led to two serious problems because of fossil fuels, including coal, oil, and gas: emissions from combustion and energy shortage, as shown in Figure 8.10.

8.6.1 CHALLENGES

To handle the energy challenges, one alternative is to encourage the usage of renewables for power generation instead of fossil energy consumption. However, owing to the intermittent nature of renewables in power generation, there is a dire need for sustainable energy storage and conversion devices. It has been discussed that hydrogen could be transmitted by pipeline and ease of refueling during long-distance driving. Fuel cells are a well-admired technology, having the highest efficiency, reliability, and zero emissions. However, for the realization of hydrogen economy,

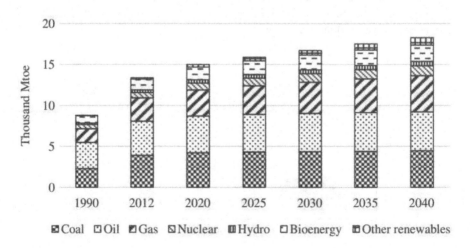

FIGURE 8.10   Changes in world primary energy demand by fuels (Birol et al. 2014).

various technical barriers have been hindering the implementation of fuel cells since their invention. These technical factors include the issues of high cost, especially of the expense of materials and the meager long-term durability of the fuel cell technology. The degradation and contamination of electrode materials are the key obstacles in the way to the commercialization of this technology.

Multiple advanced technologies are involved in the sustainability of the hydrogen economy during various processes, including production, transportation, storage, and utilization, as given in Table 8.4.

Regarding fuel cell technology, stationary fuel cells and FCVs would be common methods to store and utilize hydrogen energy. However, to witness the application of hydrogen fuel cells, the costs of manufacture, comprehensive energy efficiency, and durability are expected to be improved.

Another critical problem for the popularization of the hydrogen economy is the convenience of the supply chain to meet the intensifying demands and appropriate infrastructural services. Overall cost is one of the major questions, which tends to hinder the progress. According to the cost estimated by Tappan Bose and Pierre Malbrunot, although the cost of the feedstock is cheap (gasifying coal and biomass cost 2.6 $/GJ, steam methane reforming from natural gas is 9.3 $/GJ, and electrolysis costs 17.8 $/GJ), the whole production cost increases to 1530 $/GJ, while the net cost will increase to 3550 $/GJ, of which the unit cost for storage, transportation, and distribution is around 20 $/GJ. The major cost is contributed during distribution because to suppress hydrogen embrittlement, a nickel pipeline is required. Furthermore, the efficiency of getting hydrogen back to electricity is only about 55%, and in the case of cogeneration, it may increase to 90% (Bockris 2013). However, cost reduction depends upon the economy of scale in the future, and the increase in production supply will certainly play a leading role in bringing down the production costs.

The manufacturing cost of fuel cells includes the equipment capital, materials, labor, component fabrication, design, and assembly, which are essential in the overall manufacturing of fuel cells and stacks. Since the major applications of fuel cells include PEMFCs, the cost challenges related to them are discussed in this section. According to recent estimates, the cost of an 80-kW fuel cell stack for automobile

---

TABLE 8.4

Challenges in the Sustainability of the Hydrogen Economy

Area of Advancement	Challenges
Automobile	Longer-term durability for commercial applications
	Feasibility for mass production
	Cost reduction with small volumes
Household	Market acceptance and popularization
Hydrogen production, storage, and transportation	Standards and regulations for large-scale transportation
	Intercountry transportation
Hydrogen stations	Cost-effective development
	Standards and regulations
	Safety and security concerns

applications costs around \$73/kW. In the other study, Ahluwalia et al. (2015) produced an 80-kW$_{net}$ Argonne PEM fuel cell stack, and Yang (2013) examined the costs of a PEM fuel cell stack and indicated that the entire cost of the stack is about \$30 k/W. The PEM electrode (anode, cathode, and catalyst layer) itself represented a major part of the whole cost of the stack, about 51%. The stack assembly constitutes about 7% of the overall cost. The rest of the elements (bipolar plates and seals) cover the remaining percentage of the cost, as shown in Figure 8.11a. On the other hand, in the case of a fuel cell stack system of 80 kW$_{net}$, the cost would increase to \$59 k/W. Half of this cost is represented by the stack, with fuel, air, and thermal management the other factors that should be considered, as they add to the overall cost in the fuel cell stack system shown in Figure 8.11b.

The fuel cell cost would not be the major factor in end-user acceptance. Nevertheless, the cost of maintenance and repair is essential for end-user acceptance and stack service, which is almost neglected in considering the overall cost of the fuel cell stack. The failure of any fuel cell component due to long-term operation and cyclic conditions would result in cell failure and hence also the stack. Generally, the disassembly of the entire stack is required to replace a failed part of the stack, and this adds 100% of the cost of balancing and conditioning the stack and stack system (J. Wang 2015, 2017). As represented in Figure 8.11, such stack conditioning can represent around 22% of the entire cost of the stack system. This means that 22% of the cost of assembly, stack balancing, and conditioning is added for every repair, since these are essential treatments for every fuel cell stack system. Therefore, the cost of the whole stack is increased for every single failure of the fuel cell's components because the maintenance and repair cost can exceed 60% of the entire cost of the fuel cell system. The failure of fuel cell components mainly occurs due to the degradation and poisoning of the fuel cell components, which affects the durability and reliability of the fuel cell stack system. Thus, it can be concluded that durability and reliability are an effective strategy in significantly decreasing the cost of

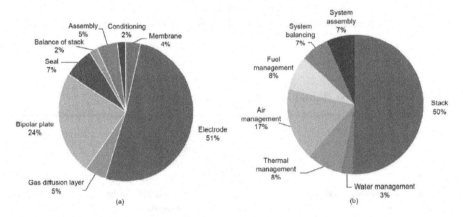

FIGURE 8.11 A pie chart to show the cost division of various components of an 80-kW$_{net}$ PEMFC for (a) stack and (b) stack system.

Adapted from J. Wang, Wang, and Fan (2018).

the entire fuel cell system and enhancing acceptance by end users of the system (J. Wang, Wang, and Fan 2018).

8.6.2 HYDROGEN AND FUEL CELLS: FUTURE MARKET AND STRATEGIES

As described, the main applications of hydrogen fuel cells are in the transport sector, which mainly depends on future targets and policy framework (McNicol, Rand, and Williams 2001). These policies, as well as investments, substantially differ from country to country. The current and future national targets for hydrogen and fuel cell technology uptake in four prominent countries are summarized in Table 8.5.

The UK and Germany are not lagging in the advancement of hydrogen technology. Both countries have defined a target of 100% zero emission vehicles (ZEVs) by the year 2040.

8.6.3 INTERNATIONAL SUPPORT AND CONSENSUS

There are several driving forces behind the global race of hydrogen and fuel cell technology. For instance, the need for air quality improvement due to transportation is the main operating principle that runs the US policy, implying that there are no national objectives for utilizing fuel cells in fixed applications. In Japan, energy security is the main reason to promote hydrogen to support national industries and lessen environmental impacts by developing three stages to make Japan a hydrogen society by (1) encouraging FCEVs and hydrogen production, (2) integrating and developing supply chains of hydrogen into the energy system by 2030, and (3) establishing a supply of zero-carbon hydrogen by 2040. Similarly, China has made a policy to reduce the issues of urban air quality by boosting economic growth via fabricating hydrogen fuel cells as a part of the MC2025 (Made in China 2025) policy. Even France and the UK have policies to terminate the sale of diesel and petrol-based cars from 2040 and the Netherlands from 2030. Norway also plans to end the sale of diesel cars from 2025 and replace them with hydrogen passenger cars. Despite all this, further policies, development, research projects, and collaborations are required to widen the awareness of hydrogen and fuel cell technology to increase acceptance by the public. A stable framework of policy and a clear long-term vision are the most important

TABLE 8.5
Summary of Goals Regarding Hydrogen Technology Roadmap in Various Countries

Country	Fuel Cell Vehicles		Hydrogen Stations	
	2025	2030	2025	2030
China	50, 000	1 M	1000	
Japan	200, 000	800, 000	320	900
South Korea	100, 000	630, 000	210	520
US	3.3 M			

elements from a social perspective. The GDP per capita is another crucial factor that is directly related to the market interest in hydrogen and fuel cells. Japan, Korea, China, the USA, and few European countries are enhancing the market value of hydrogen and fuel cells by investing more and more in this technology. For FCEVs, the US is leading with 2750 FCEVs sold as of 2017, which is more than the combined figure of FCEVs sold in Japan and Europe. The Fuel Cell Vehicle Technology Roadmap 2016 (FCVTR-2016) by China gained a target of 5000 FCEVs in 2020, with more FCEVs (in millions, shown in Table 8.5) in 2030 (Tlili et al. 2019; Staffell et al. 2019). Remarkably, FCEVs are recently entering daily life in China. One project initiated by the Ministry of Science and Technology, the Ministry of Finance, the National Development and Reform Commission, and the Ministry of Industry and Information plans to launch 10,000 new energy vehicles in ten pilot cities each year involving bus, taxi, and post areas. With agreement on the significance of the transition toward a hydrogen economy, the major countries and regions have begun to cooperate and negotiate with each other to overcome existing barriers.

8.7 CONCLUSION

In light of prior research and evolution, it is revealed that the hydrogen economy, once implemented, would play an essential role in sustainable development. As a result of this technology diffusion, there would be vibrant development in the supply chain as well. It is an undeniable truth that the cost of hydrogen production, transportation, and application is still higher than conventional technologies, which requires an effort to establish a unanimous policy. In accordance with this, various developed countries have figured out clear directions and development plans while taking into consideration the existing challenges that need to be overcome for the realization of a hydrogen economy.

However, due to the consensus on the opportunities and challenges of a hydrogen economy, the strategies and policies of technologically advanced countries are quite similar. Comparatively, Japan has demonstrated more determination in this area by investing in research and development, which has helped it own the highest number of patents around the globe. By contrast, the motive of the US government is to develop breakthrough technologies with the potential to gather profit in return. On the other side, the EU represents a steady and stable strategy between targets and investment, whereas China exhibits a huge market for hydrogen during this energy transition. More sincere but collaborative international efforts are indispensable for achieving a hydrogen economy and utilizing its benefits and advantages in a true sense.

REFERENCES

Abe, J. O., A. P. I. Popoola, E. Ajenifuja, and O. M. Popoola. 2019. "ScienceDirect Hydrogen Energy, Economy and Storage: Review and Recommendation." *International Journal of Hydrogen Energy* 44 (29): 15072–15086. https://doi.org/10.1016/j.ijhydene.2019.04.068.
Ahluwalia, R. K., X. Wang, and J.-K. Peng. 2015. "Fuel Cells Systems Analysis." *US Department of Energy Hydrogen and Fuel Cells Program 2011 Annual Merit Review and Peer Evaluation Meeting.* US Department of Energy.

Birol, Fatih, Laura Cozzi, Amos Bromhead, Tim Gould, and Marco Baroni. 2014. *World Energy Outlook2014*. www.oecd-ilibrary.org/energy/world-energy-outlook-2014_weo-2014-en.

Bockris, John O. M. 2002. "The Origin of Ideas on a Hydrogen Economy and Its Solution to the Decay of the Environment." *International Journal of Hydrogen Energy* 27 (7–8): 731–740. https://doi.org/10.1016/S0360-3199(01)00154-9.

Bockris, John O. M. 2013. "The Hydrogen Economy: Its History." *International Journal of Hydrogen Energy* 38 (6): 2579–2588. https://doi.org/10.1016/j.ijhydene.2012.12.026.

Bossel, Ulf, Baldur Eliasson, and Gordon Taylor. 2003. "The Future of the Hydrogen Economy: Bright or Bleak?" *Cogeneration and Competitive Power Journal* 18 (3): 29–70. https://doi.org/10.1080/15453660309509023.

British Petroleum Report. 2018. "67 Th Edition Contents Is One of the Most Widely Respected." *Statistical Review of World Energy* 40. www.bp.com/content/dam/bp/business-sites/en/global/corporate/pdfs/energy-economics/statistical-review/bp-stats-review-2018-full-report.pdf.

Castello, P., E. Tzimas, and P. Moretto. 2005. "Techno-Economic Assessment of Hydrogen Transmission & Distribution Systems in Europe in the Medium and Long Term." *European Commission, Joint*, no. March. http://scholar.google.com/scholar?hl=en&btnG=Search&q=intitle:Techno-economic+assessment+of+hydrogen+transmission+&+distribution+systems+in+Europe+in+the+medium+and+long+term#0.

Eberhardt, J. J. 2002. "Fuels of the Future for Cars and Trucks. Energy Efficiency and Renewable Energy." *2002 Diesel Engine . . .* 43 (5657): 5657–5678. www.fischer-tropsch.org/DOE/DOE_reports/Eberhardt, J.J/DOE-CONF-08-25-02/DOE-CONF-8-25-02.pdf.

Ferreira, J., J. Leitão, A. Monteiro, M. Lopes, and A. I. Miranda. 2017. "National Emission Ceilings in Portugal—Trends, Compliance and Projections." *Air Quality, Atmosphere and Health* 10 (9). https://doi.org/10.1007/s11869-017-0496-6.

Gerlici, Juraj, Mykola Gorbunov, Kateryna Kravchenko, Olga Prosvirova, Tomáš Lack, and Vladimír Hauser. 2018. "Assessment of Innovative Methods of the Rolling Stock Brake System Efficiency Increasing." *Manufacturing Technology* 18 (1). https://doi.org/10.21062/ujep/49.2018/a/1213-2489/MT/18/1/35.

Iordache, Ioan, Adrian V. Gheorghe, and Mihaela Iordache. 2013. "Towards a Hydrogen Economy in Romania: Statistics, Technical and Scientific General Aspects." *International Journal of Hydrogen Energy* 38 (28): 12231–12240. https://doi.org/10.1016/j.ijhydene.2013.07.034.

Lyle, Chris. 2018. "Beyond the Icao's Corsia: Towards a More Climatically Effective Strategy for Mitigation of Civil-Aviation Emissions." *Climate Law*. https://doi.org/10.1163/18786561-00801004.

McNicol, B. D., D. A. J. Rand, and K. R. Williams. 2001. "Fuel Cells for Road Transportation Purposes—Yes or No?" *Journal of Power Sources* 100 (1–2). https://doi.org/10.1016/S0378-7753(01)00882-5.

Midilli, A., M. Ay, I. Dincer, and M. A. Rosen. 2005. "On Hydrogen and Hydrogen Energy Strategies I : Current Status and Needs." *Renewable and Sustainable Energy Reviews* 9 (3): 255–271. https://doi.org/10.1016/j.rser.2004.05.003.

Mori, D., and K. Hirose. 2009. "Recent Challenges of Hydrogen Storage Technologies for Fuel Cell Vehicles." *International Journal of Hydrogen Energy* 34 (10): 4569–4574. https://doi.org/10.1016/j.ijhydene.2008.07.115.

Muthukumar, M., N. Rengarajan, B. Velliyangiri, M. A. Omprakas, C. B. Rohit, and U. Kartheek Raja. 2020. "The Development of Fuel Cell Electric Vehicles—A Review." *Materials Today: Proceedings*. https://doi.org/10.1016/j.matpr.2020.03.679.

Nejat Veziroglu, T. 2012. "Conversion to Hydrogen Economy." *Energy Procedia* 29: 654–656. https://doi.org/10.1016/j.egypro.2012.09.075.

Pudasainee, Deepak, Vinoj Kurian, and Rajender Gupta. 2020. *Coal: Past, Present, and Future Sustainable Use. Future Energy: Improved, Sustainable and Clean Options for Our Planet*. Elsevier Ltd. https://doi.org/10.1016/B978-0-08-102886-5.00002-5.

Rusman, N. A. A., and M. Dahari. 2016. "A Review on the Current Progress of Metal Hydrides Material for Solid-State Hydrogen Storage Applications." *International Journal of Hydrogen Energy* 41 (28): 12108–12126. https://doi.org/10.1016/j.ijhydene.2016.05.244.

Schüth, F. 2009. "Challenges in Hydrogen Storage." *European Physical Journal: Special Topics* 176 (1): 155–166. https://doi.org/10.1140/epjst/e2009-01155-x.

Scipioni, A., A. Manzardo, and J. Ren. (Eds.). 2017. *Hydrogen Economy: Supply Chain, Life Cycle Analysis and Energy Transition for Sustainability*. Academic Press.

Staffell, Iain, Daniel Scamman, Anthony Velazquez Abad, Paul Balcombe, Paul E. Dodds, Paul Ekins, Nilay Shah, and Kate R. Ward. 2019. "The Role of Hydrogen and Fuel Cells in the Global Energy System." *Energy and Environmental Science*. https://doi.org/10.1039/c8ee01157e.

Sun, Yahui, Chaoqi Shen, Qiwen Lai, Wei Liu, Da Wei Wang, and Kondo Francois Aguey-Zinsou. 2018. "Tailoring Magnesium Based Materials for Hydrogen Storage through Synthesis: Current State of the Art." *Energy Storage Materials* 10 (January): 168–198. https://doi.org/10.1016/j.ensm.2017.01.010.

Tlili, Olfa, Christine Mansilla, David Frimat, and Yannick Perez. 2019. "Hydrogen Market Penetration Feasibility Assessment: Mobility and Natural Gas Markets in the US, Europe, China and Japan." *International Journal of Hydrogen Energy* 44 (31). https://doi.org/10.1016/j.ijhydene.2019.04.226.

Veziroğlu, T. Nejat, and Sümer Şahin. 2008. "21st Century's Energy: Hydrogen Energy System." *Energy Conversion and Management* 49 (7): 1820–1831. https://doi.org/10.1016/j.enconman.2007.08.015.

Wang, Junye. 2015. "Barriers of Scaling-up Fuel Cells: Cost, Durability and Reliability." *Energy* 80. https://doi.org/10.1016/j.energy.2014.12.007.

Wang, Junye. 2017. "System Integration, Durability and Reliability of Fuel Cells: Challenges and Solutions." *Applied Energy*. https://doi.org/10.1016/j.apenergy.2016.12.083.

Wang, Junye, Hualin Wang, and Yi Fan. 2018. "Techno-Economic Challenges of Fuel Cell Commercialization." *Engineering*. https://doi.org/10.1016/j.eng.2018.05.007.

Wang, Shuangyin, and San Ping Jiang. 2017. "Prospects of Fuel Cell Technologies." *National Science Review* 4 (2). https://doi.org/10.1093/nsr/nww099.

Wietschel, Martin, and Michael Ball. 2009. The Hydrogen Economy: Opportunities and Challenges. *The Hydrogen Economy: Opportunities and Challenges*. Vol. 9780521882163. https://doi.org/10.1017/CBO9780511635359.

Yang, Yong. 2013. *PEM Fuel Cell System Manufacturing Cost Analysis for Automotive Applications*. Austin Power Engineering LLC.

9 Buildings for Energy and Environmental Sustainability

Muhammad Asif

9.1 INTRODUCTION

Buildings have historically been a critical part of civilizations. Buildings strongly influence the broader socio-economic dimensions of modern societies. Owing to their energy and environmental footprint, the role of buildings from a perspective of sustainable development has become ever more important. In the transition to a low-carbon, resilient, and sustainable society, buildings play a dominant role in the use of energy and are among the largest sources of greenhouse gas (GHG) emissions in most countries [1, 2]. Buildings consume a lot of resources, which is one of the main factors in the environmental impact of buildings [3–5]. The energy and environmental footprint of buildings begins with the production of materials they consume. Large quantities of raw materials, including wood, steel, concrete, and glass, are needed for both building construction and operation. During their extraction and production/refinement, these minerals can contribute to a range of issues, including emission of pollutants and greenhouse gases, habitat loss, deforestation, and soil erosion. Furthermore, considerable energy, mostly produced from fossil fuels, is used in the manufacture and transportation of building materials. Buildings are closely linked with energy use, land use, cities, and the industrial sector [6, 7].

Given its significant share in energy and resource consumption, the building sector can play a significant role in enhancing energy and environmental sustainability. Buildings that are sustainable work to reduce their negative effects on the environment, enhance occupant health and happiness, and support long-term economic viability. The International Panel on Climate Change (IPCC) asserts that significant and quick adjustments must be made in four key global systems—energy, land use, cities, and industry—if the world is to come close to meeting the climate change targets outlined in the Paris Agreement [8]. The Sustainable Development Goals (SDGs) of the United Nations (UN) have a clear understanding of the importance of buildings. While SDG 11 focuses on creating sustainable cities and communities, several other SDGs have a strong relationship with the building industry. To further encourage the development and use of low-carbon, renewable, and energy-efficient technologies in buildings, government policy support is essential. Recent trends show that the industry is making voluntary promises to increase efficiency while governments around the world are giving extensive support (including enabling policies, information

DOI: 10.1201/9781032715438-9

distribution, training, and capacity building). Large-scale energy efficiency improvements have historically been brought about by a combination of legislation, market-based instruments, incentives, capacity building, and information dissemination, especially when supported by comprehensive national policies and targets [9].

Buildings are significantly improving their energy consumption patterns and efficiency standards, especially in the developed nations, against the backdrop of the global push for sustainability. Both the technological fronts are the subject of efforts. While robust governmental frameworks are enforcing stricter norms and standards for building performance, technological breakthroughs are inventing efficient ways to further the cause. Energy management, conservation, and renewable energy technologies are among the sustainable and low-carbon solutions driving the building industry's sustainability drive.

9.2 ENERGY AND ENVIRONMENTAL FOOTPRINT OF BUILDINGS

Energy is used throughout a building's life, from construction through decommissioning. During the lifespan of a structure, energy is used both directly and indirectly. Building construction, operation, maintenance, repair, and demolition all fall under the category of direct energy use, whereas the manufacture of building materials and the setting up of machinery fall under indirect energy use [10, 11]. Overall, buildings account for around 38% of the global energy consumption. The largest portion of a building's life cycle energy usage occurs during operation for requirements such as lights, appliances, systems, heating, and cooling. The operations of buildings account for 30% of global final energy consumption and 26% of global energy-related emissions (8% being direct emissions in buildings and 18% indirect emissions from the production of electricity and heat used in buildings) [9]. The operational phase's energy consumption might range between 40% and 90% depending on a number of variables, such as climate and user behavior [12, 13].

This energy use has numerous negative effects on the ecosystem. The first is that a sizable part of energy is produced using non-renewable resources, which causes the depletion of finite resources like oil, coal, and natural gas. Second, burning fossil fuels for energy results in the emission of greenhouse gases into the atmosphere, most notably carbon dioxide (CO_2). Because of these emissions, there will be more frequent and severe weather events, higher sea levels, and changes to ecosystems as a result of climate change and global warming.

Buildings' role in greenhouse gas emissions is one of the most critical issues in terms of their environmental impact [14]. A significant portion of the world's CO_2 emissions are brought on by buildings. Both operational emissions (emissions generated while the building is being used, such as heating and cooling) and embodied emissions (emissions connected with the building's construction and upkeep) are included in this. Besides emitting GHGs, buildings also pollute air and water. For instance, burning fossil fuels for heating or energy production in structures can result in the release of airborne pollutants such sulfur dioxide (SO_2), nitrogen oxides (NO_x), and particulate matter, which can have a negative impact on air quality and human health. Buildings can also contribute to urban runoff, which introduces pollutants from parking lots, rooftops, and other surfaces into rivers and streams. Chemicals, heavy metals, and other contaminants that endanger aquatic ecosystems and pose a risk to human health may be present in this discharge. Buildings have an important role to play in the global drive for zero-carbon energy transition, as shown in Figure 9.1 [15, 16].

Large volumes of waste are produced during building construction and destruction. This waste consists of packaging materials, construction waste, and old furniture and fixtures. In landfills, a large portion of this garbage contributes to soil

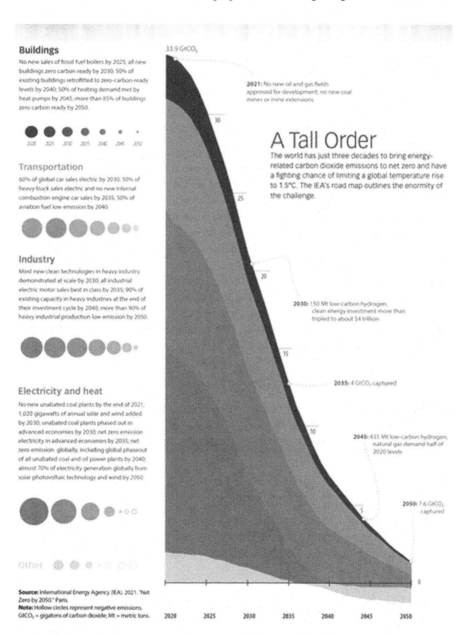

FIGURE 9.1 Role of buildings in zero carbon energy transition.

pollution and landfill congestion. It's crucial to reduce waste production during construction projects. Deconstruction, which entails painstakingly tearing down a structure to recover usable components, is one method that can greatly cut down on waste. Incorporating circular economy design ideas can also promote material reuse, recycling, and repurposing, minimizing the environmental impact of waste production.

Buildings' environmental footprint affects not only the outside environment but also indoor air quality and human health. Inadequate ventilation, the use of hazardous building materials, and the presence of indoor pollutants like radon, mold, and volatile organic compounds (VOCs) can all contribute to poor indoor air quality. Respiratory troubles, allergies, and other health problems can result from exposure to these contaminants.

9.3 BUILDINGS AND SUSTAINABLE DEVELOPMENT GOALS

Buildings strongly influence lifestyles, as they offer housing, workplaces, and social areas. But they also significantly contribute to initiatives to promote global sustainability. The Sustainable Development Goals of the United Nations offer a framework for addressing a number of international issues, including poverty, inequality, climate change, environmental degradation, and economic growth.

Buildings are essential to reaching the Sustainable Development Goals since they contribute to numerous aspects of sustainable development. Their influence ranges from combating climate change and supporting ethical production and consumption to eliminating poverty and reducing inequality. Buildings can be leveraged to promote the global agenda for a more sustainable, egalitarian, and prosperous future by adopting sustainable building practices, investing in green infrastructure, and placing a priority on inclusive and resilient communities. Some of the most relevant SDGs with regard to buildings are as follows.

9.3.1 SDG 7: Affordable and Clean Energy

Buildings use a large quantity of energy. Access to affordable, clean energy is a requirement of SDG 7. By using energy-efficient architecture, renewable energy sources, and efficient appliances to lower energy use and dependency on fossil fuels, sustainable buildings can support this objective.

9.3.2 SDG 9: Infrastructure, Industry, and Innovation

Infrastructure development is greatly aided by the construction sector. Promoting sustainable industrialization and infrastructure development, sustainable building materials, cutting-edge construction techniques, and resilient infrastructure helps to achieve SDG 9.

9.3.3 SDG 11: Sustainable Cities and Communities

SDG 11 highlights the significance of inclusive, resilient, and sustainable cities and communities. In order to create livable and environmentally responsible cities, sustainable building practices and responsible urban planning are essential components of urban development.

9.3.4   SDG 12: RESPONSIBLE PRODUCTION AND CONSUMPTION

The fundamental tenets of sustainability are responsible consumption and production. Buildings and construction use a lot of resources and produce a lot of garbage. Sustainable building methods, such as recycling materials and cutting waste, support SDG 12's goals.

9.3.5   SDG 13: ACTION ON CLIMATE

A major portion of greenhouse gas emissions comes from buildings. SDG 13 is concerned with urgent climate action. Energy-efficient building designs and the incorporation of renewable energy all help to lower emissions and combat climate change.

9.3.6   SDG 15: LAND-BASED LIFE

Responsible construction and urban growth depend on sustainable land use planning. The protection of terrestrial ecosystems and the reduction of habitat fragmentation are both emphasized by SDG 15. Both urban and rural communities can use sustainable building techniques to help them realize these goals.

9.3.7   SDG 17: PARTNERSHIPS TOWARDS GOALS

Collaboration between governmental entities, business stakeholders, members of civil society, and local communities is frequently necessary to create sustainability. SDG 17 emphasizes the value of partnerships in achieving shared sustainability goals in the building and construction industry.

Even while these SDGs are particularly pertinent to construction, it's crucial to understand that all 17 SDGs are interconnected and that progress made toward one objective frequently contributes to progress made toward others. As a result, improving sustainability in the building sector can have profoundly positive effects on many other sustainability domains.

9.4   ENERGY CONSERVATION AND MANAGEMENT

Energy efficiency through energy conservation and management (ECM) is a critical part of sustainability in the energy and buildings sectors. Due to rising energy prices, environmental concerns, and the requirement to lower greenhouse gas emissions, energy conservation and effective energy management in buildings have taken on a crucial role. Energy conservation and management, also referred to as energy management, is a process of decreasing the quantity of energy used while achieving a similar level of output. Energy conservation and management techniques not only lower energy costs but also improve occupant comfort and well-being and environmental sustainability. Energy management in a fundamental approach is quite similar to any effective project management. Some of the objectives of ECM can be:

- Improve energy efficiency and reduce energy use
- Purchase energy at a lower overall price
- Reduce environmental footprint
- Adjust operations to save on energy prices through the better choice of tariff options

To industrial and commercial organizations, ECM can be further helpful to:

- Reduce the impacts of curtailment
- Achieve strategic targets
- Develop and maintain effective monitoring and reporting of energy usage

Energy efficiency has an important role to play in the global efforts for zero-carbon energy transition, as shown in Figure 9.2 [17]. For its critical role, energy efficiency is one of the cornerstones of energy and environmental frameworks around the world [18–20]. The investment in energy efficiency technologies is also growing. Reports suggest that investments in building energy efficiency have gone up by unprecedented levels, rising by 16% in 2021 over 2020 levels to USD 237 billion [21]. Effective energy management starts with conducting energy audits and benchmarking. These evaluations point out places where energy conservation measures can be put in place. In order to get insight into a building's relative efficiency, benchmarking entails comparing its energy performance to industry norms and those of comparable buildings. Real-time energy monitoring and metering systems can be used to precisely measure energy use, spot abnormalities, and make educated decisions to reduce the consumption of energy. These methods can assist in identifying areas that require improvement and validate the efficacy of adopted strategies.

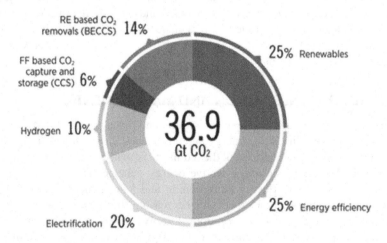

FIGURE 9.2 Major contributing factors to zero carbon energy transition.

Long-term energy management relies heavily on the incorporation of energy-efficient design principles into the initial planning for a building or the retrofitting of existing buildings. The use of passive design techniques, material choice, and building orientation can all have a big impact on energy use. Energy management systems (EMSs) can be used to provide a centralized interface for managing and improving different building systems. EMSs can perform data analysis, automate energy-saving techniques, and give building operators useful information. Demand response programs, which aid buildings in using less energy during periods of high demand, can also be facilitated by EMSs.

It is essential to involve building inhabitants in energy conservation measures. Energy use can be significantly reduced by informing residents about energy-saving techniques, giving them feedback on their energy use, and promoting a culture of sustainability.

9.5   RENEWABLE ENERGY IN BUILDINGS

Renewable energy is one of the most critical and central aspects of the sustainability drive in buildings. It can help buildings on a number of fronts, including but not limited to reducing environmental footprint, enhancing energy diversity and security, energy self-sufficiency, and saving on overall energy costs. Renewable technologies such as solar photovoltaic (PV) solar water heating, wind turbines, geothermal energy, and hydropower can provide energy to buildings that is both clean and sustainable. Communities and building owners, especially in developing regions, are becoming more and more motivated to become energy independent by producing their own power. They can lessen their dependency on centralized electricity grids thanks to renewable energy installations. Renewable energy systems offer backup power and improve energy resilience in areas vulnerable to power outages or severe weather. The adoption of renewable energy technology is being sparked by this appeal.

To encourage the use of renewable energy in buildings, governments all over the world are providing incentives, subsidies, and restrictions. Investments in renewable energy systems are encouraged by these laws, which further fuels market expansion. Renewable energy is also the foundational block of the unfolding energy transition. Buildings are increasingly using renewable energy as a result of the global transition to cleaner energy sources to tackle climate change. Owners of both residential and commercial buildings are being prompted by energy transition initiatives to look at renewable energy solutions.

The share of renewable energy in the total energy consumption in buildings has grown from 10.7% in 2009 to 14.7% in 2019 as shown in Figure 9.3 [22]. Over this period, while solar and geothermal heat has recorded a gain, bioheat has seen a slight decline. The most significant gain is in terms of renewable electricity, which mainly comes from solar PV.

Solar PV is the fastest-growing renewable technology in the building sector across the world [23–25]. PV can make a significant contribution towards reducing the energy and environmental footprint of buildings. Helped by features like scalability, ease of use, and declining price, PV has become the predominant renewable technology for application in buildings. Estimates suggest that rooftop PV can

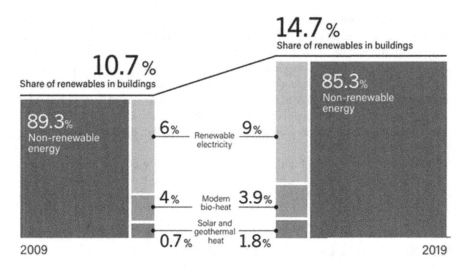

FIGURE 9.3   Growth in share of renewable energy in buildings.

help meet 25% to 49% of national electricity requirements in countries around the world. When it comes to the application of solar energy in buildings, photovoltaic has been by far the most versatile and successful technology. Small and building-related applications have played a key role in the progress of solar PV throughout the world. Most of the leading countries in installed capacity of PV have extensively used the technology in the building sector [26]. Conducive policies like feed-in-tariff and net-metering have been successfully implemented around the world. Rooftop PV is estimated to represent over 40% of the world's total PV installed capacity. Rooftop PV, due to factors like quick deployment and low levelized cost of electricity, is also regarded as important in achieving Sustainable Development Goal 7 [27, 28].

9.6   SUSTAINABLE BUILDINGS: CHALLENGES

The building sector faces a number of major challenges in its quest for sustainable buildings as the world's attention on sustainability grows. These challenges are of many types, ranging from planning and building to running and maintaining. Some of the major challenges are discussed as follows.

9.6.1   HIGH CAPITAL COST

Buildings that are sustainable face a number of difficulties, including the notion of high initial expenditures. The use of environmentally friendly materials, renewable energy sources, and energy-efficient technologies can be more expensive than traditional alternatives when it comes to sustainable construction. Even while these expenditures may lead to long-term reductions in operating and maintenance costs, many stakeholders may find the initial costs prohibitive, particularly if there are no strong incentives or legal requirements for sustainability.

9.6.2 RETURN ON INVESTMENT

The difficulty of obtaining a sufficient return on investment (ROI) for sustainable building elements is related to the problem of high starting costs. If they anticipate a lengthy payback period, investors and developers may be reluctant to invest in sustainable initiatives. To overcome this obstacle and promote wider adoption of green construction practices, it is crucial to quantify the financial advantages of sustainability, such as decreased energy costs and higher property value.

9.6.3 LACK OF AWARENESS

A major problem is the lack of knowledge and instruction on sustainable building techniques. Many building industry professionals may be unaware of all the advantages and strategies linked to sustainability. For architects, engineers, contractors, and other stakeholders to be knowledgeable about sustainable design and construction practices, there should be greater chances for educational programs, training, and certification.

9.6.4 REGULATORY COMPLEXITIES

It might be difficult to navigate the complex regulatory environment connected to sustainability. It can be challenging for developers and builders to manage the criteria for sustainable construction due to regional differences in building rules, zoning laws, and environmental norms. The adoption of sustainable practices can be facilitated by streamlining and harmonizing these regulations.

9.6.5 LIMITED AVAILABILITY OF SUSTAINABLE MATERIALS

In some areas, the supply of environmentally friendly building materials may be constrained. Costs associated with logistics and shipping may rise if specific vendors are required to obtain sustainable materials. This problem can be solved by promoting the local manufacture of eco-friendly materials and increasing their accessibility.

9.6.6 INTEGRATION OF RENEWABLE ENERGY SOURCES AND ENERGY EFFICIENCY

Sustainable buildings must have high levels of energy efficiency, although this can be difficult to do. Buildings must be carefully planned and expertly designed to maximize natural lighting, ventilation, and insulation while minimizing heat gain or loss. Additionally, because of technical, governmental, and financial considerations, integrating renewable energy sources like solar panels can be challenging.

9.6.7 OCCUPANT BEHAVIOR

Sustainable building design and technology can only go so far in accomplishing sustainability goals. An important factor in waste reduction, energy and water conservation, and general building performance is occupant behavior. It's a never-ending struggle to alter tenant behavior through instruction, rewards, and feedback methods.

9.6.8 Performance and Maintenance

Sustainable structures must maintain their great performance throughout their lifespan. Energy-efficient systems might be difficult to maintain over time because maintenance procedures may need to be changed and resources may need to be set aside for this reason.

9.6.9 Resilience and Climate Adaptation

It is becoming more crucial than ever to increase resilience to climate change and extreme weather. Buildings that are sustainable must be built to endure climate-related problems, including increasing sea levels, more frequent storms, and rising temperatures. This calls for thorough preparation and financial investment in climate-resilient design and construction methods.

9.6.10 Infrastructure Planning and Urban Design

When included in sustainable urban planning and infrastructure, sustainable buildings are most successful. However, it might be difficult to coordinate the several stakeholders, such as local governments, developers, and utilities. sustainable construction techniques. To have the greatest impact, sustainable building practices must be in line with more comprehensive urban development plans.

9.7 DISCUSSION AND CONCLUSIONS

It is crucial to consider how buildings' environmental footprints might be reduced. Buildings affect many different environmental and socioeconomic variables, as well as resource use, energy use, trash production, and greenhouse gas emissions. Adopting sustainable building practices, putting an emphasis on energy efficiency, reducing waste, and minimizing the environmental impact of buildings at every stage, from design and construction to operation, are crucial for promoting sustainability and combating climate change. In order to reduce energy consumption, operational expenses, and environmental implications, buildings must use effective energy conservation and management approaches. The integration of renewable energy sources, smart building systems, and data-driven energy management are becoming more and more feasible and cost effective as buildings continue to change as a result of technological and architectural improvements. Buildings may greatly contribute to global energy efficiency targets, reduce carbon emissions, and offer occupants comfortable and sustainable spaces by implementing a holistic strategy that combines energy-saving measures with proactive energy management tactics. Energy-efficient buildings have a bright future ahead of them as technology develops and environmental consciousness rises, ushering in a time of more resilient and sustainable built environments.

Renewable energy is an integral part of sustainable buildings, offering a range of benefits to buildings beyond. The incorporation of renewable energy technology promotes a resilient and sustainable future by providing benefits for the environment and the economy, including reduced greenhouse gas emissions, improved air quality, and the generation of jobs. Renewable energy usage in buildings is becoming more and more feasible and cost efficient as technology develops. These technologies enable building owners, communities, and individuals to have a beneficial impact on the environment while gaining economic and quality-of-life advantages by combining them with energy-efficient building design and successful energy management practices. One of the most important steps towards a more sustainable and prosperous future is the adoption of renewable energy in buildings. It is crucial to take action to reduce greenhouse gas emissions from buildings. Communities all across the world are already feeling the effects of climate change, which is causing more frequent and severe heatwaves, storms, droughts, and wildfires. We can make a big contribution to international efforts to curb global warming and lessen the effects of climate change by decreasing the carbon footprint of buildings. To provide high interior air quality and protect residents' health and well-being, buildings must be designed with enough ventilation systems, low-VOC or VOC-free materials, and measures to avoid moisture and mold formation.

Although there are substantial obstacles to achieving sustainable structures, they are not insurmountable. Stakeholders in the construction sector are more driven than ever to meet these problems as the value of sustainability and the urgent need to solve environmental and societal issues become more widely understood. To advance sustainable building practices, collaboration between architects, engineers, developers, policymakers, and communities is crucial.

REFERENCES

1. UNEP, *2020 Global Status Report for Buildings and Construction*, United Nations Environment Program, 2020.
2. H. Qudrat-Ullah, M. Asif, *Dynamics of Energy, Environment, and Economy: A Sustainability Perspective*, 1st ed., Springer, 2020. https://doi.org/10.1007/978-3-030-43578-3_1.
3. A.S. Mahmoud, M. Asif, M.A. Hassanain, M.O. Babsail, M.O. Sanni-Anibire, Energy and economic evaluation of green roofs for residential buildings in hot-humid climates, *Buildings*. 7 (2017) 30. https://doi.org/10.3390/buildings7020030.
4. F. Alrashed, M. Asif, Prospects of renewable energy to promote zero-energy residential buildings in the KSA, *Energy Procedia*. 18 (2012) 1096–1105. https://doi.org/10.1016/j.egypro.2012.05.124.
5. A.H.A. Dehwah, M. Asif, M.T. Rahman, Prospects of PV application in unregulated building rooftops in developing countries: A perspective from Saudi Arabia, *Energy Build*. 171 (2018) 76–87. https://doi.org/10.1016/j.enbuild.2018.04.001.
6. M. McGrath, Final call to save the world from "climate catastrophe", *BBC*, 2018. www.bbc.com/news/science-environment-45775309.
7. F. Alrashed, M. Asif, Challenges facing the application of zero-energy homes in Saudi Arabia: Construction industry and user perspective, *Proceedings of the ZEMCH 2012 International Conference*, 2012, pp. 391–398.
8. M. Asif, *Handbook of Energy Transitions*, CRC Press, 2022, ISBN: 978-0-367-68859-2
9. IEA, *Energy Efficiency 2019*, 2019. www.iea.org/reports/energy-efficiency-2019.
10. I. Sartori, A.G. Hestnes, Energy use in the life cycle of conventional and low-energy buildings: A review article, *Energy Build*. 39 (2007) 249–257. https://doi.org/10.1016/J.ENBUILD.2006.07.001.

11. L.F. Cabeza, L. Rincón, V. Vilariño, G. Pérez, A. Castell, Life cycle assessment (LCA) and life cycle energy analysis (LCEA) of buildings and the building sector: A review, *Renew. Sustain. Energy Rev.* 29 (2014) 394–416. https://doi.org/10.1016/J.RSER.2013.08.037.
12. J. Hong, X. Zhang, Q. Shen, W. Zhang, Y. Feng, A multi-regional based hybrid method for assessing life cycle energy use of buildings: A case study, *J. Clean. Prod.* 148 (2017) 760–772. https://doi.org/10.1016/J.JCLEPRO.2017.02.063.
13. L. Guan, M. Walmsely, G. Chen, Life cycle energy analysis of eight residential houses in Brisbane, Australia, *Procedia Eng.* 121 (2015) 653–661. https://doi.org/10.1016/J.PROENG.2015.08.1059.
14. Kh. Nahiduzaman, A. Al-Dosary, A. Abdallah, M. Asif, H. Kua, A. Alqadhib, Change-agents driven interventions for energy conservation at the Saudi households: Lessons learnt, *J. Clean. Prod.* 185 (2018) 998–1014.
15. A. Stanley, Net Zero by 2050, *International Monetary Fund*, September 2021. www.imf.org/en/Publications/fandd/issues/2021/09/infographic-series-net-zero-2050-IEA-report.
16. W. Ahmed, M. Asif, A critical review of energy retrofitting trends in residential buildings with particular focus on the GCC countries, *Renew. Sust. Energ. Rev.* 144 (2021) 111000. https://doi.org/10.1016/j.rser.2021.111000.
17. IRENA, *World Energy Transitions Outlook 2022*, International Renewable Energy Agency, 2022.
18. M. Asif, *The 4Ds of Energy Transition: Decarbonization, Decreasing Use, Decentralization, and Digitalization*, Wiley, 2022, ISBN: 978-3-527-34882-4.
19. M. Asif, *Handbook of Energy and Environmental Security*, Elsevier, 2022, ISBN: 978-0-128-24084-7.
20. W. Ahmed, M. Asif, F. Alrashed, Application of building performance simulation to design energy-efficient homes: Case study from Saudi Arabia, *Sustainability.* 11(21) (2019), 6048. https://doi.org/10.3390/su11216048.
21. UN, CO2 emissions from buildings and construction hit new high, leaving sector off track to decarbonize by 2050: United Nations, *Press Release*, 9 November 2022. www.unep.org/news-and-stories/press-release/co2-emissions-buildings-and-construction-hit-new-high-leaving-sector.
22. REN21, *Renewables Gobal Status Report 2022*, IRENA.
23. H. Khan, M. Asif, M. Mohammed, Case study of a nearly zero energy building in Italian climatic conditions, *Infrastructures.* 2(4) (2017) 9. https://doi.org/10.3390/infrastructures2040019.
24. A. Pandey, M. Asif, Assessment of energy and environmental sustainability in South Asia in the perspective of the Sustainable Development Goals, *Renew. Sust. Energ. Rev.* 165 (September 2022) 112492.
25. B. Ghaleb, M. Asif, Assessment of solar PV potential in commercial buildings, *Renew. Energ.* (2022). https://doi.org/10.1016/j.renene.2022.01.013.
26. A. Alazazmeh, A. Ahmed, M. Siddiqui, M. Asif, Real-time data-based performance analysis of a large-scale building applied PV system, *Energy Rep.* 8 (November 2022), 15408–15420.
27. B. Ghaleb, M. Asif, Application of solar PV in commercial buildings: Utilizability of rooftops, *Energy Build.* 257 (15 February 2022) 111774. https://doi.org/10.1016/j.enbuild.2021.111774.
28. S. Josji, S. Mittal, P. Holloway, P. Shukla, B. O'Gallachoir, J. Glynn, High resolution global spatiotemporal assessment of rooftop solar photovoltaics potential for renewable electricity generation, *Nat. Commun.* 12 (2021) Article number 5738.

10 The Push for Renewable Energy Adoption in Africa
Challenges and Prospects

Joan Nyika, Megersa Olumana Dinka

10.1 INTRODUCTION

Clean and affordable energy access underpins Sustainable Development Goal (SDG) 7 and seeks to alleviate the challenges of using dirty, non-renewable energy and low availability of energy to many communities of the world (Barau et al. 2020; Li et al. 2022). According to Hillerbrand (2018), adequate and accessible energy is crucial in achieving all the other SDGs owing to its role in reversing poverty, facilitating advancements in industrialization, generation of food, water supply, quality health and education. Additionally, adopting clean energy could enhance climate change adaptation and mitigation. Since the 2015 launch of the SDGs, milestones have been met towards adopting renewable energy and making it locally available to users. A forecast by the International Energy Agency (IEA 2019) confirmed this supposition, stating that in the period between 2019 and 2024, renewable energy use, particularly solar photovoltaics and wind energy, will expand by 50% globally. Similarly, renewable energy use was projected to rise from 14% in 2015 to 63% in 2050 (Gielen et al. 2019). Although the progress is remarkable, some authors are of the view that achievements are lagging behind the set timelines for energy efficiency (Hillerbrand 2018; Nerini et al. 2018).

Africa is endowed with a variety of energy sources, including wave, tidal, geothermal, wind, solar, hydropower, coal, natural gas and oil, that are unevenly distributed. As of 2018, the continent had 33 exajoules of energy, whose sources were as shown in Table 10.1 (IRENA and AFDB 2022). Despite this availability, the continent has the greatest challenge in access to energy services in the world evident from the long queues at gas stations and the high frequency of power outages (Sambo 2016). Considering that conventional energy sources, mainly natural gas and oil, are finite in nature and their expanded use is associated with global warming and climate change, adopting and transiting to renewable energy use for Africa is a priority to ensure sustainable environmental and economic growth (Samoita et al. 2020). In line with these ambitions, it is evident that some African countries are making progress to enhance off-grid connections towards expanded use of renewable energy (Pilot et al. 2019). The achievements, however, have been sluggish as a result of limited economic support to implement and run renewable energy projects amidst rising energy demands from the growing population (Gielen et al. 2019; Barau et al. 2020).

DOI: 10.1201/9781032715438-10

TABLE 10.1

Percentage Share of Different Sources of Energy in Africa as of 2018 (IRENA and AFDB 2022)

Source of Energy	Percentage Share
Biofuel and waste	43
Oil	23.3
Natural gas	16
Coal	15
Electricity and heat	2.4
Nuclear	0.4

According to Bishoge et al. (2020), renewable energy implementation in Africa is hindered by inadequate human, financial and technical resources and sociopolitical barriers, as well as weak regulatory and institutional frameworks in the region.

The challenge also has been in striking a balance on how the transition to renewable energy use from conventional energy sources will be conducted and how the challenges associated with the process will be bypassed. Usually, transitions from non-renewable to renewable energy forms are accompanied by enormous challenges whose resolution at present is limited to technological options despite the paradigmatic nature of energy systems that are sociotechnical (Bolton and Foxon 2015). In such a system, a change in energy supply and distribution induces changes in the demand and use, hence the need to use low-tech or no-tech solutions to address the energy sector issues. Energy transition also has effects on human lives. For instance, the preference for renewable energy sources induces fluctuations in the transfer of finances and the decentralization of energy systems, among other societal and institutional changes (Hillerbrand 2018). To better understand the road to adoption and use of renewable energy in Africa, this chapter will explore the continent's renewable energy potential and the determinants of growth as well as challenges and future prospects of adopting renewable forms of energy in detail.

10.2 THE CHALLENGE OF ACCESS TO ENERGY IN AFRICA

Access to energy is a fundamental need for economic growth and improved welfare of African countries. The access is, however, limited, according to statistics by the IEA (2017) that reported a crisis in energy access, particularly in sub-Saharan Africa (SSA). In the region, four of ten residents had access to electricity in 2016, compared to nine of every ten persons globally. Additionally, sixteen of the twenty countries with the highest energy deficiency were from SSA, and the region accounted for a majority, at 57%, of all global electricity deficiencies, which directly affected 609 million Africans. IEA (2017) further projected that the population affected by lack of electricity in Africa would rise to 654 million by 2030 unless mitigation measures were taken up. Northern Africa countries, in addition to Cape Verde, Ghana, South Africa, Gabon, Mauritius and Seychelles, have the highest access to electricity,

TABLE 10.2

Energy Landscape for the Regions of Africa (IRENA 2015)

Region	Population (millions)	GDP in billion/yr ($)	% Access to Electricity (% population)	Electricity per Capita (KWh)
North Africa	175	1936	98%	1574
West Africa	327	1310	47%	188
East Africa	303	646	23%	91
Central Africa	115	227	25%	167
Southern Africa	177	1100	43%	2061

compared to other regions where less than 80% of the population has no access to any form of power. In more than ten countries, only a population <20% has access to electricity. Agreeing with the statistics by IEA, Sambo (2016) observed that more than 80% of the population living in SSA had no access to electricity. Similarly, Njoh et al. (2019) highlighted that Africa has an electricity famine, with only 24% of the population having access to power due to declining per capita electricity in the continent. Recent statistics by the World Bank (2021) showed that only 36 and 60% of the African population had access to electricity and clean fuel and cooking technologies, respectively. The energy landscape of the continent based on its five regions is shown in Table 10.2 (IRENA 2015).

Apart from challenges in access, Africa also contends with issues on duration, affordability, reliability and quality of energy sources. Access to electricity on the continent has a prohibitive cost. The cost, in addition to limited access for rural dwellers and intermittent supply in informal settlements of urban areas, impedes the full use of energy resources to grow the economy. For instance, only 1% of the rural population in Niger, Liberia, Guinea Bissau, Guinea, Burkina Faso, South Sudan, Djibouti, Chad, Central African Republic and Democratic Republic of Congo had access to electricity in 2016 (IEA 2017). According to Bishoge et al. (2020), only 62% of the 1.3 billion African people have access to electricity, and the demand of power totals 700 terawatt-hours, where 70% of this total is accounted for by South Africa and north Africa. In addition, limited access to power resulted in US$60.52 billion losses, equivalent to 4% gross domestic product losses, in 2016 (African Progress Panel 2017). The energy challenge of Africa is exacerbated by population increases; poor plans for power distribution, particularly in rural areas; and informal settlements and an overdependence on fuelwood (IEA 2017).

Apart from non-renewable energy resources, Africa is also endowed with conventional energy sources, including nuclear, coal, natural gas and crude oil, as shown in Figure 10.1 (Danlami and Islam 2015). The oil reserves of the continent, mainly found in Nigeria, Libya, Algeria and Angola, are the second largest globally after those of the Middle East and account for more than 11% of global oil production. However, much of the oil is exported, since Africa only consumes 4% of global oil. The proven reserves of natural gas, mainly in West African countries,

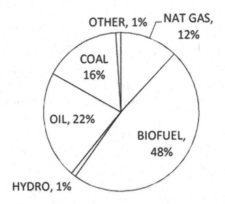

FIGURE 10.1 The share and variety of renewable and non-renewable energy resources found in Africa (IRENA 2015).

is 14.7 trillion m³, while those of coal and nuclear are approximately 31.7 billion tons and 700, 000 tons, respectively (Barau et al. 2020). Just like the oil, most of these non-renewable energy sources are exported, leaving Africa deprived of energy despite being a key producer.

10.3   ENERGY AND SUSTAINABILITY IN AFRICA

Despite having diverse non-renewable energy sources, Africa is one of the continents considered highly vulnerable to climate change due to drivers such as water scarcity, low preparedness to deal with extreme climatic events, overreliance on subsistence agriculture, population growth and attendant anthropogenic activities (Nyika 2020). The situation with regard to energy security has been worsened by the COVID-19 pandemic, which has upended previous gains on universal energy access on the continent (Li et al. 2022). IRENA and AFDB (2022) made similar suggestions, citing the need to revamp the energy sector through the use of renewable energy sources that are sustainable and deal with the post-pandemic effects on industrial development. While exploration of non-renewable forms of energy in developing African countries is driven by poverty, the need for economic development and fossil fuel vagaries of developed nations, potential resource depletion and resultant air pollution following exploration of the energy forms necessitates the adoption of renewable energy technologies (Fouche and Brent 2019). The adoption of renewable energy makes a positive contribution to sustainable development by reducing the apparent overdependence on fossil fuels as well as creating new employment (Bishoge et al. 2020). According to Hafner et al. (2018), renewable energy is an essential asset for developing African countries, and as the continent becomes tech-savvy and industrialization advances, such power sources are becoming affordable.

The economic vulnerability of Africa, influenced by fluctuations in imported fuels, would equally be reduced by adoption of renewable energy sources. Kuamoah (2020) agreed with the suggestion, claiming that the potential of renewable energy in Ghana will be high if such power sources are fully exploited and conventional energy sources are shunned. Similarly, South Africa has shown interest in the uptake of renewable energy due to its eco-friendly nature. Practices that encourage feed-in tariffs, waste-to-energy, decentralized power generation and economic development pillared on low-carbon and biomass-to-energy are now becoming common in the country (Fouche and Brent 2019). Kenya has made considerable reductions in its budgetary allocations for oil imports and over-dependence on wood biomass following adoption of renewable energy technology such as solar photovoltaics and hydropower, which, according to Kiplagat et al. (2011), is progress towards a better economy. In six east African countries, adoption of renewable energy and particularly solar power was found to improve the human development index and gross domestic product in addition to diversifying energy choices, creating employment and reducing countries' foreign dependency (Chisika and Yeom 2021). It is for these reasons that many African countries are pushing for the expansive use of renewable energy towards a smooth transition to their preference compared to conventional energy sources that are expensive, exhaustive and unclean (Sambo 2016; Pilot et al. 2019). In fact, IRENA (2020) praised the sustainable nature of renewable energy sources as the pillar of resilient societies and economies in the near future.

10.4 DETERMINANTS OF RENEWABLE ENERGY GROWTH AND ADOPTION

Evidently, the uptake of renewable energy technologies in Africa is on the rise, though at a sluggish rate. Environmental advocacy for reduced carbon emissions, overdependency on imported energy sources and the volatile prices of oil and gas are key influencers of the transition to renewable energy (Nyika et al. 2020). In this section, the factors that drive the adoption of and high preference for renewable energy are discussed under three categories: 1) political, 2) socioeconomic and 3) country-specific factors, as suggested by Aguirre and Ibikunle (2014). The specific factors in each category are summarized in Figure 10.2.

10.4.1 POLITICAL FACTORS

Political drives are the most influential drivers of adoption and growth of renewable energy, according to Kilinc-Ata (2016). Political factors drive energy security, institutional variables and public policies. Unlike conventional energy sources, renewable energy technologies are considerably expensive, and the two cannot compete unless there are supporting policies for the latter (Aguirre and Ibikunle 2014). Similarly, other authors have reported that public policies instigate and drive renewable energy adoption (Maria and Bernauer 2014; Stadelmann and Castro 2014; Kilinc-Ata 2016). Public policy regulations that encourage renewables include green certificates, feed-in tariffs, research and development, direct investment, quota policies and subsidies. Authors such as Polzin et al. (2015) advocate for the establishment of long-term policies and frameworks to grow renewable energy and emphasize the need to make them technology specific. Such advancement makes formulated policies compatible

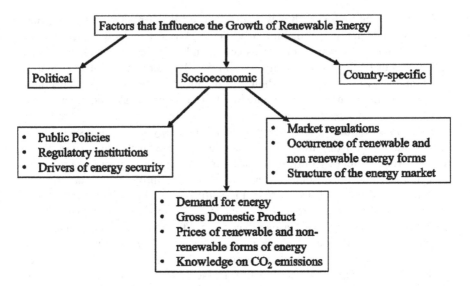

FIGURE 10.2 Factors that influence the growth in use of renewable energy resources in Africa.

with the objectives of domestic and international political institutions (Stadelmann and Castro 2014). Although such policies differ in administration, eligibility, size, structure and application, their goal to increase renewable energy capacity is common (da Silva et al. 2018).

Institutions that shape policy outcomes by influencing operations of political actors directly or indirectly are also key in the growth of renewable energy technology. For instance, political ideologies and organizations and their affiliated inter- and intra-party rivalries influence the adoption levels for environmental policies such as the Kyoto Protocol, whose aim was to reduce greenhouse gas emissions by promoting the uptake of renewable energy technologies (Stadelmann and Castro 2014). Such commitments alongside regional and global institutions such as the African Union European Union and United Nations and their member countries encourage deployment of renewables by setting targets of emission reductions (da Silva et al. 2018). At the national level, energy security through deployment of renewables is recommended to create energy independence and deconcentrate energy origins and sources (Augutis et al. 2014; Lucas et al. 2016). Therefore, countries are encouraged to push for energy security in their national agendas since it reduces overreliance on power imports and improves the trade balance (Aguirre and Ibikunle 2014).

10.4.2 Socioeconomic Factors

Socioeconomic factors that influence the growth of renewable energy include the energy demand, a country's gross domestic product, prices of non-renewable energy sources and carbon dioxide emissions (da Silva et al. 2018). The reduction of carbon emissions has been a priority in the adoption of renewable energy towards realizing

TABLE 10.3

Projected Energy Reductions with Introduction of Renewable Energy Sources (Gielen et al. 2019)

Sector	Use of Energy as of 2050 (GT/year)
Building	32
Transport	25
Heating processes	18
Power production	13
Industrial processes	8

sustainability (Rustemoglu and Andres 2016). A number of authors support the suggestion that concerns about rising carbon emissions are drivers to adopt renewables (Omri and Khuong 2014; Rafiq et al. 2014). Using 35 GT annually as the reference case in 2050, Gielen et al. (2019) supported the adoption of renewable energy sources projecting that their introduction was set to reduce carbon dioxide emissions by 94% in the year 2050. The specific sectors that will be affected are as shown in Table 10.3. The energy consumption of a given country is directly proportional to economic development and population growth. The effects of a rising population in Africa, the need to develop economically and the challenge of access to electricity necessitate renewable energy technologies use to enhance energy security sustainably (Ackah and Kizys 2015; da Silva et al. 2018). Aguirre and Ibikunle (2014) were, however, of a different view, suggesting that increased demand for energy among African developing countries encourages the use of conventional energy sources that are considered cheap. Similarly, Pfeiffer and Mulder (2013) associated increased power demand with slow adoption of renewables in preference to fossil fuels.

The prices of conventional energy sources also influence the uptake of renewable energy. Prices for the latter have been falling, with advancements in technology allowing more uptake. Renewables, however, cost more compared to conventional energy sources, a trend which could be attributable to lack of full costing of the latter and non-valuing of the consequences of their use, according to Foster et al. (2017). Despite being major sources of greenhouse gas emissions associated with climate change, fossil fuels, oil and gas exploration investments are on the rise (Abas et al. 2015). Authors such as Reboredo (2015) are of the view that renewable adoption could be encouraged by increasing the prices of conventional energy sources and vice versa. The economic potential of a country measured by the gross domestic product per capita determines the growth of renewables. In this context, wealthy countries with high GDPs have greater potential to adopt renewables since they can afford the technologies and enforce associated policies promoting such energy sources and vice versa (Ohler and Fetters 2014). Additionally, wealthy countries can respond to energy prices that come with transiting to renewable energy use. This is possible through advanced technologies and innovations in renewable energy infusions that result in their reduced costs and via government support of energy development projects, as Kumar and Agarwala (2016) noted.

10.4.3 COUNTRY-SPECIFIC FACTORS

Specific countries have diverse market regulations on renewables and potentials that dictate their uptake or otherwise. The occurrence of natural renewable energy sources such as strong winds, waterfalls and solar irradiation is positively correlated with high uptake of renewables since these resources need to be of adequate quality and quantity to compete with conventional energy sources (da Silva et al. 2018). According to Aguirre and Ibikunle (2014), countries endowed with wind, biomass and solar energy sources show higher renewable potential, unlike those without. Similarly, Stadelmann and Castro (2014) observed that countries with the natural resources to produce renewables are likely to take them up compared to those with limited resources. The structure of energy markets also influences the uptake of renewables. In markets with high regulatory exposure, low uptake of renewables was observed (Chassot et al. 2014). This trend is attributable to policy risk aversion amidst capital investors in such markets that have no autonomy provisions for potential investors. Pfeiffer and Mulder (2013) had a different viewpoint that the enactment of regulatory and economic instruments favors the diffusion of renewables. Nesta et al. (2014) positively correlated the innovation of renewables to liberalization of energy markets. This is because renewable energy production is small in size, hence the decentralization of energy sector.

10.5 THE POTENTIAL OF RENEWABLE ENERGY IN AFRICA

Africa has great potential to enhance the access and distribution of energy to its population, which would enable industrialization and agricultural production and reduce the high poverty levels by creating jobs towards inclusive economic development (Nyika et al. 2020). The expansive geographic landscape of the continent makes distribution and access to the national grid exorbitant, particularly among the rural population. Owing to this challenge, cost-efficient renewable energy technologies come in handy as solutions to rural electrification in Africa.

Specific renewable energy technologies have been adopted in Africa at high rates. For instance, the solar photovoltaic installed capacity of the continent was rated at 1334 megawatts in 2014, which was ten times higher than 2009 at only 127 MW (IRENA 2015). Countries such as South Africa and Kenya led in the use of solar energy at 780 and 60 MW, respectively, in the same year. Wind energy installation capacity for the continent in 2013 was 1463 MW, but the figure increased by 999 MW in 2014 to total 2462 MW (IRENA 2015). The growth is exponential, with countries such Morocco, South Africa and Egypt leading in installations. East Africa, through the development of the 300-MW capacity Turkana wind project, also recorded tremendous uptake of this technology (IRENA 2015). Africa also has a hydropower potential of 1,584,670 gigawatt hours per year (GWh/yr), though only 123,538 of this total was being generated in 2013, with east and southern Africa leading in generation capacity compared to other African regions (IRENA 2015). Despite having a 15-GW geothermal potential, Africa had installed only 606 MW of this total, and the majority, at 579 MW, was from Kenya (IRENA 2015). Wood fuel, biofuel and biomass residue also contributed to the energy potential of Africa by contributing more than 15 exajoules (EJ) of energy for heating and cooking, and the potential is projected to grow significantly by 2030 (IRENA 2015). These statistics show the high energy potential of the continent with a variety of renewable energy resources, which are currently underexploited.

According to Radebe (2018), renewable energy technologies enhance the productive capacity of more than 60% of Africans, particularly in Kenya, Democratic Republic of Congo, Ethiopia, Nigeria and South Africa. The cost of access to electricity from renewable energy in Africa has reduced significantly over the last decade and as such can compete with conventional energy sources such as oil, gas and coal. According to the Overseas Development Institute (ODI 2016), the cost of installing solar systems at the domestic level dropped to $350, down from $1,000, in the last half a decade. Consequently, African residents can afford this reliable energy source at a more decentralized level. Although many African countries such as Ghana, Chad, Madagascar and Nigeria have great renewable energy potential, it is largely untapped despite the likelihood of such technology to transform livelihoods towards realization of sustainable development (Tiyou 2016). According to the United Nations Development Programme (UNDP) Regional Bureau for Africa (RBA) (2018), tapping renewable energy for Africa would be transformative for the following reasons:

i. It would develop the agricultural sector, which is the mainstay of the continent's economy
ii. It would mitigate climate change in addition to creating employment opportunities
iii. It would improve the delivery of health and educational services to citizens
iv. The provision of sustainable, reliable and affordable energy would be realized if access is expanded sufficiently

For these reasons, an energy revolution via adoption and optimization of renewable energy on the continent would alleviate poverty and lead to realization of the SDGs.

Africa and in particular SSA has had a rise in population, which has created imbalances between the demand and supply for electricity. The World Bank (2017) noted that the populace without access to electricity rose to 609 million in 2014, up from 500 million in 2000 in SSA. Equally, the electrification pace due to renewable energy technologies tripled since 2012, which resulted in a demand reduction by 588 million in 2016. Enhancing renewable use in the long term is expected to further reduce the electricity demand and allow access in rural areas where conventional energy connection and distribution are expensive and, in some countries, commercially unviable due to geographical challenges (UNDP-RBA 2018). Although the annual increase in uptake of renewable energy in Africa is increasing at 12.2% compared to the global average of 8.8%, installations, including power plants using renewables to produce power, are still low, at less than 2% compared to Asia at 4% (UNDP 2018). The trend alludes to the massive potential of the continent to take up such forms of energy for a sustainable future. The continent's governments, in partnership with UNDP and the Global Environmental Facility (GEF), are intensifying knowledge transfer, funding and infrastructural setups to expand the access and use of renewable energy products (UNDP 2017). The aim is not only to improve access to electricity but also to mitigate climate change and reduce indoor pollution affiliated with the use of conventional energy sources.

10.6   EFFORTS TOWARDS EFFICIENT ENERGY USE IN AFRICA

For access to a variety of clean and renewable energy in Africa, implementation of energy management measures at industrial, commercial and residential levels should be prioritized (Aliyu et al. 2018). Energy management focuses on optimal energy use and shifting to cleaner renewable energy. According to Aliyu et al. (2015), energy management's purposes are to save energy costs and conserve finite natural resources and in climatic protection. Effective energy management results in efficient use of power especially from renewable sources and replacement of high energy-consuming equipment with less-consumers. However, such initiatives must be accompanied by monitoring and evaluation to rate their performance in reference to the preset targets (Adesola and Brennan 2019). Through energy efficiency, the global energy demand can be met with a cost-efficient, fast and environmentally friendly approach. It is for this reason countries are urged to pursue policies aimed at energy efficiency through renewables in the long term despite low fuel prices to offer sustainable energy and environmental management (Cruz-Lovera et al. 2017). Some of the strategies Africa could adopt for energy efficiency include:

   i. Insulating windows of buildings and replacing gas and oil furnaces with incandescent or solar lamps to save on energy costs.
   ii. The use of efficient motors, pumps, boilers and heating systems that can be powered by renewables such as hydropower, solar or wind energy in the industrial sector.
   iii. The use of electric cars that are gas and diesel efficient in addition to carpooling to reduce greenhouse gas emissions from the transportation sector.
   iv. At the residential level, technologies such as solar photovoltaics, fuel cells, smart metering and microcombine cogeneration can be adopted in dishwashers, washing machines, water heaters and refrigerators to save energy and transit to the use of renewables (Aliyu et al. 2015, 2018).
   v. Demand management by initiating programs, policies and actions that reduce energy consumption and encourage renewable energy use. Incentives for using renewables can be given to encourage energy and environmental conservation.
   vi. Installation of water heater controls, lighting controls, street light controls that turn off after exceeding the defined output to encourage energy conservation.
   vii. African countries should enter a partnership with international players such as UNDP to fund and mobilize resources that facilitate the use of renewables such as clean cooking stoves, plant biomass for electricity and solar photovoltaics. These partnerships are underway in countries such as Kenya, Ghana, Ethiopia, Togo, Chad, South Africa and Nigeria and have resulted in enhanced use of renewables (UNDP-RBA 2018).

Africa's new energy infrastructure based on increased use of renewables and the preexisting electricity generation system both need to incorporate energy efficiency for economic competitiveness and sustainability in the sector (Aliyu et al. 2018). The advance can be strengthened by appropriate regulatory and policy frameworks to be effective. Policy and regulatory instruments create an enabling environment to intensify renewable energy resource uptake and harness the non-renewable energy sector for intensified economic growth but with environmental caution as one of the priorities in the continent (Qudrat-Ullah and Nevo 2021).

10.7 CHALLENGES AND PROSPECTS IN TRANSITIONING TO RENEWABLE ENERGY IN AFRICA

10.7.1 Challenges

Although the potential for renewable energy in Africa is high, the infrastructure to distribute, transport, transform and efficiently use the power is limited and sometimes lacking. According to Sambo (2016), poor services and delivery mechanisms hinder the sustainability of energy services in the continent. National grid and energy pipelines of a centralized nature do not reach the majority rural populace and cost more among the poor if access is granted. In urban informal settlements, distribution networks are overstretched, and as such, electricity access is characterized by outages and even fuel supply shortages (Nyika 2020). The poor energy infrastructure results from insufficient funding from both the private and public sectors on renewable projects, as noted by Baumli and Jamasb (2020). Sy and Copley (2017) shared similar sentiments and reported that in 2013, only $8 billion was used for Africa's energy infrastructure despite the total demand being approximately $63 billion to provide adequate power access. Half of the cost was incurred by domestic public financing, while the rest was from external financing.

Developing an effective and competitive energy system demands highly skilled individuals for specialized areas such as power distribution, transmission and generation, which Africa lacks (Ohler and Fetters 2014). The specialized areas require extensive planning, modelling and analysis of energy systems at the regional, subregional and national levels. Consequently, the energy sector of the continent is highly dependent on foreign specialists, which increases the overall cost of energy. In particular, all renewable energy aspects in Africa have limited skilled manpower (Sambo 2016). Similarly, Amir and Khan (2021) noted that the lack of skilled laborers, which has worsened with the emergence of the COVID-19 pandemic, is one of the greatest challenges in the enhanced use of renewable energy in Africa.

Africa's diversity of countries in aspects of size and resource distribution often hinders cooperation and integration during delivery of both renewable and non-renewable energy forms. The trend is attributable to the inability and limitations of individual countries to mobilize huge financial investments required to enhance energy infrastructure and complete such projects (Sambo 2016). Even among the private-sector institutions trying to deploy renewable energy sources such as solar photovoltaics in a decentralized approach, financial inputs have been the greatest impediment (Li et al. 2022). With the challenges, regional integration and cooperation are indispensable. In west Africa, for instance, the introduction of the West Africa Power Pool (WAPP) to enhance cooperation of regional countries towards expansive electricity access has recorded tremendous success (Akinyemi et al. 2019). East Africa (EAPP) and South Africa (SAPP) power pools could also facilitate better regional cooperation for building energy infrastructure and access for all through synergized efforts (Nalule 2019). Regional integration towards energy security reduces the costs affiliated with power distribution and alleviates transborder challenges on energy trade and development in poor countries (Africa Energy Outlook 2014). In southern and eastern Africa, the Africa Clean Energy Corridor (ACEC) that seeks to expand the deployment and harnessing of renewable energy potential in the regions to close the electricity gap through regional cooperation has made progress in the provision of cheap alternative and clean power compared to previous overdependence on fossil fuels (Saadi et al. 2015). Although such progress is notable, poor governance and non-operational power pools hinder optimal realizations of regional cooperation towards energy security through renewable technology on the continent (Sambo 2016).

10.7.2 Prospects and Recommendations for Enhanced Renewable Energy Deployment

Evidently, Africa is suffering from large-scale and endemic energy poverty that can be reversed if optimal use and deployment of renewable energy technology, particularly hydropower and solar, occurs. With the enhanced use of renewables, the quality of life for Africans and their economy will improve significantly, as energy security will enable better provision of services and healthy business environments locally, regionally and internationally (Bishoge et al. 2020; Chisika and Yeom 2021). Although most African countries are endowed with more than one type of renewable energy system, installations are at a small scale for demonstration or pilot projects and mainly by donor agencies or countries. To optimize the use of renewables, there is need to shift to large-scale installations and the use of such energy forms in the mainstream electricity grid to enhance power access (Nyika et al. 2020). The availability of a variety of renewable energy forms in most African countries translates to huge potential and a market for renewables in the continent. However, the ability of renewables to meet the energy needs of Africa will be determined by the efforts made to optimize, finance and deploy such technologies.

To promote large-scale use of renewables, Africa must weigh the total energy consumption and supply versus the available or explorable energy resources. Additionally, using forecasting and modeling tools is imperative to produce short-, medium- and long-term scenarios on energy use and supply and ensure sustainable economic security. According to Wang et al. (2019), forecasting on energy provides a reference point for sustainable planning in addition to providing data to support distribution optimization. For optimal renewable energy transfer and knowledge sharing, African countries should engage in international cooperation amongst themselves and with developed countries. Such cooperation at regional and international levels will facilitate the pooling of power among African countries, resulting in energy efficiency from reduced costs and conditions of investments and low pollution if the associated resources are renewable (Blimpo and Cosgrove-Davies 2019). The members of the continent should initiate energy distribution strategies following projections on power demand. The strategies should be aimed at meeting the SDGs using an energy mix that has high preference for renewable energy technologies (Wang et al. 2019). The need to review individual country and regional energy laws to promote an equitable supply of energy in rural and urban areas and ensure the former takes up renewable energy in a more decentralized system to enhance expanded power access is indispensable. The initiative can be realized using effective regulatory frameworks on best practices in the energy sector and by enacting policies on management, distribution and access to power, particularly renewables. Such strong and well-coordinated policies are attributable to the widespread use of solar photovoltaics in Morocco, Kenya and Rwanda, as noted by Samoita et al. (2020).

10.8 CONCLUSION

Africa has high renewable energy potential from a variety of sources, including hydropower, solar, geothermal, wind, biofuel and biomass sources. However, the ability of the continent to meet the energy demand of its populace is far below a level that can prompt significant socio-economic growth. The situation is attributable to poor energy planning and political and legislative barriers that impede optimal exploitation of renewable energy technologies in preference to conventional energy sources. This chapter emphasizes the need to deploy renewables at large scale in the continent to close the preexisting energy gap, particularly for the rural population and residents of informal settlements in the continent. Renewable energy, in addition to meeting the power demands of Africa, will result in energy decentralization, create employment, reduce overdependence on energy imports and reverse environmental pollution affiliated with the use of fossil fuels. The chapter suggests several measures to enhance the uptake of renewable energy resources. These include:

  i. International cooperation with developed countries for knowledge and technology transfer on renewable energy technology.
  ii. Introduction, enactment and enforcement of energy policies that promote equitable energy access and practical incentives such as feed-in tariffs to encourage the uptake of renewables.
 iii. Introduction of regional or national energy supply regulatory frameworks and strategies to enable smooth phasing out of conventional energy sources in preference to renewable technologies.
  iv. Use of modeling and forecasting tools to plan energy demand and supply in the short- and long-term periods to prevent economic crises due to energy poverty.
   v. Mobilization of finances via national, regional and international avenues to fund renewable energy–associated projects and enable transmission, transformation and distribution of power, even in rural and marginalized areas.

REFERENCES

Abas, N., A. Kalair, and N. Khan. 2015. Review of fossil fuels and future energy technologies. *Futures* 69: 31–49.

Ackah, I. and R. Kizys. 2015. Green growth in oil producing African countries: A panel data analysis of renewable energy demand. *Renewable and Sustainable Energy Reviews* 50: 1157–1166.

Adesola, S. and F. Brennan. 2019. Introduction to energy in Africa: Policy, management, and sustainability. In: *Energy in Africa*. S. Adesola and F. Brennan (eds.). Cham: Palgrave Macmillan.

Africa Energy Outlook. 2014. *A Focus on Energy Prospects in Sub-Saharan Africa*. World Energy Outlook Special Report, International Energy Agency.

African Progress Panel. 2017. *Lights, Power, Action: Electrifying Africa*. Available from www.africaprogresspanel.org/policy-papers/lights-power-action-electrifying-africa (accessed August 20, 2021).

Aguirre, M. and G. Ibikunle. 2014. Determinants of RE growth: A global sample analysis. *Energy Policy* 69: 374–384.

Akinyemi, O., U. Efobi, E. Osabuohien, and P. Alege. 2019. Regional integration and energy sustainability in Africa: Exploring the challenges and prospects for ECOWAS. *African Development Review* 31(4): 517–528.

Aliyu, A., A. Bukar, J. Ringim, and A. Musa. 2015. An approach to energy saving and cost of energy reduction using an improved efficient technology. *Open Journal of Energy Efficiency* 4: 61–68.

Aliyu, A., B. Modu, and C. Tan. 2018. A review of renewable energy development in Africa: A focus in South Africa, Egypt and Nigeria. *Renewable and Sustainable Energy Reviews* 81(2): 2502–2518.

Amir, M. and S. Khan. 2021. Assessment of renewable energy: Status, challenges, COVID-19 impacts, opportunities and sustainable energy solutions in Africa. *Energy and Built Environment* 3(3): 348–362.

Augutis, J., L. Martišauskas, R. Krikštolaitis, and E. Augutienơ. 2014. Impact of the renewable energy sources on the energy security. *Energy Procedia* 61: 945–948.

Barau, A., A. Abubakar, and I. Kiyawa. 2020. Not there yet: Mapping inhibitions to solar utilization by households in African informal urban neighborhoods. *Sustainability* 12: 840.

Baumli, K. and T. Jamasb. 2020. Assessing private investment in African renewable energy infrastructure: A multi-criteria decision analysis approach. *Sustainability* 12: 9425.

Bishoge, O., G. Kombe, and B. Mvile. 2020. Renewable energy for sustainable development in sub-Saharan African countries: Challenges and the way forward. *Journal of Renewable and Sustainable Energy* 12(5): 052702.

Blimpo, M. and M. Cosgrove-Davies. 2019. *Electricity Access in Sub-Saharan Africa: Uptake, Reliability and Complementary Factors for Economic Impact*. International Bank for Reconstruction and Development/The World Bank, Washington, DC.

Bolton, R. and T. Foxon. 2015. Infrastructure transformation as a socio-technical process-Implications for the governance of energy distribution networks in the UK. *Technological Forecasting and Social Change* 90(Part B): 538–550.

Chassot, S., N. Hampl, and R. Wustenhagen. 2014. When energy policy meets free-market capitalists: The moderating influence of worldviews on risk perception and RE investment decisions. *Energy Research and Social Science* 3: 143–151.

Chisika, S. and C. Yeom. 2021. Enhancing sustainable development and regional integration through electrification by solar power: The case of six East African states. *Sustainability* 13: 3275.

Cruz-Lovera, C., A. Perea-Moreno, J. Cruz-Fernandez, J. Alvarez-Bermejo, and F. Agugliaro. 2017. Worldwide research on energy efficiency and sustainability in public buildings. *Sustainability* 9: 1294.

Danlami, A. and R. Islam. 2015. An analysis of the determinants of households' energy choice: A search for conceptual framework. *International Journal of Energy Economics and Policy* 5: 197–205.

Da Silva, P., P. Cerqueira, and W. Ogbe. 2018. Determinants of renewable energy growth in sub-Saharan Africa: Evidence from panel ARDL. *Energy* 156: 45–54.

Foster, E., M. Contestabile, J. Blazquez, B. Manzano, M. Workman, and N. Sha. 2017. The unstudied barriers to widespread renewable energy deployment: Fossil fuel price responses. *Energy Policy* 103: 258–264.

Fouche, E. and A. Brent. 2019. Journey towards renewable energy for sustainable development at the local government level: The case of Hessequa municipality in South Africa. *Sustainability* 11: 755.

Gielen, D., F. Boshell, D. Saygin, M. Bazilian, N. Wagner, and R. Gorini. 2019. The role of renewable energy in the global energy transformation. *Energy Strategy Reviews* 24: 38–50.

Hafner, M., S. Tagliapietra, and L. de Strasser. 2018. *Prospects for Renewable Energy in Africa. In: Energy in Africa.* Springer Briefs in Energy. Springer, Cham.

Hillerbrand, R. 2018. Why affordable clean energy is not enough. A capability perspective on sustainable development goals. *Sustainability* 10: 2485.

International Energy Agency (IEA). 2017. Energy access outlook 2017—From poverty to prosperity. *World Energy Outlook Special Report*, IEA, Paris. Available from www.iea. org/reports/energy-access-outlook-2017 (accessed August 16, 2021).

International Energy Agency. 2019. *Renewables 2019 Market Analysis and Forecast from 2019 to 2024.* Available from www.iea.org/renewables2019/ (accessed May 14, 2021).

International Renewable Energy Agency, IRENA. 2015. *Africa 2030: Roadmap for a Renewable Energy Future.* IRENA, Abu Dhabi. www.irena.org/rema

IRENA. 2020. *Post-COVID Recovery: An Agenda for Resilience, Development and Equality.* International Renewable Energy Agency, Abu Dhabi.

IRENA and African Development Bank (AFDB). 2022. *Renewable Energy Market Analysis: Africa and Its Regions.* ABU Dhabi and Abidjan.

Kilinc-Ata, N. 2016. The evaluation of renewable energy policies across EU countries and US states: An econometric approach. *Energy for Sustainable Development* 31: 83–90.

Kiplagat, J., R. Wang, and T. Li. 2011. Renewable energy in Kenya: Resource potential and status of exploitation. *Renewable and Sustainable Energy Reviews* 15(6): 2960–2973.

Kuamoah, C. 2020. Renewable energy deployment in Ghana: The hype, hope and reality. *Insight on Africa* 12(1): 45–64.

Kumar, R. and A. Agarwala. 2016. Renewable energy technology diffusion model for technoeconomics feasibility. *Renewable and Sustainable Energy Reviews* 54: 1515–1524.

Li, D., J. Bae, and M. Rishi. 2022. Sustainable development and SDG-7 in sub-Saharan Africa: Balancing energy access, economic growth and carbon emissions. *The European Journal of Development Research*: 1–126.

Lucas, N., G. Francés, and E. González. 2016. Energy security and renewable energy deployment in the EU: Liaisons dangereuses or virtuous circle? *Renewable and Sustainable Energy Reviews* 62: 1032–1046.

Maria, L. and T. Bernauer. 2014. Explaining government choices for promoting renewable energy. *Energy Policy* 68: 15–27.

Nalule, V. 2019. *Energy Poverty and Access Challenges in Sub-Saharan Africa: The Role of Regionalism.* Gewerbestrasse, Switzerland: Palgrave Macmillan.

Nerini, F., J. Tomej, L. To, I. Bisaga, P. Parikh, M. Black, A. Borrion, C. Spataru, V. Broto, and G. Anandarajah. 2018. Mapping synergies and trade-offs between energy and the sustainable development goals. *Nature Energy* 3: 10–15.

Nesta, L., F. Vona, and F. Nicolli. 2014. Environmental policies, competition and innovation in renewable energy. *Journal of Environmental Economics and Management* 67(3): 396–411.

Njoh, A., S. Etta, I. Ngyah-Etchutambe, L. Enomah, H. Tabrey, and U. Essia. 2019. Opportunities and challenges to rural renewable energy projects in Africa: Lessons from the Esaghem village, Cameroon solar electrification project. *Renewable Energy* 131: 1013–1021.

Nyika, J. 2020. Climate change situation in Kenya and measures towards adaptive management in the water sector. *International Journal of Environmental Sustainability and Green Technologies* 11(2): 34–47.

Nyika, J., A. Adediran, A. Olayanju, O. Adesina, and F. Edoziuno. 2020. *The Potential of Biomass in Africa and the Debate on Its Carbon Neutrality* [Online First]. London: IntechOpen.

ODI (Overseas Development Institute). (2016). *Accelerating Access to Electricity in Africa with Off-Grid Solar: The Impact of Solar Household Solutions*, by Kat Harrison, Andrew Scott and Ryan Hogarth. Available from www.odi.org/sites/odi.org.uk/files/odi-assets/publications-opinionfiles/10229.pdf (accessed August 24, 2021).

Ohler, A. and I. Fetters. 2014. The causal relationship between renewable electricity generation and GDP growth: A study of energy sources. *Energy Economics* 43: 125–139.

Omri, A. and D. Khuong. 2014. On the determinants of renewable energy consumption: International evidence. *Energy* 72: 554–560.

Pfeiffer, B. and P. Mulder. 2013. Explaining the diffusion of RE technology in developing countries. *Energy Economics* 40: 285–296.

Pilot, B., M. Muselli, P. Poggi, and J. B. Dias. 2019. Historical trends in global energy policy and renewable power system issues in Sub-Saharan Africa: The case of solar PV. *Energy Policy* 127: 113–124.

Polzin, F., M. Migendt, F. A. Taube, and P. von Flotow. 2015. Public policy influence on renewable energy investments—A panel data study across OECD countries. *Energy Policy* 80: 98–111.

Qudrat-Ullah, H. and C. Nevo. 2021. The impact of renewable energy consumption and environmental sustainability on economic growth in Africa. *Energy Reports* 7: 3877–3886.

Radebe, J. 2018. *Media Statement by the Minister of Energy on the Independent Power Producer Programmes*. South Africa, 8 March 2018. Available from https://www.energy.gov.za/files/media/pr/2018/MediaStatement-on-the-Independent-Power-Producer-Programmes-08032018.pdf (accessed January 29, 2023).

Rafiq, S., H. Bloch, and R. Salim. 2014. Determinants of renewable energy adoption in China and India: A comparative analysis. *Applied Economics* 46: 2700–2710.

Reboredo, J. C. 2015. Is there dependence and systemic risk between oil and renewable energy stock prices? *Energy Economics* 48: 32–45.

Rüstemoglu, H. and A. R. Andrés. 2016. Determinants of CO_2 emissions in Brazil and Russia between 1992 and 2011: A decomposition analysis. *Environmental Science and Policy* 58: 95–106.

Saadi, N., A. Miketa, and M. Howells. 2015. Africa clean energy corridor: Regional integration to promote renewable energy fueled growth. *Energy Research and Social Science* 5: 130–132.

Sambo, A. 2016. Renewable energy development in Africa: Issues, challenges and prospects. In: *Renewable Energy in the Service of Mankind*. A. Sayigh (ed.). Volume II. Cham, Switzerland: Springer.

Samoita, D., C. Nzila, P. Ostergaard, and A. Remmen. 2020. Barriers and solutions for increasing the integration of solar photovoltaic in Kenya's electricity mix. *Energies* 13: 5502.

Stadelmann, M. and P. Castro. 2014. Climate policy innovation in the South—Domestic and international determinants of RE policies in developing and emerging countries. *Global Environmental Change* 29: 413–423.

Sy, A. and A. Copley. 2017. *Closing the Financing Gap for African Energy Infrastructure: Trends, Challenges and Opportunities.* Policy Brief, Brookings Institution.

Tiyou, T. 2016. *The Five Biggest Wind Energy Markets in Africa: Renewable Transformation Challenge of Renewable Energy Focus.* Available from www.renewableenergyfocus. com/view/44926/the-five-biggest-wind-energy-markets-in-africa (accessed August 21, 2021).

UNDP. 2018. *UNDP Africa Regional Bureau for Africa's Country Offices' Case Studies.* Available from https://www.undp.org/africa/publications/undp-regional-programme-africa-annual-report-2018 (accessed August 23, 2021).

UNDP-Regional Bureau for Africa (RBA). 2018. Transforming lives through renewable energy access in Africa: UNDP's contributions. *UNDP Africa Policy Brief* 1(1): 1–32.

United Nations Development Programme (UNDP). 2017. *Sparking Sustainable Development: Mini-Grids as a Tool for Energy Access. Bureau of Policy and Programme Support.* UNDP, New York. Unpublished.

Wang, L., L. Zhan, and R. Li. 2019. A prediction of the energy demand trend in the Middle Africa: A comparison of MGM, MECM, ARIMA and BP models. *Sustainability* 11: 2436.

World Bank. 2017. *State of Electricity Report 2017.* Available from https://documents1.world-bank.org/curated/en/364571494517675149/pdf/114841-REVISED-JUNE12-FINAL-SEAR-web-REV-optimized.pdf (accessed August 23, 2021).

World Bank. 2021. *World Development Indicators.* Available from https://databank.world-bank.org/source/world-development-indicators (accessed May 23, 2022).

11 Energy and the Environment
A Dynamic Partnership

Guller Sahin

11.1 INTRODUCTION

Significant environmental impact occurs during the production, processing, distribution and consumption of energy resources. These movements include major land use changes resulting from fuel cycles, such as coal, biomass and hydropower, which each have an impact on both the natural and human environment (Holdren and Smith, 2000). Therefore, the dynamic partnership between energy and the environment shows that both issues should be handled with a holistic approach and evaluated together. The main solution to environmental problems arising from the consumption of hydrocarbons and fossil energy sources containing high amounts of carbon is the use of clean energy sources. Research within the scope of energy and the environment focuses on the production, storage and efficient use of energy and natural resources and the evaluation of the interaction between the environment and energy technologies in order to create environmentally friendly and renewable energy sources (SDM, 2021). Within this context, in this research, which examines the dynamic partnership between energy and the environment, the relationship between carbon dioxide (CO_2) emissions, which account for about three quarters of greenhouse gas (GHG) emissions, and energy sources is evaluated within the scope of Denmark, the United States of America (USA) and China case studies.

According to the Intergovernmental Panel on Climate Change (IPCC), human activities are estimated to have caused global warming of about 1° C above pre-industrial times. If these activities continue to increase at the current rate, global warming is expected to reach 1.5° C between 2030 and 2052 (IPCC, 2018). If the current dependence on fossil fuels continues, it is predicted that fossil fuels will further increase global warming and deplete the existing carbon budget in the next decade (IPCC, 2021). Population projections indicate that an increase of 2 billion in population over the next few decades is almost inevitable. In such a scenario, it is predicted that the demand for both energy and other resources will increase; therefore GHG emissions will increase until at least 2060 (Riahi et al., 2017). As a result, it is necessary to reduce expected increases in global average temperatures and contain global carbon emissions for less dramatic climate consequences. At the same time, it is assumed that even if GHG emissions are reduced in the next few decades, global warming will continue for at least another century (Canuto, 2021).

DOI: 10.1201/9781032715438-11

The Paris Agreement, in which the framework of the post-2020 climate change regime was determined, was accepted at the 21st Conference of Parties (COP21) held in 2015, and the 'carbon neutral' target became one of the priorities of the countries with the agreement; 192 countries agreed to keep the global temperature rise caused by GHG emissions below the 2°C (preferably 1.5°C) threshold over pre-industrial times in the long term and to reduce cumulative GHG emissions by more than 80%. To keep the temperature rise below the 2°C threshold, a limit of 1.17 gigatons (Gt) of CO_2 has been set for the amount of CO_2 until 2100, called the 'global carbon budget' (Qin et al., 2021; Kueppers et al., 2021; Le Quéré et al., 2015; Wang et al., 2021b). Attention was drawn to the fact that countries should be responsible in the fight against climate change within the framework of the principle of 'common but differentiated responsibilities and relative capabilities'. With the agreement, a framework was established to determine implementation methods regarding national contributions, mitigation, adaptation, loss/damage, financing, technology development and transfer, capacity building, transparency and due diligence in the fight against climate change (MFA, 2021). In addition, necessary measures were identified to stay within the current carbon budget, including increases in energy efficiency, employment of renewable energy sources in the energy sector, carbon capture and storage technologies and a reduction of dependence on fossil fuels (Handayania et al., 2020).

In the Emissions Gap Report, published by the United Nations Environment Program (UNEP, 2021), it is quite clearly stated that countries have to strengthen their Paris climate commitments. The report shows that new national climate commitments are combined with other mitigation measures and that the world is on track for a global temperature rise of 2.7°C by the end of the twenty-first century. It is explained that even if all the existing unconditional commitments of the Paris Agreement are met, temperatures will rise, leading to a wider and more devastating climate impact. It is stated that countries should halve their annual cumulative GHG emissions over the next eight years for the 1.5°C threshold. It also highlights the importance of decarbonizing the energy sector and electrifying end uses for a successful transition to a low-carbon society (Kueppers et al., 2021; UNEP, 2021).

Global energy demand continues to grow with the effect of increasing welfare and raising living standards. Significant inequalities persist in energy consumption and access to energy. The current structure of the global energy system has the following trends (BP, 2020; Pata, 2021):

- Fossil energy sources, being the cause of environmental degradation and endangering human health, increase orientation towards renewable energy sources.
- The energy demand structure changes over time due to the increasing share of renewable energy and the decreasing role of fossil fuels, which is offset by the increasing role of electricity.
- A transition to lower-carbon energy systems tends to restructure the energy system, thanks to a more diverse mix of energy sources, more localized energy markets and increased levels of integration, greater consumer access and competition.

- Oil demand is expected to decline within thirty years. The scale and pace of this decline is estimated to be due to increased efficiency and electrification of road transport.
- The natural gas (NG) outlook is more flexible than oil. The role of NG demand in supporting fast-growing emerging economies by decarbonizing and reducing their dependence on coal is supported by being a near-zero carbon energy source when combined with carbon capture and storage.
- It is assumed that the energy source will grow fastest in thirty years, in line with significant increases in the development of and investment in renewable energy, wind and solar energy capacity.

Figure 11.1 reflects the global primary energy demand structure for the period 1995–2020 in the light of the information explained previously. In this direction, approximately 83% of the primary energy need is met from fossil-based fuels. It is seen that the global primary energy demand decreased by 4.5% in 2020 compared to 2019, and this decrease is the first decrease in energy consumption after the 2009 Global Financial Crisis. In 2020, consumption decreased in all energy sources except for renewable energy (9.7%) and hydropower (1%). At the same time, oil continues to have the largest share (31.2%) in the energy mix; it can be seen that coal is the second-largest fuel source and constitutes 27.2% of the total primary energy

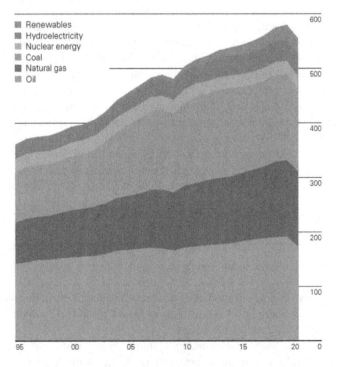

FIGURE 11.1   Global primary energy demand structure: 1995–2020.

Source: BP (2021)

consumption. Moreover, the share of NG (24.7%) and renewable energy (5.7%) in the global energy mix is increasing (BP, 2021).

Emissions from fossil energy use have doubled in the last thirty years, quadrupled in the last sixty years, and have seen an almost twelvefold increase in the twentieth century. CO_2 emissions, which were 0.2 Gt in 1850, are an extremely small percentage of approximately thirty-seven Gt CO_2 emissions in 2021. While the vast majority of CO_2 emissions emanate from the burning of fossil fuels, human activity, such as land use, land use changes and deforestation, has also contributed significantly to the cumulative total. Changes in land use and forestry increased cumulative CO_2 emissions by about one third between 1850 and 2021, and fossil fuels and cement by about two thirds. Cumulative emissions between 1850 and 2021 account for about 86% of the carbon budget for a chance to stay below the 1.5° C threshold (Evans, 2021; GCP, 2021).

Figure 11.2 shows that cumulative CO_2 emissions from energy consumption fell 6.3% in 2020, the largest ever drop, which was due to the Covid-19 pandemic. This reduction, which is about 2 Gt of emissions, is about five times greater than the decline after the global financial crisis. This was the largest reduction in emissions since the Second World War. It is also noteworthy that the carbon density of the energy mix decreased by 1.8%. However, CO_2 emissions in the atmosphere are returning to pre-epidemic levels, so GHG concentrations continue to rise. CO_2 emissions fell more than energy demand in 2020, as the pandemic affected oil and coal demand more than any other energy source. Despite the stated decrease, the amount of emissions from energy consumption is around 31.5 Gt, which is the highest average annual concentration of emissions in the atmosphere so far. At the same time, the amount of emissions in 2020 is about 50% higher than when the Industrial Revolution started. In 2021, global energy-related emissions are projected to increase by 4.8% as economies recover and demand for fossil-based energy sources increases. The increase in emissions of more than 1500 million tons (Mt) is expected to be the largest increase since the carbon-intensive economic recovery from the global financial crisis (BP, 2021; IEA, 2021; UNEP, 2021; Wang et al., 2021b).

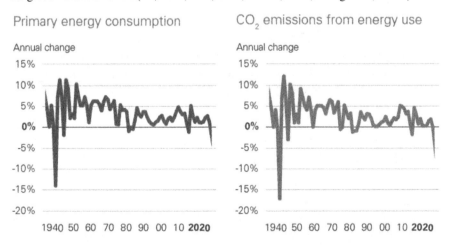

FIGURE 11.2   Global energy consumption and CO_2 emissions from energy use: 1940–2020.

Source: BP (2021)

It is estimated that if annual CO_2 emissions remain at current levels, there is a 50% probability that from 2022 the 1.5° C threshold will be exhausted within ten years (Evans, 2021). Therefore, efforts to limit emissions are among the most important issues on the world agenda, as discussed at the twenty-sixth United Nations Climate Change Conference (COP26). At the summit, it was noted that countries should halve their emissions by 2030 in order to limit the global temperature increase to 1.5° C (Samar, 2021). It was stated that not only states but also companies, financial institutions and city governments, which are non-state actors, need to take action to combat climate change. There were higher targeted commitments on methane, coal and deforestation than in previous COPs (Eryar Ünlü, 2021).

11.1.1 IMPORTANCE OF RESEARCH

It is widely accepted that irreversible long-term effects occur if the global temperature rise is above 1.5° C. To avoid this possible situation, large-scale transformations in energy systems are imperative. The historical trend in energy resources shows that the world will continue to be dependent on fossil fuels in the short term. In this context, 63–78% of the global primary energy demand is still expected to be met from fossil fuels by 2030. Due to the urgency of climate change and international commitments, the need for multifaceted research, primarily to reduce CO_2 emissions in the fossil fuel supply chain, explains the importance of the research.

In light of the aforementioned, the evaluation of the countries that present exemplary performance in the current situation in the production and use of energy in a cleaner way, in the management of environmental policies and in the formation of rational policies for the future, has a guiding quality. In this context, another reason attributed to the importance of the research is to make macro policy recommendations from the micro-policies applied at the country level for the dynamic relationship between energy and the environment.

11.1.2 AIM OF THE RESEARCH

The aim of this research is to make policy recommendations at a global level from case studies from Denmark, the USA and China, which have implemented successful policies in the dynamic partnership between energy and the environment. For the aim, a theoretical framework is adopted within the scope of document analysis based on statistical data.

11.2 DENMARK CASE STUDY

Having become the largest oil producer in the European Union after Brexit, Denmark is a country that wants to be a role model in energy transition. In this direction, it carries out its policies based on energy efficiency-oriented energy saving and renewable energy sources as much as possible. It aims to have all of its energy demand based on renewable energy by 2050. During the transition from fossil fuel energy sources to renewable energy sources, it was the country that reduced its annual CO_2 emissions the most (4.4%) between 2009 and 2019 (BP, 2021). In this context, the transition to a low-carbon energy economy is evaluated in the Denmark case study from the perspective of CO_2 emissions.

Denmark had only one energy source in the early 1970s with a self-sufficiency rate of less than 2%. Before the 1973–74 oil crisis, the energy system was almost 100% dependent on fossil fuels, especially oil imports. Petroleum (93%), an important source of energy consumption, relied entirely on imports. This situation, together with the oil crises of 1973–74 and 1979–80, caused significant economic problems and brought the country into a crisis. This problematic situation has created an open attitude in parliament towards alternative ideas in the fight against the energy crisis. In this context, it has revealed alternative ideas, such as renewable energy creating employment, reducing imports and providing relatively fixed prices on energy. As a result, in addition to the establishment of the NG grid, an energy program supporting renewable energy was initiated and an energy tax policy was adopted (Veenman et al., 2019; Wang et al., 2019).

The oil crises led to the implementation of four national energy plans, the main goals of which are to reduce dependence on imported oil and ensure sustainable development (Eser and Polat, 2015; GWM, 2004; Veenman et al., 2019; Yücel et al., 2021):

- 'Energy Plan 1976': After 1977, fossil energy resources were taxed and domestic energy demand was reduced. The 'Energy Act' of 1979 introduced an energy tax on oil, allowing greater fuel flexibility, with municipalities empowering local governments to implement district heating plans. In addition, the first incentive application was made in the form of granting 30% of the wind turbine cost.
- 'Energy Plan 1981': The potential role of renewable energy is enhanced by introducing subsidies and/or tariff guarantees to support wind power, district heating and conversion of existing combined heat and power plants to biomass.
- 'Energy 2000': Incentives given to wind energy applications were gradually reduced after 1989, and after 1993, fixed price guarantees were applied in varying amounts, depending on the type of source. Emphasizing the role of biomass fuel, a target of 20% reduction in CO_2 emissions in the 1988–2005 period was set. In addition, it was determined for combined heat and power to develop national oil and gas resources in the North Sea, improve oil and gas self-sufficiency rates, increase national NG use and generate 10% of electricity from wind turbines by 2005.
- 'Energy 21': In 1997, individuals who joined wind energy cooperatives were given tax refunds for income from wind energy production. By 2005, the aim was to obtain 12–14% of energy from renewable energy sources, with the aim of increasing this rate to 35% by 2030 and further increasing it by 1% each year thereafter.

After the 2000s, the country focused strongly on renewable energy with different incentive practices, aiming to diversify its energy supply and to enable the public to benefit from wind energy projects (Bohnerth, 2015). The amendments made in favor of producers with the Law on the Promotion of Renewable Energy Resources in 2008 showed its effect in the form of an increase in wind energy production in 2009 (Yücel et al., 2021). In line with the 2020 targets, various incentive mechanisms were developed in line with the aim of increasing energy production with renewable resources. Fixed price guarantee, premium guarantee and subsidies were among the important incentive mechanisms. As a tax incentive, a law was enacted to facilitate individual investment in renewable energy and to regulate the conditions of individual investment, effective from 1 January, 2011. As of 1 July, 2013, a subsidy fund was created to meet the 2020 targets. With the said fund, producers who turn to renewable energy production were supported between 45 and 65% of their investment costs (Eser and Polat, 2015).

The country's net-zero commitment and advances in renewable energy have placed the country at the forefront of energy transition. With the Climate Law of 2019, it aims to reduce GHG emissions by 70% compared to 1990 levels by 2030 and to meet at least half of the total energy consumption with renewable energy sources. It states that it will gradually close its coal power plants, provide 90% of its district heating from non-fossil sources and end the sale of gasoline and diesel cars. At the same time, it also has a political agreement targeting renewable energy to supply all electricity and 55% of total consumption. It aims to make all of the gas injected into its grid renewable by 2035. The country, which invests heavily in renewable energy, especially wind energy, has decided to take a serious step towards becoming carbon neutral and has committed to achieving net-zero emissions by 2050. It also reduced the tax it pays for electrical heat as part of its plan to become fossil fuel free, offered financial incentives for investments in district heating and cooling infrastructure, and allowed direct citizen participation in renewable energy projects (IEA, 2021; IRENA, 2021). At the same time, it has announced that it will end all offshore oil and NG activities in the North Sea by 2050, with the decision taken by its parliament.

In Denmark, which does not have nuclear energy sources, the main renewable energy sources are wind energy and biomass. It is the country that benefits most from wind energy in the world (Şen, 2017). Backed by a flexible local power system and a high level of interconnection, the country is recognized as a global leader in both integrating variable renewable energy and providing a highly reliable and secure electricity grid. The country also offers great potential for large-scale use of combined heat and power plants with heat storage capacity and increased use of wind power, efficient integration of heat and electricity systems (IEA, 2021).

Examining the data in Table 11.1 shows the developments in primary energy consumption in the period 2009–2020. Total primary energy consumption in 2020 was 0.59 exajoules (EJ); a 13% decrease compared to 2019. In the same year, the rate of change in oil, NG and coal consumptions decreased by 17.5%, 20% and 14.3%, respectively, compared to the previous year. In this context, the highest rate of decrease among fossil fuels was in NG, successor oil consumption. Total renewable energy consumption was 0.21 EJ; the highest rate among this consumption was wind energy (0.15 EJ).

Figure 11.3 explains the annual amount of CO_2 emissions between 1843 and 2020. Emissions, which were 59.22 Mt in the 1973 oil crisis, seem to have been the

TABLE 11.1

Primary Energy Consumption (EJ): 2009–2020

Energy type	2009	2010	2011	2012	2013	2014	2015	2016	2017	2018	2019	2020
Primary energy	0.78	0.82	0.78	0.72	0.75	0.72	0.70	0.71	0.70	0.70	0.68	0.59
Oil	0.35	0.36	0.35	0.32	0.32	0.32	0.33	0.32	0.32	0.32	0.32	0.26
Natural gas	0.16	0.19	0.16	0.15	0.14	0.12	0.12	0.12	0.12	0.11	0.10	0.08
Coal	0.17	0.16	0.14	0.10	0.14	0.11	0.07	0.09	0.07	0.07	0.04	0.03
Solar	0.00	0.00	0.00	0.00	0.00	0.01	0.01	0.01	0.01	0.01	0.01	0.01
Wind	0.06	0.07	0.09	0.10	0.10	0.12	0.13	0.12	0.13	0.12	0.14	0.15
Geo, biomass and other	0.03	0.04	0.04	0.04	0.04	0.04	0.04	0.04	0.06	0.05	0.05	0.05

Source: Prepared by the researcher using the BP Statistical Review of World Energy 2021 Report.

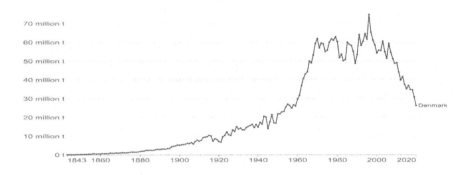

FIGURE 11.3 Annual CO_2 emissions from the burning of fossil fuels (Mt): 1843–2020.

Source: Ritchie and Roser (2020)

highest in 1996 (74.85 Mt) within the historical path. It can be seen that the emissions decreased approximately threefold in 2020 (26.19 Mt) compared to 1996. It can also be seen that the biggest decrease in the amount of emissions was in 1997 (9.40 Mt), and in 2020, emissions decreased by 4.70 Mt compared to the previous year.

Denmark is known as a leader in renewable energy, especially in the field of wind energy, which accounted for 56% of electricity production in 2020. In addition, bioenergy plays an important role in the country's power mix. It has significantly reduced CO_2 emissions due to the decarbonization of the electricity sector. Renewable energy consumption is the reason its emissions from electricity generation have decreased by more than 70% since 1990. In 2020, 26.19 Mt of CO_2 emissions were produced. This is a decrease of approximately 18% compared to 2019. This reduction in emissions was the result of the Covid-19 pandemic, which caused dramatic emissions reductions around the world. However, the downward trend of emissions continued before the epidemic in the country, and emissions fell by about 60% over the period 1996–2019 (EPI, 2020; Ritchie and Roser, 2020; Tiseo, 2021).

11.3   THE USA CASE STUDY

The importance of natural gas in the energy mix is related to its environmental value as it is a relatively clean energy source, emitting fewer pollutants compared to other fossil fuels. In this context, NG is a relatively more efficient and cleaner energy source than other fossil fuels due to its high unit heat value, low exhaust pollution and low cost characteristics (Aune et al., 2004; Burnham et al., 2012; De Gouw et al., 2014; Li et al., 2020; UCS, 2014). As a result of the combustion of natural gas, approximately 50% less CO_2 emissions are emitted per unit of energy compared to coal and oil (De Gouw et al., 2014). Historically, electricity generation from natural gas has been shown to help reduce GHG emissions. At the same time, NG provided a valuable complement to intermittent renewable generation technologies, such as solar and wind power (Tanaka and O'Neill, 2018). However, it is important to emphasize that although natural gas remains a critical fuel source in the current energy mix, it is not a purely clean energy source due to it being a fossil fuel (Russo, 2021; UCS, 2014).

In recent years, NG has been considered a 'bridge fuel' for a low-carbon future, with the hope of lower-cost and larger-term short-term emissions reductions (Tanaka and O'Neill, 2018). It is estimated that oil and NG resources will continue to make up half of the energy market by 2040. Therefore, the use of natural gas as a transitional source in the early stages of the low-carbon conversion is extremely important (Lu et al., 2020). Given the reduced carbon budget, NG is seen as an alternative for low emissions targets, thanks to the significant expansion in natural gas infrastructure and the simultaneous improvement in the competitiveness of renewable energy after 2010 (Woollacott, 2020). NG demand is predicted to rise on the axis of cost-competitiveness and reliability of low-carbon alternatives (Tanaka and O'Neill, 2018).

While the world is transitioning to a low-carbon energy economy, NG is distinguished from oil and coal by being a bridge fuel. The country with the highest NG fuel type in primary energy production (32.93 EJ) and consumption (29.95 EJ) globally is the USA (BP, 2021; Lu et al., 2020). In this context, the transition process to a low-carbon energy economy is evaluated in the USA case study from the NG perspective.

The structure of the US NG market has undergone significant changes over the past three decades. Among these changes, the role of the deregulation period brought to the market by the Natural Gas Wellhead Inspection Law in 1989 is important. With the deregulation in question, although the focus was generally on price mechanisms and the supply side of the market, there were also significant changes on the demand side (Hou and Nguyen, 2018). Since 2000, NG production has resulted in an increasing share of natural gas in the national energy mix, thanks to advances in drilling technology and advances in hydraulic fracturing. While the share of NG production among all fuel sources was the highest after 2010, consumption has been the second most important share after petroleum products in the energy market since 2006. Forecasts indicate that the leading role of natural gas in the energy mix will continue into the future (De Gouw et al., 2014; EIA, 2021).

On the supply side, NG was predominantly extracted from gas wells in the past, and this conventional form accounted for 79% of the total supply. However, in recent years NG production has given way to unconventional forms. The products found in shale gas and coal bed wells especially have increasingly become the main source of the gas market. In 2016, 48% of the total NG production was provided in non-traditional ways. Commercial and residential users were the main users, with more than 50% of the market share, while spending from electricity generators increased dramatically, accounting for more than one third of total NG consumption (Hou and Nguyen, 2018). After the start of the shale gas revolution, NG became the most important source of primary energy generation with a market share of 31.8% in 2017 (Rubaszek and Uddin, 2020). NG consumption contributed to a growing share of over 40%, and growth in NG consumption dominated the energy market in 2018 (Wang et al., 2021c).

The country's NG market is characterized by strong domestic production, a dense pipeline network and developed consumption markets. National NG production and consumption show an increasing trend, and this production and consumption is expected to increase. Since the beginning of the twenty-first century, the gap between production and consumption has been narrowing (Dong et al., 2018; Ruester and Neumann, 2008). The United States was a leading producer and consumer of oil and NG during the late nineteenth and twentieth centuries. The practice of directional drilling and fracturing made it possible to access previously locked oil reservoirs, which was originally called unconventional production. Most of the production increases after 2005 were the result of horizontal drilling and fracking techniques, especially in shale, sandstone, carbonate and other tight geological formations. The country produced more NG than the natural gas it consumed in 2020. Steady and rapid production growth in natural gas, driven by technological innovation and market forces, accounted for the 36,000 million cubic feet produced in 2020. About 65% of US oil and 85% of dry NG production were from tight formations. Dry NG production in 2020 was approximately 10% more than total NG consumption. Production was lower than in 2019, largely due to a drop in drilling activity as a result of lower NG and oil prices, driven by the decline in demand due to the Covid-19 pandemic. In addition to falling energy demand and increasing distribution of renewable resources, coal has also suffered from a loss of competitiveness relative to natural gas (BP, 2021; EIA, 2021; Field and Derwent, 2021).

When Table 11.2 is examined, it can be seen that the total primary energy consumption in 2020 was approximately 88 EJ and that there was a 7.7% decrease compared to 2019. It is observed that in 2020, oil, NG and coal consumptions were 32.54 EJ, 29.95 EJ and 9.20 EJ, respectively, and that the rate of change compared to the previous year decreased by 12.6%, 2.3% and 19.1%, respectively. In this context, it is explained that the highest rate of decrease among fossil fuels is in the consumption of coal, followed by oil. In particular, the rate of decrease in coal consumption attracts attention. NG consumption increased by about 3.2% in the period 2009–2019, with most of its energy needs being met from oil and natural gas. The highest consumption rate in renewable energy consumption in 2020 was experienced in wind energy (3.03 EJ). Despite an increase in renewable energy sources, more than 70% of the country's energy comes from fossil fuels.

Since the Industrial Revolution, at least 2,500 billion tons of CO_2 has been released into the atmosphere, mostly due to human-induced activity. More than 500 billion

TABLE 11.2

Primary Energy Consumption (EJ): 2009–2020

Energy type	2009	2010	2011	2012	2013	2014	2015	2016	2017	2018	2019	2020
Primary energy	89.88	92.91	92.05	89.62	92.04	92.99	92.09	91.96	92.26	95.64	94.90	87.79
Oil	35.05	35.61	34.91	34.11	34.70	34.94	35.66	35.92	36.28	37.14	37.13	32.54
Natural gas	22.23	23.33	23.70	24.77	25.45	26.00	26.77	26.97	26.64	29.58	30.57	29.95
Coal	19.74	20.88	19.70	17.42	18.08	18.04	15.58	14.26	13.87	13.28	11.34	9.20
Solar	0.02	0.03	0.04	0.08	0.15	0.27	0.36	0.50	0.70	0.84	0.96	1.19
Wind	0.70	0.90	1.13	1.32	1.56	1.68	1.75	2.08	2.31	2.46	2.66	3.03
Geo, biomass and other	0.69	0.70	0.71	0.71	0.74	0.77	0.76	0.75	0.75	0.73	0.68	0.68

Source: Prepared by the researcher using the BP Statistical Review of World Energy 2021 Report.

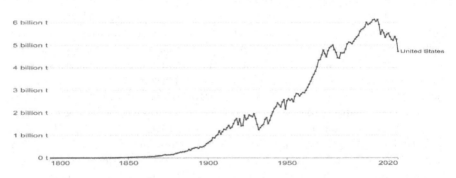

FIGURE 11.4 Annual CO_2 emissions from the burning of fossil fuels: 1800–2020 (Bt).

Source: Ritchie and Roser (2020)

tons of this amount was produced by the USA. The specified amount corresponds to 20% of the cumulative CO_2 rate (Samar, 2021). The USA is currently the country with the highest carbon emissions per capita and is the second-largest emitter of cumulative CO_2 emissions after China. At the same time, it has cumulatively produced more CO_2 than any other country throughout history. The citizens of the country have a carbon footprint of about three times the global average (Gabbatiss, 2021).

In Figure 11.4, the course of annual CO_2 emissions over the 1800–2020 period is monitored. In this context, it can be seen that emissions generally follow a decreasing trend after peaking at 6.13 billion tons (Bt) in 2007 in the said time period. Compared to 2019, emissions decreased by approximately 11.6% in 2020, with a reduction rate of 119.67 Bt.

The country, which is one of the world's largest emitters of CO_2, has implemented a series of measures to reduce the amount of increasing emissions (Dong et al., 2018). In this context, the USA will reduce its emissions by at least 50% by 2030 compared to 2005 levels, aiming for half of new vehicles being electric by 2030, and it promises a carbon-neutral target by 2050 (BBC, 2021). The reduction in NG prices as a result of the development of hydraulic fracturing technology has brought environmental and economic benefits. In this context, the capacity to generate electricity from natural gas has expanded significantly, helping to reduce carbon and other emissions from coal-fired electricity generation at nearly twice that of coal (EPA, 2019; Woollacott, 2020). In its statement of international public support for the US clean energy transition, it declared that it will support the clean energy transition and end international investment in fossil fuels in the energy sector by the end of 2022. With the COP26 Global Methane Commitment, it aims to reduce methane emissions by at least 30% by 2030 compared to 2020 (Eryar Ünlü, 2021). It has also announced two new draft laws on the Environmental Protection Agency, which will reduce methane losses from new and existing oil and gas pipelines and the Ministry of Transport to reduce potentially dangerous leaks from NG pipelines (Haber7, 2021). In its environmental plan, the country aims to expand the use of green energy. In addition, a total of $150 billion in rewards has been created for energy companies that have abandoned fossil fuels and switched to clean fuels (BBC, 2021).

11.4 CHINA CASE STUDY

The most important areas of difficulty faced by countries regarding the energy sector are the following: increasing energy demand, energy supply security problem, global warming, pollution, public health threats, energy poverty, exhaustion of non-renewable energy sources, dependency on fossil fuels and risk of volatility in fossil fuel prices. With the transition to renewable energy sources, the aim is to reduce these problems and to encourage clean production around the world. Therefore, the decarbonization opportunities offered by renewable energy sources are attracting more and more attention. China ranks first with its share of increase (1 EJ) in renewable energy production in 2020. In this context, the transition process to a low-carbon energy economy is evaluated in a case study of China from the perspective of renewable energy sources.

China has become the world's largest emitter of CO_2 and energy consumer, accounting for about 31% of cumulative emissions. The main reason for this is the rapid industrialization and increasing urbanization rates of the country after its reform and opening up in 1978. At the same time, the country already has a large population. In the aftermath of the Covid-19 pandemic, China has shown strong vitality in restoring social order and economic recovery. It is expected that the economy will tend to grow and the population will continue to increase until at least 2030. It is assumed that the simultaneous growth of both the economy and the population will further stimulate an increase of CO_2 emissions in the long run (BP, 2021; Shi et al., 2022).

More than a hundred countries signed the Kyoto Protocol in 1997, and leading developed countries planned CO_2 reduction targets to protect humanity from the threat of global warming. In the IPCC Fifth Report of 2013, it was announced that CO_2 emissions from the burning of fossil fuels are the main cause of global warming and climate change. During this period, China did not take any responsibility for emissions reduction. The rapid economic development experienced after 1978 made the country the second-largest economy in the world and, accordingly, the country's CO_2 emissions increased rapidly (Hsu et al., 2021; Yin et al., 2015). Because of this, the country has been subjected to considerable pressure in international climate change negotiations. As a result, the country has adopted short-term and medium-term targets to reduce CO_2 emissions and a wide-ranging set of policies that contribute to achieving these targets. These policies are partly shaped by certain goals, such as promoting economic growth, reducing local air pollution and developing strategic industries (Sandalow, 2019).

The Chinese economy has undergone a period of rapid growth, which is heavily dependent on energy consumption, especially of fossil fuels. In this context, fossil fuel-based energy consumption has increased rapidly, and this has led to a significant decrease in environmental quality. Therefore, non-renewable energy consumption was a direct cause of high CO_2 emissions. In this context, priority was given to the development of renewable energy in order to both maintain sustainable development and economic growth and to reduce environmental impact. Accordingly, both production and consumption of renewable energy have grown in recent years, and this growth is expected to continue in the future (Chen et al., 2019; Wang et al., 2021a).

China's coal-based energy consumption structure has great effects on energy security, energy independence and environmental pollution. Renewable energy is essential for the country to build a safe, independent and low-carbon energy system. The depletion of fossil energy sources and pressure to save energy and reduce emissions has forced the country to develop renewable energy (Lin and Zhu, 2019). Strong political initiatives accelerated the country's transition to renewable energy sources since 2015. In the energy structure, targets were set to increase electricity production from renewable energy sources. Due to environmental pressure and self-imposed climate targets, it has greatly increased the use of renewable energy sources, accelerating the development of solar and wind energy (Li and Haneklaus, 2021). For example, while renewable energy consumption almost tripled in the period 1990–2018, the share of conventional energy consumption decreased from 95% to 86% in the same time span (Wang et al., 2021a). The installed power of hydroelectric, wind energy, photovoltaic energy production and biomass energy generation reached 338 million, 154 million, 102 million and 13.3 million kW, respectively, ranking first in the world. At the same time, although the country is currently experiencing a rapid development of renewable energy, the development of renewable energy faces a number of problems due to market failures and various factors. First, because of the high up-front cost, the development of renewable energy requires a high initial investment and a large amount of government subsidies. To achieve sufficient market competitiveness in renewable energy, the level of renewable energy technological innovation should be further encouraged. Second, although it has relatively high technological research and development capability in hydropower, nuclear power and thermal energy, wind power, biomass and suchlike have weaker innovation capability. Third, despite the rapid development in renewable energy technological innovation, there are large differences between states (Lin and Zhu, 2019).

In China, the main driver of massive fossil fuel consumption has been stunning economic growth. Approximately 84% of the total primary energy consumption in the country is met by fossil fuels, and 57% of the total is due to coal consumption

TABLE 11.3

Primary Energy Consumption (EJ) in China: 2009–2020

Energy type	2009	2010	2011	2012	2013	2014	2015	2016	2017	2018	2019	2020
Primary energy	97.53	104.29	112.54	117.05	121.38	124.82	126.53	128.63	132.80	137.58	142.03	145.46
Oil	16.69	18.99	19.68	20.63	21.54	22.39	24.24	25.06	26.20	27.06	27.94	28.50
NG	3.25	3.92	4.87	5.43	6.19	6.78	7.01	7.54	8.69	10.22	11.10	11.90
Coal	70.58	73.22	79.71	80.71	82.44	82.49	80.94	80.21	80.59	81.11	81.79	82.27
Solar	0.00	0.01	0.02	0.03	0.08	0.22	0.36	0.60	1.06	1.58	2.00	2.32
Wind	0.26	0.46	0.69	0.96	1.27	1.46	1.69	2.18	2.74	3.27	3.61	4.14
Geo, biomass and other	0.20	0.23	0.26	0.28	0.34	0.42	0.49	0.56	0.72	0.84	1.00	1.20

Source: Prepared by the researcher using the BP Statistical Review of World Energy 2021 Report.

(Li and Haneklaus, 2021). According to Table 11.3, total primary energy consumption in 2020 was 145.46 EJ, an increase of 3.43 EJ compared to 2019. China, one of the few countries where energy consumption increased in 2020, recorded the largest increase at 2.1%. Total coal production was 80.91 EJ, while coal consumption was 82.27 EJ. Coal was followed by oil (28.5 EJ) and NG (11.9 EJ), respectively. As the world's largest primary energy consumer, China was the largest contributor (1.0 EJ) to growth in renewable energy in 2020. There was an increase of 28.9% in renewable energy consumption in the 2009–2019 period. The highest consumption amount among renewable energy sources occurred in wind energy (4.14 EJ).

Industrialization in the country took place at a rapid pace, and the per capita CO_2 emissions amount, which was 1.5 metric tons in 1980, increased to 4.46 metric tons in 2005 due to increasing fossil fuel consumption. The country has important renewable energy resources and aims to increase the ratio of renewable energy consumption in total energy consumption in order to reduce environmental pollution. Since coal is an energy source that is consumed more than other energy sources, the Renewable Energy Law was enacted in 2005 to promote renewable energy. A series of renewable energy policies were then proposed, providing guarantees for the further development of renewable energy consumption. The rapid decline in the prices of renewable energy technologies was a positive sign for the use of renewable resources. At the same time, policies have been developed to support and encourage the use of renewable energy, such as the policy of subsidies for renewable resources. Increasing subsidies for renewable energy sources is expected to contribute to increased demand and therefore sustainability (Sharma et al., 2022; Chen et al., 2019).

Since 2009, the country has responded to increasing energy and environmental problems by reducing carbon emissions and expanding the use of alternative energy sources. As part of the Paris Agreement, it committed to reduce CO_2 emissions intensity by 60–65% compared to 2005 and to increase the proportion of non-fossil energy in primary energy to 20% by 2030 (Shi et al., 2022; Yi, 2016). Under the US-China Joint Announcement on Climate Change, it promised to achieve the highest CO_2 emissions reduction by 2030 (Chen et al., 2019). The country has recently become a global leader in renewable energy and has started to invest heavily in renewable energy technologies (Jaganmohan, 2021). In 2016, it established the Global Energy Interconnection Development Cooperation Organization. This has greatly improved renewable energy generation, represented by wind power and photovoltaic power generation. In this context, it has become the country with the largest installed capacity of wind energy and photovoltaic power generation in the world. Renewable energy production technology has developed in the country, and while the cost of energy production has steadily decreased, market competitiveness has increased. This has provided an important basis for large-scale applications of renewable energy and larger interconnections of electricity grids (Huang, 2020). The country's experience with public policy practices has shown that energy regulation policies focus on energy intensity and CO_2 emissions indicators, financial incentives, access to green technologies, regulatory reform and industry restructuring that are all useful practices for facilitating the development of the renewable energy industry and coping with environmental degradation (Chen et al., 2020).

From a historical perspective, China's status as the world's leading emitter of CO_2 is relatively recent. In the nineteenth and twentieth centuries, the amount of emissions was relatively small, but in the early twenty-first century, it started to increase in parallel with the economic development of the country and exceeded emissions generated by the USA. China's cumulative emissions since the start of the Industrial Revolution were about half that of the United States (Sandalow, 2019). Over past decades, environmental regulation and policies have been implemented to reduce coal use and improve air quality, with the country becoming a world leader in renewable energy and nuclear power installation (Li and Haneklaus, 2021).

According to Figure 11.5, the total amount of CO_2 emissions, which was 6.3 Bt in 2010, increased to 10.67 Bt in 2020, accounting for approximately 31% of global CO_2 emissions. The rate of increase in CO_2 emissions, which was 6.4% in the 2004–2014 period, decreased to 2.4% annually in the 2009–2019 period. The rate of increase in emissions was 0.6% in 2020 (BP, 2021; Li and Haneklaus, 2021).

China, the country with the highest CO_2 per capita after the USA, has the highest cumulative emissions (BBC, 2021). The country's continued economic growth trend

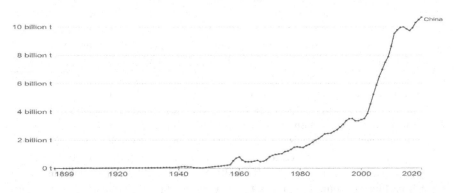

FIGURE 11.5   Annual CO_2 emissions from the burning of fossil fuels: 1899–2020 (Bt).

*Source:*Ritchie and Roser (2020)

may inevitably lead to more carbon emissions in the future. At the same time, in accordance with the Paris Agreement, the country is expected to control CO_2 emissions by converting the industrial structure to more sustainable energies. It promotes cleaner technologies to achieve carbon neutrality goals (Qin et al., 2021). It sees the development of renewable energy as a key measure to implement the energy supply revolution strategy. Its strategy, based on local multiple supply security, highlights the country's focus on building an energy supply system to develop non-coal energy (Yu et al., 2019). In this context, it strives to increase investment in the renewable energy industry, particularly in solar and wind energy. Local incentives have enabled the renewable industry to reduce employment barriers, reduce socio-economic risks and help boost sustainable development for the future. In terms of installed generation capacity, new solar and wind installations have continued to grow rapidly since 2008. Wind and solar power accounted for approximately 63% of the new power generation capacity facilities. In 2015, Germany's installed solar capacity exceeded its base and had the largest solar market globally, with both new installations and total capacity (Li and Haneklaus, 2021).

11.5   POLICY IMPLICATIONS AT THE MACRO LEVEL

A significant reduction target in emissions requires fundamental, rapid and large-scale systematic transformation to fully decarbonize the global energy system. All countries and sectors have different roles to play in decarbonizing to achieve global climate neutral goals. At the same time, the timing and pace of emissions reduction differs according to countries' conditions, such as their dependence on fossil fuels, processes in energy transition, socio-economic and political conditions and their capacity to reduce emissions.

11.5.1   POLICY IMPLICATIONS FROM THE DENMARK CASE STUDY

Presenting a successful example of the relationship between energy and the environment, Denmark teaches important lessons in the integration of variable renewable energy. In an overall assessment within the framework of the country's goal of achieving an energy system independent of fossil fuels by 2050, it states that coal consumption has greatly decreased over the last thirty years and that this decrease is marginal compared to the changes in oil and natural gas. It points out that despite the decrease in the amount of fossil fuels, these fuels are still important energy consumption components. The importance of large oil fields in oil production and the use of renewable energy are increasing (Karabiber and Xydis, 2019; Sällh et al., 2014).

The approach in the Denmark energy model is to establish a link between sectors and systems that create synergies among themselves rather than focusing on sectoral components and concepts. The three main elements of the model are the development of energy efficiency, renewable energy and the electricity distribution system. Public-private cooperation supports innovation and investment. The main reason Denmark has become an exemplary country, in line with its goal of reducing CO_2 emissions, is that it produces energy using renewable energy sources instead of fossil-based sources. In this context, the country has reduced its GHG emissions by 20% due to an increase in the use of renewable energy resources in agriculture and transportation and a decrease in the use of petroleum-based resources (Yücel et al., 2021).

When the Denmark case study is evaluated in terms of policy implications at the macro level, it shows the importance of various incentive practices, such as fixed price guarantees and premium guarantees and subsidies, particularly in renewable energy. Policies such as diversifying energy supply, facilitating individual investment in renewable energy, enabling citizens to benefit from wind energy projects, supporting the investment costs of manufacturers who turn to renewable energy production and providing tax refunds can be presented among the policy recommendations. In particular, with high taxes on oil and coal, the competitive advantage can be transformed in favor of renewable resources.

11.5.2 POLICY IMPLICATIONS FROM THE USA CASE STUDY

NG is replacing coal in many countries to help countries achieve medium-term climate reduction targets in the short term. The USA provides a successful example of this. Anticipating that natural gas will play a leading role in the energy mix in the near future, NG can play an important role in short-term climate mitigation targets. In the long run, in the absence of carbon capture and storage, fossil energy sources can be replaced by renewable and nuclear generation to meet climate targets. In this context, it is predicted that NG demand will vary significantly depending on the decarbonization path, renewable technology deployment, energy efficiency improvement and climate legislation (Guo and Hawkes, 2019). As one of the strategic transition energies towards a low carbon economy, the lower cost of natural gas and its feature of being a bridge fuel can be stated as a macro-level policy implication for countries' short-term goals of reducing CO_2 emissions.

The increase in demand for fossil energy sources affects costs arising from negative externalities, such as air pollution and CO_2 emissions. NG continues to be a critical energy source, both economically and environmentally. In the initial phase of the clean energy transition, the resource stock and characteristics of relatively less polluting natural gas enable natural gas to serve as an important transition fuel. In this context, it is thought that understanding the supply and demand structure of the NG market will help design an effective energy planning system where natural gas can reserve market space as a valuable transition fuel (Joshi, 2021).

Even if the net emissions impact of NG infrastructure expansion is negligible, it can still offer cost-effective emissions reductions under climate policies in the short term. The rapidly increasing capacity, infrastructure requirements and other features of NG generation can provide significant short-term and medium-term synergies with expanding renewable energy generation (Woollacott, 2020).

11.5.3 POLICY IMPLICATIONS FROM THE CHINA CASE STUDY

Emissions reduction actions by China, the world's highest CO_2-emitting country, will greatly contribute to ameliorating global problems related to climate change. The country is using an integration of various renewable energy sources and systems to reduce the harmful consequences of high CO_2 emissions (Hsu et al., 2021). The move to promote the use of renewable energy in the energy mix is designed to reduce reliance on higher-polluting emissions from fossil fuel use. Renewable energy sources have historically contributed significantly to clean electricity in GHG production and the reduction of CO_2 emissions worldwide (Li and Haneklaus, 2021).

The country has adopted several targets for the deployment of renewable energy by 2020. These targets are substantial, both in terms of total installed capacity and the expected contribution of renewable energy to total electricity generation. An important goal of developing renewable energy is to reduce CO_2 emissions and reliance on imported energy by decoupling the increasing use of fossil energy from economic growth over the next few decades. The emphasis on renewable energy is also designed to increase the country's competitiveness as a leading global supplier of clean and low-cost renewable energy technologies. The targets for renewable energy deployment form part of a broader set of energy and climate policies. National targets were set for the reduction of energy and carbon intensity, as well as the contribution of primary energy from non-fossil sources. These broad targets have been supported by measures targeting increases in wind, solar and biomass electricity generation types (Qi et al., 2014).

The macro-level policy implications of the Chinese case study are that renewable energy is an important solution to reduce negative environmental impact (Wang et al., 2021a). The development of renewable energy is an effective way to reduce cumulative GHG emissions, improve the ecological environment and optimize the energy mix. At the same time, the partnership between the environment and renewable energy is the basis for ensuring sustainability (SDM, 2021). The key to achieving harmonious and sustainable development between energy and the environment is to strengthen the development of renewable energy consumption and reduce non-renewable energy consumption (Chen et al., 2019).

11.6 CONCLUSION

GHG emissions are mostly generated by industrial activities, such as transportation, power plants and energy-intensive industries. The highest amount of these emissions is CO_2 emissions, with a rate of 75% (Khezri et al., 2022). CO_2 emissions, which increased by 257%, and other environmental pollutants, which increased by 122% compared to pre-industrial levels, have seriously triggered global warming and climate change. Since the 1970s, the ecological footprint has exceeded the planet's capacity to regenerate itself, resulting in significant adverse effects on environmental degradation, human well-being and ecosystems (Pata, 2021). It is widely accepted that, due to rapid industrialization, CO_2 emissions have risen to unprecedentedly high levels in recent decades, which has caused the global temperature to rise. The acceleration of economic activity, along with population growth and globalization, causes serious threats to the ecosystem by causing CO_2 and other GHG emissions to increase (Qin et al., 2021). The intensity of CO_2 emissions in the atmosphere reached its highest level in two million years in 2019. Concentrations of methane and nitrous oxide gases, which are important GHGs, were at their highest level in 800,000 years in 2019 (GCP, 2021; Samar, 2021).

Due to the importance and urgency of the issue, the partnership between energy and environment in this study is examined in the case studies of Denmark, the USA and China. The main conclusions obtained from the case studies indicate that natural gas can be used as a bridge fuel as a short-term solution to reduce CO_2 emissions. At the same time, since natural gas is not a completely clean energy source, it is necessary to turn to renewable energy sources for long-term solutions. Carbon capture and storage technologies, which are becoming increasingly important in the climate policies and net zero emission strategies of countries, also constitute one of the short-term solution proposals. In this context, studies conducted by the International Energy Agency and IPCC show that carbon capture is of great importance in reducing global emissions and meeting international climate targets.

REFERENCES

Aune, F.R., Golombek, R., Kittelsen, S.A.C. (2004). Does increased extraction of natural gas reduce carbon emissions? *Environmental and Resource Economics*, *29*, 379–400. https://doi.org/10.1007/s10640-004-9456-3

BBC (2021). İklim krizi: Dünyayı en çok kirleten ülkeler, karbon emisyonunu azaltmak için neler yapıyor? www.bbc.com/turkce/haberler-dunya-59088481

Bohnerth, J.C. (2015). Energy cooperatives in Denmark, Germany and Sweden—A transaction cost approach. Master thesis in Sustainable Development 243. Uppsala Universitet.

BP (2020). Energy outlook: 2020 edition. https://www.bp.com/content/dam/bp/business-sites/en/global/corporate/pdfs/energy-economics/energy-outlook/bp-energy-outlook-2020.pdf

BP (2021). Statistical review of world energy: 70th edition. https://www.bp.com/content/dam/bp/business-sites/en/global/corporate/pdfs/energy-economics/statistical-review/bp-stats-review-2021-full-report.pdf

Burnham, A., Han, J., Clark, C.E., Wang, M., Dunn, J.B., Palou-Rivera, I. (2012). Life-cycle greenhouse gas emissions of shale gas, natural gas, coal, and petroleum. *Environmental Science and Technology*, *46* (2), 619–627. https://doi.org/10.1021/es201942m

Canuto, O. (2021). The road to decarbonisation. *Capital Finance International*, 122–123. https://www.researchgate.net/publication/355945438_The_Road_to_Decarbonisation

Chen, K., Ren, Z., Mu, S., Sun, T.Q., Mu, R. (2020). Integrating the Delphi survey into scenario planning for China's renewable energy development strategy towards 2030. *Technological Forecasting & Social Change*, *158*, 120157. https://doi.org/10.1016/j.techfore.2020.120157

Chen, Y., Zhao, J., Lai, Z., Wang, Z., Xia, H. (2019). Exploring the effects of economic growth, and renewable and non-renewable energy consumption on China's CO_2 emissions: Evidence from a regional panel analysis. *Renewable Energy*, *140*, 341–353. https://doi.org/10.1016/j.renene.2019.03.058

De Gouw, J.A., Parrish, D.D., Frost, G.J., Trainer, M. (2014). Reduced emissions of CO_2, NO_x, and SO_2 from US power plants owing to switch from coal to natural gas with combined cycle technology. *Earth's Future*, *2*, 75–82. https://doi.org/10.1002/2013EF000196

Dong, K., Sun, R., Wu, J., Hochman, G. (2018). The growth and development of natural gas supply chains: The case of China and the US. *Energy Policy*, *123*, 64–71. https://doi.org/10.1016/j.enpol.2018.08.034

EIA (2021). Annual energy outlook 2021 with projections to 2050. US Energy Information Administration, Washington. https://www.eia.gov/outlooks/aeo/pdf/AEO_Narrative_2021.pdf

EPA (2019). Inventory of U.S. greenhouse gas emissions and sinks: 1990–2017. U.S. Environmental Protection Agency, Washington. https://www.epa.gov/sites/default/files/2019-04/documents/us-ghg-inventory-2019-main-text.pdf

EPI (2020). Environmental performance index 2020: Global metrics fort the environment: Ranking country performance on sustainability issues. EPI 2020.pdf.

Eryar Ünlü, D. (2021). Türkiye COP26'da 4 taahhüde imza attı. www.dunya.com/kose-yazisi/turkiye-cop26da-4-taahhude-imza-atti/639989

Eser, Y.L., Polat, S. (2015). Elektrik üretiminde yenilenebilir enerji kaynaklarının kullanımına yönelik teşvikler: Türkiye ve İskandinav ülkeleri uygulamaları. *Gümüşhane Üniversitesi Sosyal Bilimler Elektronik Dergisi, 12*, 201–225.

Evans, S. (2021). Analysis: Which countries are historically responsible for climate change? www.carbonbrief.org/analysis-which-countries-are-historically-responsible-for-climate-change

Field, R.A., Derwent, R.G. (2021). Global warming consequences of replacing natural gas with hydrogen in the domestic energy sectors of future low-carbon economies in the United Kingdom and the United States of America. *International Journal of Hydrogen Energy, 46*, 30190–30203. https://doi.org/10.1016/j.ijhydene.2021.06.120

Gabbatiss, J. (2021). The carbon brief profile: United States. www.carbonbrief.org/the-carbon-brief-profile-united-states

GCP (2021). www.globalcarbonproject.org/index.htm

Guo, Y., Hawkes, A. (2019). The impact of demand uncertainties and China-US natural gas tariff on global gas trade. *Energy, 175*, 205–217. https://doi.org/10.1016/j.energy.2019.03.047

GWM (2004). Co-operative energy: Lessons from Denmark and Sweden. GWM 2014.pdf.

Haber7 (2021). COP26'da 100'den fazla ülke bu taahhüde imza attı. www.haber7.com/dunya/haber/3158015-cop26da-100den-fazla-ulke-bu-taahhude-imza-atti

Handayania, K., Filatova, T., Krozer, Y., Anugrah, P. (2020). Seeking for a climate change mitigation and adaptation nexus: Analysis of a long-term power system expansion. *Applied Energy, 262*, 114485. https://doi.org/10.1016/j.apenergy.2019.114485

Holdren, J.P., Smith, K.R. (2000). Energy, the environment, and health. In: *World energy assessment: Energy and the challenge of sustainability*. The United Nations Development Programme, The United Nations Department of Economic and Social Affairs, The World Energy Council.

Hou, C., Nguyen, B.H. (2018). Understanding the US natural gas market: A Markov switching VAR approach. *Energy Economics, 75*, 42–53. https://doi.org/10.1016/j.eneco.2018.08.004

Hsu, C.-C., Zhang, Y., Ch, P., Aqdas, R., Chupradit, S., Nawaz, A. (2021). A step towards sustainable environment in China: The role of eco-innovation renewable energy and environmental taxes. *Journal of Environmental Management, 299*, 113609. https://doi.org/10.1016/j.jenvman.2021.113609

Huang, Q. (2020). Insights for global energy interconnection from China renewable energy development. *Global Energy Interconnection, 3*(1), 1–11. https://doi.org/10.1016/j.gloei.2020.03.006.

IEA (2021). Global energy-related CO_2 emissions, 1990–2021. www.iea.org/data-and-statistics/charts/global-energy-related-co2-emissions-1990-2021

IPCC (2018). Summary for policymakers. In: *Global warming of 1.5°C. An IPCC special report on the impacts of global warming of 1.5°C above pre-industrial levels and related global greenhouse gas emission pathways, in the context of strengthening the global response to the threat of climate change, sustainable development, and efforts to eradicate poverty*. Edited by Valérie Masson-Delmotte, Panmao Zhai, Hans-Otto Pörtner, Debra Roberts, Jim Skea, Priyadarshi R. Shukla, et al. https://www.ipcc.ch/site/assets/uploads/sites/2/2022/06/SPM_version_report_LR.pdf

IPCC (2021). Climate change 2021: The physical science basis. Edited by Masson-Delmotte, Valérie, Panmao Zhai, Anna Pirani, Sarah L. Connors, Clotilde Péan, Yang Chen, et al. https://report.ipcc.ch/ar6/wg1/IPCC_AR6_WGI_FullReport.pdf

IRENA (2021). World energy transitions outlook: 1.5°C pathway. International Renewable Energy Agency. https://www.irena.org/-/media/Files/IRENA/Agency/Publication/2021/Jun/IRENA_World_Energy_Transitions_Outlook_2021.pdf

Jaganmohan, M. (2021). Global renewable energy industry—Statistics & facts. www.statista.com/topics/2608/global-renewable-energy-industry/#dossierKeyfigures

Joshi, J. (2021). Sectoral and temporal changes in natural gas demand in the United States. *Journal of Natural Gas Science and Engineering*, *96*, 104245. https://doi.org/10.1016/j.jngse.2021.104245

Karabiber, O.A., Xydis, G. (2019). Electricity price forecasting in the Danish day-ahead market using the TBATS, ANN and ARIMA methods. *Energies*, *12*(928), 1–29. https://doi.org/10.3390/en12050928

Khezri, M., Heshmati, A., Khodaei, M. (2022). Environmental implications of economic complexity and its role in determining how renewable energies affect CO_2 emissions. *Applied Energy*, *306*, 117948. https://doi.org/10.1016/j.apenergy.2021.117948

Kueppers, M., Pineda, S.N.P., Metzger, M., Huber, M., Paulus, S., Heger, H.J., et al. (2021). Decarbonization pathways of worldwide energy systems—Definition and modeling of archetypes. *Applied Energy*, *285*, 116438. https://doi.org/10.1016/j.apenergy.2021.116438

Le Quéré, C., Moriarty, R., Andrew, R.M., Canadell, J.G., Sitch, S., Korsbakken, J.I., et al. (2015). Global carbon budget 2015. *Earth System Science Data*, *7*(2), 349–396. https://doi.org/10.5194/essd-7-349-2015

Li, B., Haneklaus, N. (2021). The role of renewable energy, fossil fuel consumption, urbanization and economic growth on CO_2 emissions in China. *Energy Reports*, *7*, 783–791. https://doi.org/10.1016/j.egyr.2021.09.194

Li, Y., Chevallier, J., Wei, Y., Li, J. (2020). Identifying price bubbles in the US, European and Asian natural gas market: Evidence from a GSADF test approach. *Energy Economics*, *87*, 104740. https://doi.org/10.1016/j.eneco.2020.104740

Lin, B., Zhu, J. (2019). Determinants of renewable energy technological innovation in China under CO_2 emissions constraint. *Journal of Environmental Management*, *247*, 662–671. https://doi.org/10.1016/j.jenvman.2019.06.121

Lu, H., Ma, X., Azimi, M. (2020). US natural gas consumption prediction using an improved kernel-based nonlinear extension of the Arps decline model. *Energy*, *194*, 116905. https://doi.org/10.1016/j.energy.2020.116905

MFA (2021). Paris anlaşması. www.mfa.gov.tr/paris-anlasmasi.tr.mfa

Pata, U.K. (2021). Linking renewable energy, globalization, agriculture, CO_2 emissions and ecological footprint in BRIC countries: A sustainability perspective. *Renewable Energy*, *173*, 197–208. https://doi.org/10.1016/j.renene.2021.03.125

Qi, T., Zhang, X., Karplus, V. (2014). The energy and CO_2 emissions impact of renewable energy development in China. *Energy Policy*, *68*, 60–69. https://doi.org/10.1016/j.enpol.2013.12.035

Qin, L., Hou, Y., Miao, X., Zhang, X., Rahim, S., Kirikkaleli, D. (2021). Revisiting financial development and renewable energy electricity role in attaining China's carbon neutrality target. *Journal of Environmental Management*, *297*, 113335. https://doi.org/10.1016/j.jenvman.2021.113335

Riahi, K., van Vuuren, D.P., Kriegler, E., Edmonds, J., O'Neill, B.C., Fujimori, S., et al. (2017). The shared socioeconomic pathways and their energy, land use, and greenhouse gas emissions implications: An overview. *Global Environmental Change*, *42*, 153–168. https://doi.org/10.1016/j.gloenvcha.2016.05.009

Ritchie, H., Roser, M. (2020). CO_2 and greenhouse gas emissions. https://ourworldindata.org/co2-and-other-greenhouse-gas-emissions

Rubaszek, M., Uddin, G.S. (2020). The role of underground storage in the dynamics of the US natural gas market: A threshold model analysis. *Energy Economics*, *87*, 104713. https://doi.org/10.1016/j.eneco.2020.104713

Ruester, S., Neumann, A. (2008). The prospects for liquefied natural gas development in the US. *Energy Policy, 36*, 3160–3168. https://doi.org/10.1016/j.enpol.2008.04.030

Russo, T.N. (2021). Responsibly sourced gas: Time to change the natural gas industry's narrative. *Climate and Energy, 37*, 22–27. https://doi.org/10.1002/gas.22215

Sällh, D., Höök, M., Grandell, L., Davidsson, S. (2014). Evaluation and update of Norwegian and Danish oil production forecasts and implications for Swedish oil import. *Enegy, 65*, 333–345. http://doi.org/10.1016/j.energy.2013.11.023

Samar, K. (2021). Dünya'yı en çok hangi ülkeler kirletiyor? 1850 yılından bu yana ülkelere göre karbondioksit salımı. https://tr.euronews.com/2021/11/08/dunya-y-en-cok-hangi-ulkeler-kirletiyor-1850-y-l-ndan-bu-yana-ulkelere-gore-karbondioksit-

Sandalow, D. (2019). 2019 guide to Chinese climate policy. www.energypolicy.columbia.edu/research/report/2019-guide-chinese-climate-policy

SDM (2021). Energy and the environment. www.sdsmt.edu/Research/Research-Areas/Energy-and-the-Environment/

Şen, S. (2017). Yenilenebilir enerji üretiminde maliye politikası aracı olarak teşvikler: Seçilmiş bazı Avrupa ülkelerinin deneyimleri ve Türkiye. *Journal of Life Economics, 4*(1), 59–76. https://doi.org/10.15637/jlecon.185

Sharma, A., Dharwal, M., Kumari, T. (2022). Renewable energy for sustainable development: A comparative study of India and China. *Materialstoday: Proceedings, 60*(2), 788–790. https://doi.org/10.1016/j.matpr.2021.09.242

Shi, H., Chai, J., Lu, Q., Zheng, J., Wang, S. (2022). The impact of China's low-carbon transition on economy, society and energy in 2030 based on CO_2 emissions drivers. *Energy, 239*, 122336. https://doi.org/10.1016/j.energy.2021.122336

Tanaka, K., O'Neill, B. (2018). The Paris agreement zero-emissions goal is not always consistent with the 1.5°C and 2°C temperature targets. *Nature Climate Change, 8*, 319–324. https://doi.org/10.1038/s41558-018-0097-x

Tiseo, I. (2021). Carbon dioxide emissions in Denmark 1970–2020. www.statista.com/statistics/449517/co2-emissions-denmark/

UCS (2014). Environmental impacts of natural gas. www.ucsusa.org/resources/environmental-impacts-natural-gas

UNEP (2021). Emissions gap report 2021: The heat is on—A world of climate promises not yet delivered. https://www.unep.org/resources/emissions-gap-report-2021

Veenman, S., Sperling, K., Hvelplund, F. (2019). How future frames materialize and consolidate: The energy transition in Denmark. *Futures, 114*, 102473. https://doi.org/10.1016/j.futures.2019.102473

Wang, C.-J., Han, X., Xin, S., Liu, D.-H., Xu, M., Ma, J.-Q., et al. (2019). An empirical analysis of Denmark's energy economy and environment and its sustainable development policy. *Journal of Sustainable Development, 12*(2), 29–38. https://doi.org/10.5539/jsd.v12n2p29

Wang, J., Zhang, S., Zhang, Q. (2021a). The relationship of renewable energy consumption to financial development and economic growth in China. *Renewable Energy, 170*, 897–904. https://doi.org/10.1016/j.renene.2021.02.038

Wang, R., Assenova, V.A., Hertwich, E.G. (2021b). Energy system decarbonization and productivity gains reduced the coupling of CO_2 emissions and economic growth in 73 countries between 1970 and 2016. *One Earth, 4*(19), 1614–1624. https://doi.org/10.1016/j.oneear.2021.10.010

Wang, Z.-X., He, L.-Y., Zhao. Y.-F. (2021c). Forecasting the seasonal natural gas consumption in the US using a gray model with dummy variables. *Applied Soft Computing, 113*, 108002. https://doi.org/10.1016/j.asoc.2021.108002

Woollacott, J. (2020). A bridge too far? The role of natural gas electricity generation in US climate policy. *Energy Policy, 147*, 111867. https://doi.org/10.1016/j.enpol.2020.111867

Yi, S. (2016). Can the Chinese economy take on a renewable energy revolution? https://environment-review.yale.edu/can-chinese-economy-take-renewable-energy-revolution-0

Yin, J., Zheng, M., Chen, J. (2015). The effects of environmental regulation and technical progress on CO_2 Kuznets curve: An evidence from China. *Energy Policy*, *77*, 97–108. https://doi.org/10.1016/j.enpol.2014.11.008

Yu, S., Zheng, Y., Li, L. (2019). A comprehensive evaluation of the development and utilization of China's regional renewable energy. *Energy Policy, 127*, 73–86. https://doi.org/10.1016/j.enpol.2018.11.056

Yücel, U., Özdemir, E., Ayaz, M. (2021). Yenilenebilir enerji kaynaklarından üretilen elektrik enerjisi teşvik yöntemlerinin incelenmesi. *Düzce Üniversitesi Bilim ve Teknoloji Dergisi*, *9*, 774–790. https://doi.org/10.29130/dubited.774963

12 Climate Change
Challenges and Responses for Business

Nugun P. Jellason

12.1 INTRODUCTION

The world population continues to increase and is projected to reach 9.8 billion people by 2050 (Collins and Page 2019). This will trigger more demand for food and other consumables, leading to increased production (Forbord and Vik 2017). To meet these demands, businesses will more than double their existing energy use, increase natural resource depletion such as forests with implications for environmental conservation (Alaze et al. 2021). The global energy portfolio has been dominated by fossil fuels (e.g., coal, oil and natural gas) despite the developments in renewables and other forms of non-fossil energy (Diczfalusy 2014; Arnold 2011). Burning fossil fuels to power economies has led to climate change, thereby endangering the Earth (Doppelt and McDonough 2010; Arnold 2011). Demand for fossil fuels is projected to grow by 19% in Organisation for Economic Co-operation and Development (OECD) countries and by 85% in non-OECD countries (Arnold 2011). The wide margin of demand is due partly to the economic growth trajectory of non-OECD countries (Arnold 2011).

Globally, climate change remains a significant concern for businesses, civil society and governments in the 21st century and constitutes a challenge to business management and organisation (Kolk and Pinkse 2004; Okereke, Wittneben, and Bowen 2011). Available evidence increasingly suggests human-induced carbon dioxide (CO_2) emissions as the cause of climate change (IPCC 2007; Stern 2006). To tackle the increasing human-induced climate change, businesses have a critical role to play by transforming their processes, culture and policies for a low-carbon future (Bumpus et al. 2014; Doppelt and McDonough 2010). Business-to-business (B2B) transactions have been reported as significant channels for high carbon emissions through the supply chains (Lye 2014). These emissions will need to be cut down to allow businesses to effectively compete in a constantly changing market. In line with responsible business culture, international businesses must innovate and act as global citizens with responsibilities for solving the climate change challenge (Bumpus et al. 2014).

In addition to the need for environmental stewardship and intergenerational equity (Arnold 2011), the Covid-19 pandemic has added a layer to the risks and uncertainties around international business environments. Increased Covid-19 vaccine

DOI: 10.1201/9781032715438-12

rollouts in developed economies and the economic recovery measures being implemented will likely increase energy demand and by implication CO_2 emissions above the current levels (IEA 2021). There is an opportunity and indeed a need for recovery measures to be sustainable. Against this backdrop, this chapter sets out to explore the challenges climate change poses to international businesses. First, the risks and uncertainties of climate change to international businesses will be explored from a human-induced climate change and internationalisation of business theoretical perspective; then the challenges experienced will be x-rayed in the context of production, operational and regulatory challenges. Further, the response strategies will be considered in line with the Sustainable Development Goals (SDGs), and finally the chapter will conclude.

12.2   CLIMATE CHANGE RISKS AND UNCERTAINTIES

In the last decade, the global economy experienced significant expansion amidst continuous climate change events. As a result of the risks and uncertainties surrounding climate change (Kunreuther et al. 2013) and economic expansion, it is likely that the opportunity to curb emissions to a 1.5–2.0° C temperature limit by 2030 will be missed (IPCC 2021). Re-occurring climate change events have inspired thinking about energy consumption and production, consumerism and transportation that has gradually made existing technologies such as internal combustion engines and coal-fired power plants obsolete (Blowfield 2013). While the changing climate constitutes an environmental liability to businesses and shareholders thereby reducing shareholders' and business value (Doppelt and McDonough 2010), it presents yet an opportunity for businesses to positively harness (Ihlen 2009; Millennium Ecosystem Assessment 2005).

As earlier highlighted, both risks and opportunities abound with the changing climate, which will likely lead to winners and losers (Adger and Nicholson-Cole 2011). Hence, the risks and uncertainties of climate change and its significance to the global business community due to growing greenhouse gas (GHG) emissions requires attention (Azarkamand et al. 2020) (Figure 12.1).

Against this backdrop, institutions must function more effectively under environments of uncertainty (Oliver 1991) and climate change mitigation plans well developed to reduce the carbon footprint of businesses to ensure the sustainability of any business (Azarkamand et al. 2020).

In an attempt to find solutions to the lingering climate crises, in 1979, the first World Climate Conference was held in Geneva, Switzerland, while in 1988, the Intergovernmental Panel on Climate Change (IPCC)—saddled with the responsibility of investigating the extent and likely impact of climate change—was established (Okereke, Wittneben, and Bowen 2011). The IPCC's first report advocated for a global agreement on climate change, leading to the signing in 1992 of the United Nations Framework Convention on Climate Change (UNFCCC). Signatories to the convention from developed countries committed to work towards reducing human-induced carbon emissions and other greenhouse gas emissions to their 1990 levels (UNFCCC article 4). The treaty was activated in 1994 and had 197 signatory parties as of 2015.

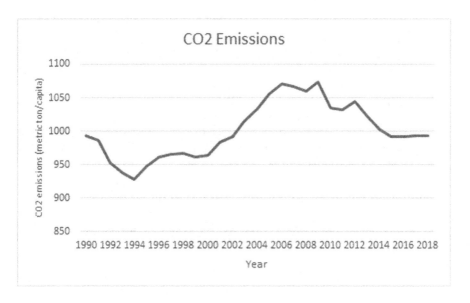

FIGURE 12.1 CO_2 emissions (metric ton/capita from 1990–2018 across the globe).

Source: World Bank data. License: Attribution-Non Commercial 4.0 International (CC BY-NC 4.0).

The World Business Council for Sustainable Development (WBCSD) high-lighted in its submission to the 1992 Earth Summit in Rio de Janeiro, Brazil, that the relationship between climate change and business was paradoxical and difficult to unpick. They argued that, on the one hand, businesses driving industrial activi-ties cause human-made CO_2 emissions (Okereke, Wittneben, and Bowen 2011). On the other hand, innovation, adaptation, green investment and clean technology are needed to achieve economic development amidst climate change. Hence, to address climate change challenges, a disruptive approach to the current industrial processes is essential (Okereke, Wittneben, and Bowen 2011). The next section will explore the challenges of climate change in the context of international business.

12.3 CHALLENGES OF CLIMATE CHANGE FOR INTERNATIONAL BUSINESSES

12.3.1 PRODUCTION CHALLENGES

International businesses are mostly operated by multinational enterprises (MNEs) working in multiple countries with a subsidiary in the host country (Collinson, Narula, and Rugman 2020; Levy and Kolk 2002). The cross-border operations involve production and the movement of people, raw materials and finished goods from one place to the other to fulfil customer and business needs. For example, before the Covid-19 pandemic, companies such as the apparel giant Levi Strauss, Marks & Spencer, Clarks, Next and other related companies outsourced the production of the merchandise sold in high-end stores in the US or Europe to factories in countries such as India, Pakistan and China that use fossil fuels such as coal energy as a source of power for the factories (Eavis and Krauss 2021). After production, the merchandise is transported via planes and ships that use fossil-based jet fuel and diesel (Eavis and Krauss 2021). These movements of goods involve the use of different forms of energy that may likely be inefficient and destructive to the environment (WBCSD 2021), hence the need to rethink the types of energy used to power the economy.

Transitioning from conventional energy use to a new efficient energy source has never been straightforward, as trade-offs are involved (Jamieson 2011). Business leaders and chief executive officers (CEOs) are often faced with the dilemma of delivering profits to investors and at the same time have a moral and social obligation to implement sustainability in a business (Epstein and Buhovac 2017). A 'win-win' strategy is essential for the corporate heads to deliver these competing demands made by international businesses. However, the trade-offs that often exist are in energy access and cost that make 'win-win' conditions challenging to attain by businesses in pursuit of profitability.

The global health pandemic of the novel Covid-19 led to a decline of 3.5% of the world gross domestic product (GDP) in 2020, and the International Monetary Fund (IMF) projects that the global GDP will expand by 6% in 2021 (Figure 12.2) due, on the one hand, to the successful vaccination program and other economic recovery programs such as Biden's stimulus package in the United States of America (USA) (IEA 2021).

The European Union (EU), on the other hand, may experience, overall, a slow recovery from the bloc's economic stimulus plan due to the severe impact of the second wave of the Covid-19 across Europe. The EU will also likely experience about 2.3% growth below the 2019 levels (IEA 2021). Although the return of industrial

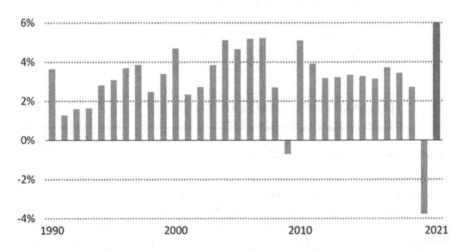

FIGURE 12.2 Annual rate of change in world GDP 1990–2021.

Source: IEA 2021. All rights reserved.

production to pre-Covid-19 levels in some sectors in the EU has been attributed to the recovery of international trade (IEA 2021); the impact of this recovery is likely to reverse the gains made from GHG emission reduction from the beginning of the pandemic due to national lockdowns and travel bans imposed to reduce the spread of the virus. There is a need to ensure that the EU recovery plans take into consideration sustainability approaches to obtain multiple benefits. The following section explores climate change impact on the operations and logistics of international business.

12.3.2 Operations, Logistics and Supply Challenges

International businesses constitute big energy users, thereby requiring a significant reduction or replacement, especially for businesses highly dependent on fossil fuels to become sustainable. Global energy demand for 2021 was predicted by the International Energy Agency (IEA) to rise by 4.6%, reversing the 2020 energy demand contraction and 0.5% above the 2019 levels (IEA 2021). About 70% of the predicted global energy demand will be in developing economies and emerging markets based on a 3.4% increase above the 2019 levels (IEA 2021). Fossil fuel (e.g., coal, oil and natural gas) demand has been projected to significantly grow in 2021, with coal alone expected to increase by about 60%, which is likely to raise CO_2 emissions by 1500 Mt (IEA 2021). Overall coal demand is likely to rise by about 4.5% in 2021, with Asia responsible for 80% of that growth. China accounts for about 50% of the growth of the Asian energy demand (Figure 12.3), as demand in the European Union (EU) and the US are also returning to historic high levels (IEA 2021).

Oil for transport is expected to increase by 6.2% in 2021; however, this increase will be about 3% below the 2019 levels due to the disruptions in road and air transport by the effects of the Covid-19 pandemic (IEA 2021). A full recovery to the pre-Covid-19 oil demand levels will require CO_2 emissions to be pushed by 1.5% above the 2019 levels (IEA 2021).

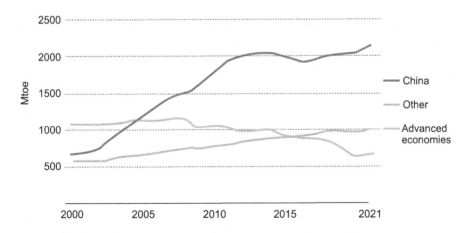

FIGURE 12.3   Coal consumption by region, 2000 to 2021.

Source: IEA 2021.

China's profile on fossil fuels consumption such as coal and oil has been on the increase. Yet, China accounts for about half of the global increase in the generation of renewable electricity, and it is on course to generate over 900 TWh from wind and solar photovoltaic (PV) by the end of 2021. These figures are almost double the figures projected for the EU (about 580 TWh) and the US (about 550 TWh) (IEA 2021).

Natural gas demand is projected to grow by 3.2% in 2021 relative to 2019 levels, with demand mostly higher from the Russian Federation, the Middle East and Asia (IEA 2021). The US, being the world's biggest market for natural gas, will experience a net decrease in demand compared to the 2020 figures due to growing natural gas prices and an increasing shift to renewables (IEA 2021). The Ukraine-Russia war has further exposed the vulnerability of the European countries' dependence on Russia's natural gas supplies. Most growth in natural gas demand globally in 2021 was from the building sectors and industry (IEA 2021).

Growth in energy usage shows an increase in renewables use by 3% in 2020 linked to the increase in electricity generation from solar PV and wind sources of approximately 330 terawatt hour (TWh) (IEA 2021). The share of renewables in total electricity generation has grown significantly in the past two years and projected to reach 30% by the end of 2021 compared to the less than 27% witnessed in 2019 (IEA 2021). However, there's room for more expansion of renewable sources of electricity by businesses to meet or surpass the 2030 emissions reduction target (Kolk and Pinkse 2004; Salite et al. 2021).

Suppliers of goods and commodities across supply chains are an integral part of businesses and are responsible for scope 3 greenhouse gas emissions that emanate from the supply chain of a firm (Hertwich and Wood 2018). Business firms have opportunities to reduce their scope 3 emissions by compelling their suppliers to adopt environmentally and socially friendly management practices in their supply chain (Epstein and Buhovac 2017). For example, L'Oréal, a beauty product manufacturer operating in France and other parts of the world, expects its suppliers to maintain operational standards as the standards found on its own business sites (Epstein and Buhovac 2017). This is based on its commitment as a signatory to the United Nations Global Compact. L'Oréal also expects its suppliers to respect International Labour Organisation (ILO) laws and local legislation on occupational health and safety, working hours and minimum wage (Epstein and Buhovac 2017). Non-compliance with these contractual standards is likely to affect all business transactions with L'Oréal.

In the last decade, the food and beverage and the metal industries were the sectors that seriously focused on scrutinising the emissions of their suppliers, whereas in other sectors, companies do not necessarily monitor their suppliers' emissions but rather select suppliers based on their existing environmental programmes, such as ISO certification (Kolk and Pinkse 2004). Cutting down emissions could be challenging for businesses, as they need to accurately measure their greenhouse gas emissions to ascertain their prevailing carbon footprint (Kolk and Pinkse 2004) and find alternative cleaner energy sources that help them remain profitable (Eavis and Krauss 2021). Where cleaner substitutes are not within reach, businesses purchase carbon credits to enable them to continue production, as their emissions are being offset by the company selling the emission allowance (Eavis and Krauss 2021). However, this approach has been said to be counterproductive to attaining net zero targets.

12.3.3 REGULATORY AND INSTITUTIONAL CHALLENGES

As climate change constitutes a challenge to international businesses, often, the drive for a sustainability strategy is from government regulation, pressure from non-governmental organisations (NGOs), competitor action or marketplace demands (Epstein and Buhovac 2017) and local- or global-level policy requirements. This drive could result in a behavioural change by businesses leading to technology or energy change or efficiency enhancement.

At the global level, regulatory frameworks and policies have been proposed at different fora and climate conventions, such as the 1992 Rio Conference that initiated the first international agreement on greenhouse gas emissions reduction; the Kyoto Protocol of the United Nations Framework Convention on Climate Change (1997) agreed at Kyoto, Japan, after countries signed to extend the Rio agreement; and the Conference of Parties (COP) to reduce greenhouse gas emissions based on equal but differentiated responsibilities (Pauw et al. 2014). The adoption of the Kyoto Protocol led to changes in corporate strategies (Grubb, Vrolijk, and Brack 2018) and encouraged the development of regulations as NGOs increased pressure on both developed and developing country governments to ratify the protocol and for companies to take necessary actions to stem the tide of global warming (Kolk and Pinkse 2004).

The global-level frameworks are ambitious, but several scholars (Andonova, Betsill, and Bulkeley 2009; WBCSD 2021) argue that they have not yet yielded the desired results. They posit that the resolutions of such conferences were non-binding, often incongruent with local development plans (Al-Saidi 2021; Silva-Chávez, Schaap, and Breitfeller 2015), hence the difficulty of implementation of the frameworks. For example, under the Trump administration, the US withdrew from the Paris Agreement of the UNFCCC, thereby threatening any gains made on the journey to a net-zero CO_2 emission world by 2050. The US action will lead to a leakage effect where gains by other parties will be reversed given the US government's global influence and resolve to continue the use of non-renewables, for example, coal, to generate power for its industries (Zhao and Alexandroff 2019). However, Urpelainen and Van de Graaf (2018) argue that the US' withdrawal from the Paris Agreement alone without any fundamental change to US federal climate policies is unlikely to lead to any further emissions. Rather, the US's non-participation in contributing to climate finance could have been a major barrier to future climate cooperation at the global level (Urpelainen and Van de Graaf 2018). The reversal of that policy by the Biden administration from its first day in office will potentially change the dynamics of future climate cooperation (Bodansky 2021).

Emerging economies such as China and India have for years declined to make commitments to cut their coal consumption as a source of energy to power their growing industries (Urpelainen and Van de Graaf 2018), as eliminating coal is challenging due to a lack of political will from governments (Zhao and Alexandroff 2019). On the contrary, it is argued that declining production and increased commitments to reduce CO_2 emission have led to a reduction in coal consumption in China (Urpelainen and Van de Graaf 2018) linked to slowed economic growth for China and suggesting that China has already met its CO_2 commitment target (Guan et al. 2018).

In the past, moral suasion was one of the strategies adopted for parties to commit to emissions reduction together with other economic instruments such as carbon trading and payments for adaptation and mitigation activities. For example, pressure by non-governmental organisations is mounted on governments to ratify the Kyoto Protocol and on businesses to act appropriately to address the planet's warming (Kolk and Pinkse 2004). However, these approaches may not be sufficient to lead to action, as lack of available and accessible climate finance could affect mitigation measures and is discussed next.

12.3.4 FINANCIAL CHALLENGES

In addition to operational costs, businesses face additional costs attributed to climate change mitigation actions. These costs include the cost of adapting systems to cope with the impact of climate change such as the cost of replacing or improving critical infrastructure and retrofitting systems for energy efficiency (Monios and Wilmsmeier 2020). Businesses will likely be able to practice climate change mitigation when they have access to extra financing opportunities or high liquidity (Amran et al. 2016).

To solve problems of climate financing, several initiatives exist for providing funding for businesses to transition and reduce GHG emissions. These include government incentives and subsidies and carbon financing from international multilateral organisations that provides a window for financing climate change adaptation and mitigation initiatives. However, these are linked to conditionalities that businesses must meet (Warhurst 2005). There is a need to shift investments to zero and low-carbon energy sources from fossil fuels by financial institutions if we are to achieve the 2050 climate change mitigation agenda (WBCSD 2021). The response strategies for managing climate change challenges for international business are explored next.

12.4 RESPONSE STRATEGIES AND FRAMEWORK FOR MANAGING CLIMATE CHANGE CHALLENGES FOR INTERNATIONAL BUSINESSES

Sustainability proponents argue sustainability should be at the core of corporate strategy (Pérez Henríquez 2014) as businesses experience negative impacts from the high frequency and intensity of extreme climate events (Amran et al. 2016). Businesses have responded to climate change in diverse ways (Begg, Van der Woerd, and Levy 2018; Sullivan 2017), as they are perceived as part of the problem, thereby requiring action (Ihlen 2009). However, progress on climate mitigation actions has been slow compared to the urgency that climate change demands (Lye 2014), hence the growing agreement about the need to accelerate climate change mitigation and adaptation strategy implementation by businesses (Amran et al. 2016).

Certain instruments have been advocated to support mitigation plans, ranging from economic instruments (such as emission trading) to command and control (regulation) (Blowfield 2013; Stern 2006). For example, climate change mitigation activities offer opportunities for the private sector to develop carbon credits that are internationally transferable for use by other countries to meet their Nationally Determined Contributions (NDCs) or other environmental commitments (UNFCCC 2021a). Evidence exists that points to progress by large businesses in transitioning to a low-carbon economy. For example, AstraZeneca, a large international pharmaceutical company operating in Europe (UK and Sweden) and the US, has reduced its absolute greenhouse gas emissions by over 68% in the last decade (Doppelt and McDonough 2010). As of 2020, AstraZeneca had 99.9% of imported electricity from renewable sources globally and is set to eliminate its direct emissions from company operations and indirect energy emissions (scopes 1 and 2 GHG emissions, respectively) by 2025 (AstraZeneca 2020).

The recent Conference of the Parties (COP 26) held in Glasgow, United Kingdom, in partnership with Italy came up with resolutions to make the 2020s a period of climate action with enhanced support (UNFCCC 2021b). Parties resolved to work towards reducing CO_2 emissions to limit the rise in global average temperatures to 1.5°. This CO_2 emission reduction will be achieved through fulfilling the pledge by developed countries to provide $100 billion annually to developing countries, downscale the use of coal for power and cut down inefficient fossil fuel subsidies (UNFCCC 2021b). The private sector is incentivised through the UNFCCC mechanism and cooperative approaches to carry out mitigation actions in various sectors and using technologies in transport, energy efficiency and reforestation initiatives (UNFCCC 2021a).

About 450 top financial enterprises, such as banks, asset owners and managers, financial service providers and insurers, with over 130 trillion worth of assets from 45 countries have joined the Glasgow Financial Alliance for a Net Zero world to mobilise private capital for third world countries (UNFCCC 2021c). Also, by November 2021, about 5,235 businesses across 110 countries had joined the Race to Zero, a global campaign led by the High-Level Champions to mobilise action towards reducing greenhouse gas emissions (UNFCCC 2021c).

Transitioning to sustainability will require transformational changes to the overall business culture. Doppelt and McDonough (2010) advocated seven levers of change for sustainability in business that are circular and non-linear, and organisations can start from any of the levers:

a) Alter the thinking, beliefs and assumptions that led to the current way of work of the organisation.
b) Alter the way the organisation plans and makes decisions by getting diverse teams together.
c) Reorient the goals, visions and guiding principles of the organisation to achieve sustainability.
d) Restructure the strategies the organisation uses to reach its goals and missions by pointing the organisation in new directions.
e) Shift information flow of the organisation towards addressing the new vision, strategies and goals to emphasise the urgency to achieve sustainability.
f) Improve the organisation's learning capacity through practice, as innovation is required to steer organisations towards sustainability.
g) Embed the new goals, visions and strategies into standard company operating policies and procedures.

These leverage points need to be adequately addressed, no matter the starting point, to ensure smooth transitioning to sustainability, underpinned by the awareness of

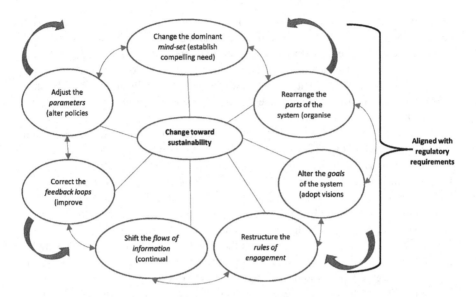

FIGURE 12.4 Leverage points for changing towards sustainability.

Adapted from Doppelt and McDonough (2010).

stakeholder expectations to change behaviour (Doppelt and McDonough 2010). An addition to these seven levers will be the alignment of the new vision, strategies and goals with regulatory requirements (Figure 12.4).

This alignment will ensure that the firm or business transition conforms to regulatory requirements and takes advantage of any incentives available. Successful transitioning is accompanied by numerous benefits and opportunities, including reduced operational costs, attraction of the best talents to work in a reputable company and attracting public attention to the company's emission footprints, thereby translating into increased brand value and competitiveness of the firm (Doppelt and McDonough 2010; Epstein and Buhovac 2017).

12.4.1 CLIMATE CHANGE GOVERNANCE

Climate change governance involves the response approaches and mechanisms for the prevention, adaptation and mitigation of the risks imposed by climate change (Jagers and Stripple 2003). Policies and regulatory frameworks to achieve these for international businesses are explored in the following sub-section.

12.4.1.1 Regulatory Frameworks and Policy Instruments for Climate Change Governance in International Business

Global challenges such as climate change are likely to be addressed through governance (Sullivan 2010; Van Tulder, Verbeke, and Strange 2014). Regulatory institutions that create standards and rules to shape the behaviours of business actors have been developed (Utting 2014). These institutions are concerned with setting environment, sustainability and governance (ESG) standards in line with social and environmental elements of business success and other aspects of 'good governance' that encompass accountability and transparency involving state and non-state actors (Utting 2014). According to the WBCSD, environmental and socioeconomic systems need to be transformed as the world faces three important challenges of climate emergency, biodiversity loss and high inequality that are interwoven, each of which can drastically impact the planet and business (WBCSD 2021). The institutions and systems operate at an international level due to the nature of climate change challenge (Levy and Kolk 2002) and help in reducing the greenhouse gas emissions of businesses (Sullivan 2010).

The introduction of voluntary standards and codes of conduct serves as an alternative to government regulations and reduces cost to taxpayers (Epstein and Buhovac 2017). For example, external certification schemes such as the Rain Forest Alliance, Fair Trade, GlobalG.A.P and several multistakeholder roundtables aim to enhance sustainable production or sourcing (Schouten, Vellema, and van Wijk 2014; Utting 2014). In some cases, government regulations when applied could be counterproductive if they do not support innovation by reducing competition (Epstein and Buhovac 2017). Ethical sourcing of raw materials for businesses by slowing deforestation will help reduce greenhouse gas emissions (Stern 2006).

International businesses face different country-specific government regulations in addition to the global-level regulatory governance arrangements for climate change mitigation and adaptation. These country-specific regulations are categorised into home and host country regulations, termed the home country effect and host country effect, respectively, and determine the strategies adopted by international businesses in response to climate change (Amran et al. 2016; Kolk and Pinkse 2004; Levy and Kolk 2002). This makes it difficult for a 'one-size-fits-all' approach to climate change mitigation to be effective for international businesses.

International oil companies formulate climate strategies based on the home country regulatory environment. For example, it is generally believed that European regulators and consumers are more environmentally conscious in their dealings compared to their American colleagues and will likely sacrifice economic for environmental benefits (Oliver 1991). Also, a company's culture and distinctive history influence its response to institutional pressures (Oliver 1991). For example, this can be seen in how European companies respond positively to the requirements of the Kyoto Protocol due to the EU's ratification of the protocol compared to US companies' attitude towards the protocol, as witnessed by the US government's initial refusal to ratify the protocol (Kolk and Pinkse 2004).

12.5   CONCLUSION

This chapter explored climate change challenges faced by international businesses and the response strategies employed to limit greenhouse gas emissions across the operations and supply chains of international businesses. Climate change affects businesses at different levels, and diverse approaches have been implemented by actors. These approaches are underpinned by regulations and policies at local and international levels, including the Kyoto Protocol (1997), the Paris Agreement (2015) and successor action plans to achieve a net-zero world.

Whilst response strategies have been advocated to promote the carbon transition through market forces (economic instruments) and policy changes, market forces will take longer to actualise; hence the need to emphasise more policies to accelerate the transition to a net-zero carbon economy (Roberts and Geels 2019). Businesses can respond to mitigate future emissions through technology change, improved efficiency or change in the source of energy/fuels used (Kirstein, Halim, and Merk 2018). Reliance on long-distance energy supply chains could be reduced to enhance decarbonisation of business sectors due to the sector's dependence on fossil fuels to power business transactions (Monios and Wilmsmeier 2020).

As countries are recovering from the impact of the Covid-19 through different economic stimuli, concerns have been raised. For example, in the 2021 Global Energy Review, the IEA highlighted the risks of driving CO_2 emissions up to new levels never attained due to a likelihood of a rebound of global economic output by about 6% in 2021 (IEA 2021). Covid-19 and the Ukraine-Russia war, instead of being setbacks, offer businesses an opportunity to incorporate sustainable approaches in their recovery plans to yield a 'win-win' outcome. Hence, the need for recovery policies to be sustainable and carbon emission-proof to avoid the rebound effect (IEA 2021).

Despite China's efforts to increase its renewable portfolio, the journey to net zero may take longer than expected (Doppelt and McDonough 2010) given its industrialisation drive, internationalisation strategy and positioning to become the world biggest economy and overtake the US (Peters et al. 2021). For example, under China's ambitious Belt and Road Initiative (BRI), it aims to build alternative secure trade routes and achieve economic interdependence with participating countries (Jie and Wallace 2021). While China's ambitions are legitimate, certain factors such as the pressure of its population on the state social security system, very high cost of production and low efficiency of factors of production compared to the US makes it difficult to outcompete the US (Beckley 2020).

As climate change is a global phenomenon and countries experience its impact differently, multinational companies must adapt business strategies to meet the realities of their host countries (Amran et al. 2016). Hence, in-depth understanding of the host country business environment and regulations (Collinson, Narula, and Rugman 2020), in line with wider global-level climate change governance frameworks and policies such as the Sustainable Development Goals (SDGs), is key to reducing risks and promoting a climate-friendly international business that is transformative. Realising the ambitions of the SDGs could unlock about US$12 trillion of economic opportunities annually by 2030 (WBCSD 2021).

REFERENCES

Adger, Neil W., and Sophie Nicholson-Cole. 2011. "Ethical dimensions of adapting to climate change-imposed risks." In *The Ethics of Global Climate Change*, edited by Denis G. Arnold, 255–271. Cambridge: Cambridge University Press.

Alaze, Anita Franziska, Saskia Karina Coomans, Persefoni Dimitsaki, Maud Alline Mol, and Matilda Smith-Cornwall. 2021. "Time for action towards a sustainable future: A policy brief for 'Green Supermarkets'." *South Eastern European Journal of Public Health (SEEJPH)*. https://doi.org/10.11576/seejph-4686.

Al-Saidi, Mohammad. 2021. "Cooperation or competition? State environmental relations and the SDGs agenda in the Gulf Cooperation Council (GCC) region." *Environmental Development* 37:100581.

Amran, Azlan, Say Keat Ooi, Cheng Yew Wong, and Fathyah Hashim. 2016. "Business strategy for climate change: An ASEAN perspective." *Corporate Social Responsibility and Environmental Management* 23 (4):213–227. https://doi.org/10.1002/csr.1371.

Andonova, Liliana B., Michele M. Betsill, and Harriet Bulkeley. 2009. "Transnational climate governance." *Global Environmental Politics* 9:52–73.

Arnold, Denis G., ed. 2011. *The Ethics of Global Climate Change*. Cambridge: Cambridge University Press.

AstraZeneca. 2020. *What Science Can Do: AstraZeneca Annual Report and form 20-F Information 2020*. Cambridge: AstraZeneca.

Azarkamand, Sahar, Alsnosy Balbaa, Christopher Wooldridge, and Rosa M. Darbra. 2020. "Climate Change—Challenges and Response Options for the Port Sector." *Sustainability* 12 (17). https://doi.org/10.3390/su12176941.

Beckley, Michael. 2020. "China's Economy Is Not Overtaking America's." *Journal of Applied Corporate Finance* 32 (2):10–23.

Begg, Kathryn, Frans Van der Woerd, and David Levy. 2018. *The Business of Climate Change: Corporate Responses to Kyoto*. Abingdon: Routledge.

Blowfield, M. 2013. *Business and Sustainability*. Oxford: Oxford University Press.

Bodansky, Daniel. 2021. "Climate change: Reversing the past and advancing the future." *American Journal of International Law* 115:80–85.

Bumpus, Adam G., Blas L. Pérez Henríquez, Chukwumerije Okereke, and James Tansey. 2014. "Carbon governance, climate change and business transformation." In *Carbon Governance, Climate Change and Business Transformation*, edited by Adam G. Bumpus, James Tansey, Blas L. Pérez Henríquez and Chukwumerije Okereke, 1–11. London: Routledge.

Collins, Jason, and Lionel Page. 2019. "The heritability of fertility makes world population stabilization unlikely in the foreseeable future." *Evolution and Human Behavior* 40 (1):105–111.

Collinson, Simon, Rajneesh Narula, and Alan Rugman. 2020. *International Business*, 8th ed. London: Pearson Education.

Diczfalusy, Bo. 2014. "What does a post-peak carbon economy look like?" In *Carbon Governance, Climate Change and Business Transformation*, edited by Adam G. Bumpus, James Tansey, Blas L. Pérez Henríquez and Chukwumerije Okereke, 12–26. London: Routledge.

Doppelt, B., and W. McDonough. 2010. *Leading Change Toward Sustainability: A Change-Management Guide for Business, Government and Civil Society*, 2nd ed. Sheffield: Routledge.

Eavis, Peter, and Clifford Krauss. 2021. "What's Really Behind Corporate Promises on Climate Change? Many Big Businesses Have Not Set Targets for Reducing Greenhouse Gas Emissions. Others Have Weak Goals." *The New York Times*, 22 February. Accessed 02/01/2022. www.nytimes.com/2021/02/22/business/energy-environment/corporations-climate-change.html.

Epstein, Marc J., and Adriana Rejc Buhovac. 2017. *Making Sustainability Work: Best Practices in Managing and Measuring Corporate Social, Environmental and Economic Impacts*. Oakland: Berrett-Koehler.

Forbord, Magnar, and Jostein Vik. 2017. "Food, farmers, and the future: Investigating prospects of increased food production within a national context." *Land Use Policy* 67:546–557. https://doi.org/10.1016/j.landusepol.2017.06.031.

Grubb, Michael, Christiaan Vrolijk, and Duncan Brack. 2018. *Routledge Revivals: Kyoto Protocol (1999): A Guide and Assessment*. London: Routledge.

Guan, Dabo, Jing Meng, David M. Reiner, Ning Zhang, Yuli Shan, Zhifu Mi, Shuai Shao, Zhu Liu, Qiang Zhang, and Steven J. Davis. 2018. "Structural decline in China's CO 2 emissions through transitions in industry and energy systems." *Nature Geoscience* 11 (8):551–555.

Hertwich, Edgar G., and Richard Wood. 2018. "The growing importance of scope 3 greenhouse gas emissions from industry." *Environmental Research Letters* 13 (10):104013.

IEA. 2021. "Global Energy Review 2021: Assessing the Effects of Economic Recoveries on Global Energy Demand and CO2 Emissions in 2021." Accessed 27/01/2024. https://www.iea.org/reports/global-energy-review-2021.

Ihlen, Øyvind. 2009. "Business and Climate Change: The Climate Response of the World's 30 Largest Corporations." *Environmental Communication* 3 (2):244–262. https://doi.org/10.1080/17524030902916632.

IPCC. 2007. "Summary for policymakers." In *Climate Change 2007: The Physical Science Basis. Contribution of Working Group I to the Fourth Assessment Report of the Intergovernmental Panel on Climate Change*, edited by S. Solomon, D. Qin, M. Manning, Z. Chen, M. Marquis, K. B. Averyt, M. Tignor and H. L. Miller. Cambridge: Intergovernment Panel on Climate Change (IPCC).

IPCC. 2021. "Climate change 2021: The physical science basis." In *Contribution of Working Group I to the Sixth Assessment Report of the Intergovernmental Panel on Climate Change*, edited by V. Masson-Delmotte, P. Zhai, A. Pirani, S.L. Connors, C. Péan, S. Berger, N. Caud, Y. Chen, L. Goldfarb, M.I. Gomis, M. Huang, K. Leitzell, E. Lonnoy, J.B.R. Matthews, T.K. Maycock, T. Waterfield, O. Yelekçi, R. Yu and B. Zhou. Cambridge: Cambridge University Press.

Jagers, Sverker C., and Johannes Stripple. 2003. "Climate Governance Beyond the State." *Global Governance: A Review of Multilateralism and International Organizations* 9 (3):385–399. https://doi.org/10.1163/19426720-00903009.

Jamieson, Dale. 2011. "Energy, ethics, and the transformation of nature." In *The Ethics of Global Climate Change*, edited by Denis G. Arnold, 16–37. Cambridge: Cambridge University Press.

Jie, Yu, and Jon Wallace. 2021. "What Is China's Belt and Road Initiative (BRI)?" *Chatham House*. Accessed 29/12/2021. www.chathamhouse.org/2021/09/what-chinas-belt-and-road-initiative-bri.

Kirstein, Lucie, Ronald Halim, and Olaf Merk. 2018. *Decarbonising Maritime Transport. Pathways to Zero-Carbon Shipping by 2035*. Paris, France: International Transportation Forum.

Kolk, Ans, and Jonatan Pinkse. 2004. "Market strategies for climate change." *European Management Journal* 22 (3):304–314.

Kunreuther, Howard, Geoffrey Heal, Myles Allen, Ottmar Edenhofer, Christopher B. Field, and Gary Yohe. 2013. "Risk management and climate change." *Nature Climate Change* 3 (5):447–450.

Levy, David L., and Ans Kolk. 2002. "Strategic responses to global climate change: Conflicting pressures on multinationals in the oil industry." *Business and Politics* 4 (3):275–300. https://doi.org/10.2202/1469-3569.1042.

Lye, Geoff. 2014. "Business as low carbon transformation driver." In *Carbon Governance, Climate Change and Business Transformatio*, edited by Adam G. Bumpus, James Tansey, Blas L. Pérez Henríquez and Chukwumerije Okereke, 55–79. London: Routledge.

Millennium Ecosystem Assessment, MEA. 2005. *Ecosystems and Human Well-Being*. Vol. 5. Washington, DC: Island Press.

Monios, Jason, and Gordon Wilmsmeier. 2020. "Deep adaptation to climate change in the maritime transport sector—a new paradigm for maritime economics?" *Maritime Policy & Management* 47 (7):853–872. https://doi.org/10.1080/03088839.2020.1752947.

Okereke, Chukwumerije, Bettina Wittneben, and Frances Bowen. 2011. "Climate change: Challenging business, transforming politics." *Business & Society* 51 (1):7–30. https://doi.org/10.1177/0007650311427659.

Oliver, Christine. 1991. "Strategic responses to institutional processes." *The Academy of Management Review* 16 (1):145–179. https://doi.org/10.2307/258610.

Pauw, Pieter, Clara Brandi, Carmen Richerzhagen, Steffen Bauer, and Hanna Schmole. 2014. *Different Perspectives on Differentiated Responsibilities: A State-of-the-Art Review of the Notion of Common but Differentiated Responsibilities in International Negotiations. Discussion Paper.* Bonn: Deutsches Institut für Entwicklungspolitik gGmbH.

Pérez Henríquez, Blas L. 2014. "The problem of climate change: Challenges and opportunities in carbon governance." In *Carbon Governance, Climate Change and Business Transformation*, edited by Adam G. Bumpus, James Tansey, Blas L. Pérez Henríquez and Chukwumerije Okereke, 27–54. London: Routledge.

Peters, Michael A., Alexander J. Means, David P. Ericson, Shivali Tukdeo, Joff P. N. Bradley, Liz Jackson, Guanglun Michael Mu, Timothy W. Luke, and Greg William Misiaszek. 2021. "The China-threat discourse, trade, and the future of Asia. A Symposium." *Educational Philosophy and Theory*:1–21. https://doi.org/10.1080/00131857.2021.1897573.

Roberts, Cameron, and Frank W. Geels. 2019. "Conditions for politically accelerated transitions: Historical institutionalism, the multi-level perspective, and two historical case studies in transport and agriculture." *Technological Forecasting and Social Change* 140:221–240.

Salite, Daniela, Joshua Kirshner, Matthew Cotton, Lorraine Howe, Boaventura Cuamba, João Feijó, and Amélia Zefanias Macome. 2021. "Electricity access in Mozambique: A critical policy analysis of investment, service reliability and social sustainability." *Energy Research & Social Science* 78:102123. https://doi.org/10.1016/j.erss.2021.102123.

Schouten, Greetje, Sietze Vellema, and Jeroen van Wijk. 2014. "Multinational enterprises and sustainability standards: Using a partnering-intensity continuum to classify their interactions." In *International Business and Sustainable Development*, edited by Rob Van Tulder, Alain Verbeke and Roger Strange, 117–139. Bingley, UK: Emerald Group Publishing Limited.

Silva-Chávez, Gustavo, Brian Schaap, and Jessica Breitfeller. 2015. "REDD+ finance flows 2009–2014." *Trends and Lessons Learned in REDD+ Countries.* Accessed 27/01/2024. https://www.forest-trends.org/wp-content/uploads/2015/11/doc_5029.pdf.

Stern, Nicholas. 2006. *The Economics of Climate Change: The Stern Review.* Cambridge: HM Treasury.

Sullivan, Rory. 2010. "An assessment of the climate change policies and performance of large European companies." *Climate Policy* 10 (1):38–50.

Sullivan, Rory. 2017. *Corporate Responses to Climate Change: Achieving Emissions Reductions Through Regulation, Self-Regulation and Economic Incentives.* Abingdon: Routledge.

UNFCCC. 2021a. "COP26 Outcomes: Market mechanisms and non-market approaches (Article 6)." *United Nations.* Accessed 27/12/2021. https://unfccc.int/process-and-meetings/the-paris-agreement/the-glasgow-climate-pact/cop26-outcomes-market-mechanisms-and-non-market-approaches-article-6#eq-2.

UNFCCC. 2021b. "The Glasgow climate pact—Key outcomes from COP26." *United Nations.* Accessed 27/12/2021. https://unfccc.int/process-and-meetings/the-paris-agreement/the-glasgow-climate-pact-key-outcomes-from-cop26.

UNFCCC. 2021c. "Momentum for global climate action continues to build." *United Nations Climate Change.* Accessed 28/12/2021. https://unfccc.int/news/momentum-for-global-climate-action-continues-to-build.

Urpelainen, Johannes, and Thijs Van de Graaf. 2018. "United States non-cooperation and the Paris agreement." *Climate Policy* 18 (7):839–851.

Utting, Peter. 2014. "Multistakeholder regulation of business: Assessing the pros and cons." In *International Business and Sustainable Development*, edited by Rob Van Tulder, Alain Verbeke and Roger Strange, 425–446. Bingley: Emerald Group Publishing Limited.

Van Tulder, Rob, Alain Verbeke, and Roger Strange. 2014. "Taking stock of complexity: In search of new pathways to sustainable development." In *International Business and Sustainable Development*, 1–22. Bingley, UK: Emerald Group Publishing Limited.

Warhurst, Alyson. 2005. "Future roles of business in society: The expanding boundaries of corporate responsibility and a compelling case for partnership." *Futures* 37 (2–3):151–168.

WBCSD. 2021. "Time to transform." In *How Business Can Lead the Transformations the World Needs*. Geneva: World Business Council for Sustainable Development (WBCSD).

Zhao, Stephen, and Alan Alexandroff. 2019. "Current and future struggles to eliminate coal." *Energy Policy* 129:511–520. https://doi.org/10.1016/j.enpol.2019.02.031.

13 International Climate Change Regimes in the 21st Century
From Stockholm to Paris

Emrah Atar, İlker Yasin Durmaz

13.1 INTRODUCTION

Political turmoil, civil wars, hunger, water wars, and international migration are issues that the 21st-century world struggles with. While not all of these problems are directly linked to global warming and climate change, the impact of global warming is more than one's anticipation of such issues. For instance, migration, traditionally associated with civil wars and economic factors, is now witnessing a rise in the number of individuals compelled to relocate due to environmental and climate-related factors. Despite the growing recognition of the significance of climate change and the imperative to cooperate globally, the resolution of this issue is hindered by political ideologies and national interests, preventing more decisive actions.

Moreover, considering the severity of the climate crisis and the possible damage it will cause to human beings and the world, the lack of intensity around the subject is worth attention for research. Nevertheless, climate change continues to focus on people at an increasing level in studies conducted in many countries; with growing awareness of the use of renewable energy and social mobility, young people have started to make their voices heard more.

In recent years, globalisation and environmental concerns have consistently emerged as focal points in world politics. In this perspective, it is evident that environmental issues have emerged as among the most crucial challenges faced by nearly every country. Collaborative efforts have been initiated to address these problems. However, the question arises: Do diverse groups of countries, with varying interests and priorities, undertake sufficient measures to effectively combat this pressing issue?

The answer to the question surrounding these issues is poised to persist on the global agenda for years to come. Examining these subjects within the context of national and international regulations makes it evident that they carry adverse economic, social, and political implications at a general level. Establishing crucial business partnerships becomes paramount at this juncture—this involves identifying issues, formulating and implementing effective policies, and adhering to them. In essence, it is widely believed that environmental challenges, now a shared global

DOI: 10.1201/9781032715438-13

concern, can only be effectively addressed through international cooperation. Consequently, numerous collaborative initiatives have been undertaken over time, particularly under the leadership of the United Nations.

Countries which embrace the fact that environmental problems are of a cross-border dimension have taken steps in the global arena to mitigate the circumstances of these problems. International mobility on ecological issues began in conjunction with the United Nations Environment Conference held in Stockholm in 1972. The process of climate change was formalised, especially with the first World Climate Conference held in 1979. It was then followed, respectively, by United Nations Framework Convention on Climate Change, Kyoto Protocol, Copenhagen Accord, Cancún Agreements, and Paris Agreement.

As can be seen, there have been many initiatives related to climate change and global warming in the international arena. The Framework Convention on Climate Change (1992), which is considered the first of the steps taken in the international arena, and the Kyoto Protocol (1997) signed within the Framework Convention, are agreements that envisage a 5.2% reduction in greenhouse gas emissions of developed countries compared to 1990. With the expiration of this Protocol in 2015, the Paris Agreement (2016) was signed by 197 countries to prevent the climate crisis. On the other hand, the United Nations 2030 Agenda for Sustainable Development can be considered a universal development pointing to a paradigm shift and foreseeing global participation in climate change.

This chapter provides a comprehensive analysis of climate change, examining the evolution of international climate change regimes from Stockholm to Paris in the 21st century. It delves into the current debates, outlining the strengths, weaknesses, opportunities, and threats (SWOT) associated with the issue. The discussion will then construct a framework detailing how the global community currently addresses the climate change dilemma and how it should ideally proceed.

13.2   GLOBAL ENVIRONMENTAL ISSUES IN BRIEF

It can be said that the history of humanity is late to fully understand the effects of environmental problems on the atmosphere, the world, and humanity. The issue of environmental problems, which has increased in importance in the last 50 years, has managed to attract the attention of many segments of our world. This situation brings with it the question of whether it is too late to solve these problems. Scientists, who focused on regional problems in the first stages, had the opportunity to realise, albeit late, that the problem was global due to studies carried out later. Over time, it should not be ignored that these problems damage not only the material elements of people but also moral values. In this direction, it will be helpful at this stage to mention what these problems are that have international effects. It is seen that environmental problems, which were previously seen as water and air pollution and mostly encountered in industrial areas, range from toxic waste to the depletion of the ozone layer; the destruction of biological richness in nature, that is, the extinction of some living species forever; climate change; and pollution of the sea and oceans (El-Kholy 2012; Hasnat, Kabir, and Hossain 2018). Since it is impossible to address all environmental problems at this stage, the chapter only draws attention to some general issues. In this part, the research will briefly touch on the main problems of population, pollution, nuclear issues, global warming/climate change, natural hazards, and loss of biodiversity, among the most critical global environmental problems.

Population: The world population continues to increase day by day. Even environmental problems, epidemics, wars, and migration cannot help stop this rapid rise. The United Nations (UN) closely monitors the world population and makes use of this foresight in future studies. Accordingly, the UN estimates that the world population will be around 11–12 billion by 2050 (Fricke et al. 2018). At this point, not only should the increase in the number of people be considered but also the rapid increase in the number of people living in cities. By 2050, faster growth in urban areas is predicted, with a rapid expansion in the number of megacities with populations of 10 to 20 million or more (Cohen 2003; Wenzel, Bendimerad, and Sinha 2007). Population pressures of this magnitude will adversely affect natural and institutional resources. Rapid population growth not only puts pressure on cities by increasing the rate of urbanisation but also causes rapid consumption of human resources provided by nature and more frequent hunger. Increasing demands for resources, including water, electricity, oil, timber, hydrocarbons and food, will require anticipating and resolving conflicts over competing users and preventing harm to people and the environment; also, high population growth and urbanisation result in increasing energy demand, economic expansion, and waste generation (Flower 2006).

Pollution: The deterioration of the ecological balance, which has arisen with increased environmental problems, has revealed an unhealthy and poor quality of life. Increasing ecological issues could cause these problems to turn into crises. Therefore, there are long-term, complex, and high-cost situations in ecological issues. Regarding pollution, air pollution is one of the most critical environmental problems. With the increase in the world population and developing technology, the start of the Industrial Revolution in the 18th century caused air quality to decrease day by day. The factories and types of equipment established with this historical event brought along many environmental problems. Although there are many air pollution causes, they can be examined in two groups: naturally occurring air pollution and human-induced air pollution (Theodore 2008; Shahadin et al. 2018). Carbon dioxide, nitrogen oxide and dioxide, and chlorofluorocarbons are pollutants caused by manufactured air pollution. In addition, different subjects such as water pollution, sound pollution, and waste pollution can be examined.

Nuclear issue: Nuclear problems can be considered a continuation of the pollution theme, or they can be separated. Broader issues of nuclear energy are emerging through the proliferation and growth of nuclear weapons, especially sustainability. The environmental effects of nuclear energy should be taken into consideration when utilising it. Environmental damage caused by nuclear energy can be evaluated in the category of transboundary damage. Activities that can cause damage by nuclear pollution are nuclear tests, nuclear power plant accidents, leaks, and radioactive waste during the transportation of dangerous materials on nuclear ships because these activities can lead to destructive results that are out of human control and irreversible (Özkan 2016).

Global warming: Global warming or climate change can be considered the long-term change of temperature and typical weather patterns in a place. Climate change is having an impact on particular places or the planet as a whole. The excessive increase in population density in the world, intense migration and urbanisation movements, and the increase in living standards can be considered the primary reasons. However, due to the excessive consumption of fossil fuels such as oil, coal, and natural gas and changes in land cover, the release of large amounts of harmful gases and particles into the atmosphere is the most critical factor causing global warming (Özmen 2009). The climatic changes that occur with global warming may cause weather forecasts to be more difficult, but they show that the expected temperature and precipitation levels can no longer be trusted, unexpected weather conditions will occur, and the atmosphere will enter a warmer environment. While the natural disasters observed as a result of the phenomenon of global warming make the living conditions of plants and animals, especially human life, complex, it is estimated that these sudden, extreme, and drastic changes in the world climate system will not leave land to be cultivated or liveable earth in 30–40 years (Akin 2017).

Natural disaster: The current century is perhaps the century in which natural disasters are experienced the most. Countries have paid a great price in the face of these unexpected disasters, especially in areas where the population is dense, which causes the size of disasters to be more significant. More than a million people were adversely affected by disasters in 2021 (McEntire 2021). The harms of such disasters that can be experienced both for our environment and humanity can be briefly summarised as follows (Pearce 2003): It is seen that they disrupt the economy and growth targets throughout the country, creating adverse effects on income distribution and increasing poverty even more. Sustainable development is greatly affected by the suspension of planned investments and the cutting of resources allocated to investments, leading to unemployment, deterioration of social balance, and sudden and uncontrolled population movements. However, ignoring the environmental problems that arise and not carrying out renovation work may cause more environmental damage. For this reason, it is necessary to carry out studies to reduce disaster risk in the management of crises caused by natural disasters.

Loss of biodiversity: Biodiversity loss continues its impact. Nature is home to millions of species and helps these species to live concerning with other and the environment. The ongoing destruction of natural habitats is resulting in the daily loss of numerous plant and animal species. Habitat destruction is primarily due to mining, agricultural use, forest monoculture, or land conversion for settlements (Flower 2006). At this point, essential steps are being taken. Particularly shortly after the 1992 Earth Summit in Rio de Janeiro, interest in understanding how biodiversity loss can affect the dynamics and functioning of ecosystems and the supply of goods and services has increased dramatically. Major international research initiatives were established, hundreds of experiments have been conducted in ecosystems all over the world, and new ecological theories have been developed and tested against experimental results (Cardinale et al. 2012).

Unfortunately, the list of damage caused by the environmental problems mentioned previously would be very long. However, at this point, some of them can be summarised as follows so that the cause-effect relationship might be understood better: nuclear accidents, stratospheric ozone degradation, increase in UV radiation, genetic resources and losses in biodiversity, decreased quality of springs and groundwater, increase in transportation and storage of hazardous wastes, climate change, deforestation, nuclear waste, decreased urban air quality, conservation of natural resources and damage to sensitive ecosystems, increasing industrial accidents, destruction of the seas as a result of pollution by direct draining or discharge, difficulties in protecting threatened species, desertification, drinking water supply and its difficulties, hygiene and quality safety in food, and soil erosion (Baykal and Baykal 2014).

13.3 CHALLENGES TO GLOBAL WARMING AND CLIMATE CHANGE

As mentioned before, the issue of global warming or climate change is at the top of the world agenda and will have terrible consequences for every part of the earth. Rising temperatures are fuelling environmental degradation, natural disasters, extreme weather, food and water insecurity, economic deterioration, and conflict and terrorism; sea levels are rising; the Arctic is melting; coral reefs are dying; oceans are acidifying; and forests are burning (United Nations 2021). Although the world is taking steps to deal with this problem, these steps are either insufficient or do not reach implementation. While a more detailed discussion of actions taken in the battle against the problem will follow in the next section, it's crucial to emphasise the significance of international efforts at this point. Effectively addressing the challenges in the global fight against warming necessitates dedicated efforts from all stakeholders, including scientists, politicians, and economists. Regrettably, the necessary leadership often lags behind, and studies at the desired level in the international arena are hindered either due to a lack of initiative or the influence of vested interests (Howe 2014).

Despite the evidence, climate change is the most complex and crucial political issue our society has ever encountered, and this is not to say that there has not been an improvement. Although our economy and people have grown, greenhouse gas emissions in the United States have stayed steady since 1990 (Hockstad and Hanel 2018). However, global greenhouse gas levels have risen since then, bringing humankind dangerously near the expected global warming levels (Olivier and Peters 2017). All these demonstrate that some steps have been taken, even if they are insufficient.

The dynamic integrated model of climate and economics (Nordhaus 1994; Heris and Rahnamayan 2020) put forward by scientists is just one of the systems for this effort to get results. It is feasible to investigate emissions, concentrations, climate change, damage, and emission controls using this method. The model can be used to calculate the costs and benefits of various techniques for slowing climate change and analyse the impact of control strategies over time. Nordhaus also looks at several possible approaches to climate change policy, including no control, economic optimisation, geoengineering, emissions and climate stabilisation, and a ten-year wait in implementing climate change legislation (1994). As can be seen, the determination, processing, implementation, and results of climate change policies are a long-term process.

To determine and analyse the problems encountered at the point of climate change, it is also necessary to determine whether co-operation can be achieved in the current situation. For this reason, it is also required to conduct a SWOT analysis, especially in the field of international governance.

The issue of global warming or climate change has had a very high acceptance and impact on society, especially in recent years. It indicates that positive steps can be taken, and the support of the public can be obtained at the point of solving this problem, which has increased awareness around the world. The possibility of burden-sharing at the local and global level and with different stakeholders also provides an affirmative effect. However, the lack of sufficient data at the point of finding a solution to the problem, the complexity of the problem, and the increase in logistical difficulties are difficulties that may be encountered. On the other hand, the gradual increase in the possibilities of building infrastructure to mitigate climate change and the improvement of capacity management in this direction increase the opportunities for the solution of the problem on an international scale. In contrast, although the urgency of the situation reveals that steps should be taken soon, targeting climate activists poses a threat to public opinion. Targeting activists who have an impact on society can be seen as a disadvantage in keeping the issue on the agenda.

TABLE 13.1
Initiatives on Climate Change in the International Arena

United Nations Environment Programme	After the Stockholm Conference in 1972, the United Nations Environment Programme (UNEP) was established within the United Nations General Assembly (UNGA) to assess environmental issues on an international scale.
World Climate Conference (WCC)	The World Climate Conference was organised in 1979 by the World Meteorological Organization (WMO) in Geneva, Switzerland. The conference aims to evaluate the climate change regime in light of scientific data and to present a projection in this context.
United Nations Framework Convention on Climate Change	The UN created an Intergovernmental Negotiating Committee (INC) to assess these concerns at an international scale. The United Nations Framework Convention on Climate Change (UNFCCC) was drafted within the INC and was opened for signature in Rio De Janeiro in 1992.
Kyoto Protocol	The Kyoto Protocol, the first legally binding agreement, was negotiated at a Third Conference of Parties (COP3) in 1997. After the ratification process, it entered into force in 2005.
Copenhagen Accord	The 15th Conference of Parties (COP15) was convened in Copenhagen to arrive at an international agreement on climate issues.
Cancún Agreements	The 16th Conference of Parties (COP 16) was held in Cancun, Mexico. Following the non-binding Copenhagen Accord, the Cancun agreements have presented comprehensive environmental governance, including the Green Climate Fund and technology mechanisms.
Paris Agreement	The Paris Agreement was held at the 21st Conference of Parties (COP21) to UNFCCC in Paris. The legally binding agreement, in which 195 countries are parties, sets forth a new international climate regime to combat climate change.

13.4 MAJOR INTERNATIONAL CLIMATE CHANGE ACTIONS, NEGOTIATIONS, AND AGREEMENTS

In the previous sections, attention was drawn to the environmental problems that negatively affect the ecosystem and the most important one among them, climate change. In this part of the chapter, treaties that have emerged as a result of necessary international joint studies handled in line with the principle of combating climate change will be discussed. In this direction, the United Nations Environment Program titles, World Climate Conference, United Nations Framework Convention on Climate Change, Kyoto Protocol, Copenhagen Accord, Cancún Agreements, and Paris Agreement are examined. The first conferences covered in the study in this chapter try to build the work scientifically. After the framework, the agreements are discussed with legally binding organisations. However, the Paris Agreement changes the responsibilities they impose on countries and foresees a voluntary struggle.

13.4.1 UNITED NATIONS ENVIRONMENT PROGRAMME

The climate crisis and environmental problems go beyond the boundaries of the sovereign territories of the states. Due to this structure, which directly affects the 'global commons', environmental initiatives accommodate the 'global environment governance approach' (Hierlmeier 2001). As of the 1970s, due to the drought extending from Africa to Asia, world public opinion put the climate issue on the agenda within this approach. As a result of the increasing warming problem and the shortage of food supplies, the United Nations set an initiative. After the Stockholm Conference in 1972, United Nations Environment Programme was established within the United Nations General Assembly to assess environmental issues on an international scale. This programme, which aims to comprehensively evaluate environmental governance, depletion of the ozone layer, climate change, and marine issues within the UN, is becoming important in terms of addressing environmental issues on an international scale for the first time. The report published by UNEP in 1972 contains the objectives of this international environmental governance approach as follows:

(a) To promote international cooperation in the field of the environment and to recommend, as appropriate, policies to this end;

(b) To provide general policy guidance for the direction and coordination of environmental programmes within the United Nations system;

(c) To receive and review the periodic reports of the Executive Director of UNEP on the implementation of environmental programmes within the United Nations system;

(d) To keep under review the world environmental situation in order to ensure that emerging environmental problems of wide international significance receive appropriate and adequate consideration by Governments;

(e) To promote the contribution of the relevant international scientific and other professional communities to the acquisition, assessment and exchange of environmental knowledge and information and, as appropriate, to the technical aspects of the formulation and implementation of environmental programmes within the United Nations system;

(f) To maintain under continuing review the impact of national and international environmental policies and measures on developing countries, as well as the problem of additional costs that may be incurred by developing countries in the implementation of environmental programmes and projects, and to ensure that such programmes and projects shall be compatible with the development plans and priorities of those countries;

(g) To review and approve the programme of utilization of resources of the Environment Fund.

(UNGA 1972)

In the context of this responsibility, the UNEP program was initially established with the membership of 58 states (Petsonk 1989). However, on 15 March 2013, by a resolution of the UN General Assembly, the Governing Council in UNEP was expanded to include all UN member states (UNGA 2013). This decision contributed to a more effective solution to the environmental problems encountered and strengthened international environmental governance. Since its establishment, UNEP has contributed to the solution of many environmental problems and has submitted a secretariat to the conventions on problems. The Vienna Convention, Montreal Protocol, Convention on International Trade in Endangered Species of Wild Fauna and Flora, and Basel Convention are a few of the basic agreements to which UNEP has contributed. Besides solutions to environmental problems, reports on the balance of development provided to developing countries also increase the importance of UNEP. Despite the framework offered by the aforementioned agreements, as Hierlmeier emphasised, UNEP has a limited capacity for solving environmental issues. Problems such as lack of authority, legitimacy issues, and the problem of financial capacity are the Achilles heel of UNEP (Hierlmeier 2001). Over and above that, the growing trade networks of international companies ignore global commons and limit UNEP's capacity for the measure. Although UNEP has a limitation of capacity, it provides a substantial ground in solving global environmental problems and played a central role in International Environment Law. Many agreements concluded under the Secretariat of UNEP have expanded the legal and social framework in ecological issues. However, the emerging themes such as lack of co-operation on environmental matters and budgetary problems prevent UNEP from establishing international environmental law. In addition, the non-binding character of its decisions also makes UNEP's global position debatable.

13.4.2 WORLD CLIMATE CONFERENCE

The World Climate Conference was organised in 1979 by the World Meteorological Organization in Geneva, Switzerland. The conference aims to evaluate the climate change regime in light of scientific data and to present a projection in this context. The organising committee consisted of scientists and representatives of international bodies from 50 different countries. According to a report by WMO, the conference has two main objectives. The first is "to review knowledge of climatic change and variability, due both to natural and anthropogenic causes", and the second is "to assess possible future climatic changes and variability and their implications for human activities"

(Kellogg 1979). In this context, the conference's primary purpose, which does not make a given policy proposal, is to evaluate the human activities that affect climate change (Agrawala 1998). The conference looks for the causes of the climate crisis in humanitarian activities. Using various modelling methods, scientists from distinctive disciplines have addressed the effects of increasing economic activities and globalisation on the climate crisis. They theorised different modelling to evaluate the phenomenon of climate change from water sources to land use. At the end of the two-week process, scientists aimed to reach a consensus on climate change.

Consequently, "WCC is a foundational period, during which scientific concern about global warming developed", as Daniel Bodansky classified it (2001). Despite this situation, the conference has been one of the cornerstones in the climate crisis discussions held in the following process. In addition, it has provided a solid basis for conferences that evaluate climate change in light of scientific data, such as the Villach Conference and the Toronto Conference.

13.4.3 United Nations Framework Convention on Climate Change

In the 1980s, increasing scientific data about human activities that caused climate change led to concern in the public agenda. Considering these scientific data, international organisations carried out a series of conferences to solve the problem. The UN created an Intergovernmental Negotiating Committee to assess these concerns at an international scale. The United Nations Framework Convention on Climate Change was drafted within the INC and was opened for signature in Rio De Janeiro in 1992. According to the report by the UN, the purpose of this Convention is the "stabilisation of greenhouse gas concentrations in the atmosphere at a level that would prevent dangerous anthropogenic interference with climate change" (United Nations 1992). The Convention emphasises effective international co-operation to reduce these anthropogenic impacts. In the context of this goal, 197 countries have become parties to the Convention. A decoupling of developing and developed countries has been made between the participating countries. With this decision, the UN has determined the principles of local and regional struggle in the climate crisis. The Convention imposes obligations on all parties to reduce greenhouse gases and provide financial and financial support for developing countries (United Nations 1992).

Despite the search for a scientific consensus on climate change in UNFCC, the definition of "dangerous anthropogenic" in Article 2 has caused controversy. As Bodansky (1993) shows, the Convention's language is far from indicating a scientific consensus. Similarly, Oppenheimer and Petsonk (2005), who consider the agreement's language in a historical and political context, argue that the definition of dangerous anthropogenic interference is problematic. One of the reasons underlying the problematisation of the definition of "dangerous" is the contextual nature of the word since the "dangerous" threshold in the humanitarian activities of the developed and developing countries that are parties to the agreement is quite different. By 1995, a Conference of the Parties was held to discuss contractual disputes with the parties. As a result of the COP held at regular intervals, an attempt was made to resolve the language ambiguities and the discussions on the categorisation of the countries. In its first decision, the Berlin Mandate, parties accept that industrialised countries should take precautions to prevent greenhouse gas emissions and developing countries act at a later stage (Kuyper, Schroeder, and Linnér 2018).

In conclusion, UNFCCC has provided a solid basis for global policymaking. The observer organisations that are parties to the Convention have contributed to this worldwide policymaking process. The critical point of the agreement is that the parties include states and sub-national organisations and non-state actors (Hickmann et al. 2021). This comprehensive approach to climate change can be evaluated within the framework of the "global environment governance" approach, which has been repeated since the establishment of UNEP.

13.4.4 KYOTO PROTOCOL

The Kyoto Protocol, the first legally binding agreement, was negotiated at the Third Conference of Parties in 1997. After the ratification process, it entered into force in 2005. According to the ultimate objective in UNFCCC Article 2, the Protocol aims to stabilise atmospheric concentrations of greenhouse gas emissions. Following goals defined by UNFCCC, the Protocol provides binding obligations for parties to the Convention, considers in detail how these obligations will be fulfilled, and includes provisions on the supervision of obligations imposed on parties are the most critical aspects of the Kyoto Protocol. Additionally, the Protocol organises a flexible mechanism for implementing the reduction obligations.

The most crucial obligation stipulated by the Kyoto Protocol is in the third article of the Protocol. Accordingly,

> the parties included in Annex I shall, individually or jointly, ensure that their aggregate anthropogenic carbon dioxide equivalent emissions of the greenhouse gases listed in Annex A do not exceed their assigned amounts, calculated according to their quantified emission limitation and reduction commitments inscribed in Annex B and following the provisions of this Article, to reduce emissions of such gases by at least 5 per cent below 1990 levels in the commitment period 2008 to 2012.
>
> (United Nations 1997)

In addition to the obligations on greenhouse gas reduction, the Protocol introduces specific policies and precautionary responsibilities to parties to the Convention. These policy proposals, which range from increased energy efficiency in related sectors of the national economy to limiting or reducing methane through waste recovery and use, guide climate change.

Despite the solid legal ground it has brought about on climate change, the Kyoto Protocol has also led to several controversies since its opening for signature in 1997. In the early 2000s, the USA raised objections to the agreement. According to President Bush, the different obligations imposed by the agreement on the countries were unacceptable. At the same time, according to the president, who noted that the cost of the agreement to the countries economically would be very high, the reduction of the mandatory order envisaged by the agreement was entirely improper. Moreover, studies that consider the political economy of Kyoto make similar points. According to these arguments, Kyoto is an essential cornerstone for climate change and contains strong question marks due to its economic cost (Barrett 1998).

13.4.5 Copenhagen Accord

After the Kyoto period (2008–2012), the 15th Conference of Parties was convened in Copenhagen to arrive at an international agreement on climate issues. The Copenhagen Accord, which was attended by representatives of developing and developed countries, emphasised "dangerous anthropogenic interference with climate change" mentioned in Article 2 of UNFCCC. Besides the underlying continuity of the Kyoto Protocol, the Copenhagen Accord recalls the mitigation and adaptation process to prevent climate change. According to the report by UNFCCC, the Accord "underline[s] that climate change is one of the greatest challenges of our time" and "emphasises our strong political will to urgently combat climate change by the principle of common but differentiated responsibilities and respective capabilities" (UN Copenhagen Accord 2009). In the Copenhagen Accord, actors also agreed "that deep cuts in global emissions are required according to science to reduce the global temperature below 2° C and take action to meet this objective consistent with science and based on equity" (UN Copenhagen Accord 2009). Another significant point in the agreement is that it provides financial support from developed countries to reduce emissions from deforestation and forest degradation (REDD). According to Hunter (2010), "REED and technology transfer may subsequently be viewed as critical building blocks in an effective, comprehensive climate regime". Despite it emphasising the importance of global environmental governance and providing technology transfer, the Accord was criticised by legal scholars. One of the criticisms was that the agreement does not have legal binding. As a result of that, scholars believe it cannot contribute to a legal framework for climate issues. The second criticism is that the agreement can be seen as a "soft law" (Ottinger 2009). Also, the Accord is insufficient to keep global warming below 2°C. According to a study, "current national emissions targets can't limit global warming to degrees 2 Celsius; they might even lock the world into exceeding degrees 3 Celsius" (Rogelj et al. 2010).

13.4.6 Cancún Agreements

The 16th Conference of Parties took place in Cancun, Mexico, following the nonbinding Copenhagen Accord. The Cancun agreements introduced a comprehensive environmental governance framework, incorporating initiatives like the Green Climate Fund and technology mechanisms. In alignment with previous agreements, it unequivocally acknowledges that climate change stands as one of the most formidable challenges faced globally. Parties to Cancun Agreements also re-emphasise the end that the global average temperature is below 2° C. Along with the emphasis on previous agreements, the Cancun agreement provides several new institutional arrangements. According to the report, parties to agreements "[decide] to establish a Green Climate Fund, to be designated as an operating entity of the financial mechanism of Convention under the Article 11 . . . to support projects, programmes, policies and other activities in a developing country" (United Nations 2011).

Additionally, parties to the agreements will establish technology mechanisms to facilitate technology development and transfer. A technology mechanism consists of the Technology Executive Committee and Climate Technology Centre and Network. The mechanism aims to enhance climate issue management for developing countries. Besides financial and institutional arrangements, the agreements created a system that includes measurement, reporting, and verification (MRV). This system aims to increase the transparency of mitigation actions and their effects. This comprehensive agreement also includes the Cancun Adaptation Framework to support developing countries. The Adaptation Framework includes "planning, prioritising and implementing adaptation actions" to "improving climate-related research and systematic observation for climate data collection, archiving, analysis and modelling to provide decision-makers at the national and regional levels" (United Nations 2011). As a result, the Cancun agreement provides decision-makers with comprehensive institutional and financial arrangements. Moreover, it shapes environmental governance bodies by creating the Green Climate Fund and technology mechanisms.

13.4.7 Paris Agreement

The Paris Agreement was held at the 21st Conference of Parties to UNFCCC in Paris. The legally binding agreement, to which 195 countries are parties, sets forth a new international climate regime to combat climate change. Following reaching a consensus on the mitigation and adaptation of climate change, many people called the Paris Agreement a historic success and promising. Ban Ki Moon, former general secretary of the UN, claims that the agreement is "a monumental triumph for people and our planet" (Falkner 2016). According to the report, it has three objectives. First is "holding the increase in global average temperature to well below 2° C above pre-industrial levels". The second is "increasing the ability to adapt to the adverse impacts of climate change". The third is "making finance flows consistent with a pathway towards low GHG emissions & climate-resilient development" (United Nations 2015).

Unlike its predecessors, the Paris Agreement "acknowledges the primacy of domestic politics in climate change and allows countries to set their level of ambition for change mitigation" (Falkner 2016). This marks a historical turn since post-Kyoto negotiations. The Paris Agreement, which highlights the principle of volunteerism taking into account countries' specific situations, heralds a new era of global environment approach. Under the "new global environment approach", it also funds $100 billion per year in financial support from 2020 through 2025 to combat greenhouse gas emissions and other climate issues in developing countries (Horowitz 2016). Furthermore, another critical issue regarding the agreement is its legally binding character. Bodansky (1993) argued that every provision in the agreement could not be considered a legally binding character. For him, the nature of the agreement, which has both mandatory and non-mandatory provisions, is one of the most important outcomes of the Paris Agreement.

Despite the agreement being a milestone in combatting climate change, it risks managing climate change issues. As mentioned, agreements differentiate responsibilities to challenge climate change between countries in the context of countries' privacy. This differentiation approach could cause problems in the mitigation and collective action process (Streck, Keenlyside, and Von Unger 2016). Consequently, the Paris Agreement is a product of negotiation, which has been an ongoing process since 1972 in Stockholm. Many countries have become parties to the agreement and have confirmed that climate change is one of humanity's most important problems.

It also funds financial support to challenge climate change and considers countries' privacy. This differentiation principle could provide more effective solutions to climate change and disentangle distributional conflicts.

13.5 INTERNATIONAL REGIMENS

The increasing awareness of climate issues has coincided with the formulation of international legal agreements. As demonstrated, all these agreements acknowledge and underscore the significance of "anthropogenic causes" in the context of climate issues and climate change. Although human beings drastically affect the environmental cycle, the solution is still in our hands. Along with the broadening literature and data, mitigation and adaptation processes have been accelerated, and thus all initiatives from UNEP to Paris Agreement put a different agenda on climate issues and policies. Following seeking consensus on climate change and its effects (the main concerns in WCC), international organisations such as the UN and other non-governmental organisations set the framework for policy-making processes (UNFCCC). According to this framework, countries that are party to UNFCCC should consider global environmental problems and re-organise their local policies. Besides the normative core of this agreement, the UNFCCC has established a mechanism which allows interaction and co-operation among parties. This organisational structure entails authoritative agreements such as the Kyoto Protocol, which is evaluated as the first legally binding agreement on climate change. As emphasised, the Kyoto Protocol is a cornerstone in terms of combating climate change and reduction of greenhouse gas emissions. Although the Protocol marks important thresholds in the climate change regime, it has been criticised regarding its economic cost, and also differentiated responsibility for combatting climate change remains questionable.

After the Kyoto period, the Copenhagen Accord was convened by developing and developed countries. In line with former agreements, the Accord emphasises the urgent need to reduce the global temperature by 2°C. Moreover, the Accord has established regulatory and supervisory mechanisms to share the burden of combatting climate change between developing and developed countries. The Accord, which has no legally binding framework, has also been criticised as a "soft law" despite all this effort. The Cancún Agreements, which were held in COP16, take concrete steps to combat climate change such as the Green Climate Fund, which is anticipated to fund developing countries' local policies regarding climate change action. After the Cancun Agreements, the Paris Agreement was held in COP21. It is accepted that Paris Agreement is a historic triumph in terms of climate change policies adopted by countries. According to the Agreement, all countries should share the burden of challenging climate change, and they should differentiate their responsibilities on climate change. Moreover, the Agreement has a legally binding character that allows the effective implementation of climate change policy.

Despite agreements having different contexts, they have highlighted "anthropogenic causes" and "economic cost". These two themes are the major issues in combatting climate change. Besides these two, the international governance approach matters as well. Agreements envisage governments and non-governmental organisations as actors in combatting climate change. However, the role of the non-governmental organisation depends on the relationship between civil society and the state. In the third world, countries could encounter problems organising non-governmental organisations. Another questionable point of these agreements is the differentiated responsibilities in combatting climate change. These agreements generally interpret these responsibilities within the economic structure. However, these responsibilities should take into consideration cultural and religious practices. To this end, agreements and international organisations have to consider the cultural background which determines the relationship between humans and nature.

13.6 CONCLUSION

Although environmental issues gained universal significance relatively recently, they have garnered the attention of stakeholders since the 1960s. The efforts of countries in the global struggle against these issues and the impact on their populations vary. It can be observed that the poorest countries are most susceptible to environmental changes. Furthermore, addressing ecological problems should not only account for the present but also consider the well-being of future generations, necessitating responsible action. From this standpoint, it is anticipated that the country groups referred to as the Global South would take the lead in policymaking. Unfortunately, as previously mentioned, these initiatives remain insufficient due to governance shortcomings.

Global warming and climate change stand as urgent environmental challenges demanding immediate attention and solutions. Adhering to appropriate policies in addressing these issues will enhance the possibility of our generation and those to come leading healthy, environmentally friendly lives, while also preserving the rich diversity of life. Recent summit meetings consistently prioritise environmental concerns, with global press and media organisations prominently featuring efforts in the fight against climate change on their front pages. This chapter delves into the United Nations' (UN) role as the leading institution in addressing these challenges. While the UN initially grappled with defining the problem in its early years, it later sought to enhance collaboration among country groups and encouraged the implementation of concrete steps to tackle the issue. Throughout, there is a persistent call for prompt resolution of the situation.

REFERENCES

Agrawala, Shardul. 1998. 'Context and Early Origins of the Intergovernmental Panel on Climate Change'. *Climatic Change* 39 (4): 605–620. https://doi.org/10.1023/A:1005315532386.

Akin, Galip. 2017. 'Küresel Isınma, Nedenleri ve Sonuçları'. *Ankara Üniversitesi Dil ve Tarih-Coğrafya Fakültesi Dergisi* 46 (2). www.dtcfdergisi.ankara.edu.tr/index.php/dtcf/article/view/1450.

Barrett, Scott. 1998. 'Political Economy of the Kyoto Protocol'. *Oxford Review of Economic Policy* 14 (4): 20–39.

Baykal, Hülya, and Tan Baykal. 2014. 'Küreselleşen Dünya'da Çevre Sorunları/ Environmental Problems in A Globalized World'. *Mustafa Kemal Üniversitesi Sosyal Bilimler Enstitüsü Dergisi* 5 (9).

Bodansky, Daniel. 1993. 'The United Nations Framework Convention on Climate Change: A Commentary'. *The Yale Journal of International Law* 18: 451.

Bodansky, Daniel. 2001. 'The History of the Global Climate Change Regime'. In *International Relations and Global Climate Change*, edited by Urs Luterbacher and Detlef F. Sprinz, 23:23–40. Cambridge, MA: The MIT Press.

Cardinale, Bradley J., J. Emmett Duffy, Andrew Gonzalez, David U. Hooper, Charles Perrings, Patrick Venail, Anita Narwani, et al. 2012. 'Biodiversity Loss and Its Impact on Humanity'. *Nature* 486 (7401): 59–67. https://doi.org/10.1038/nature11148.

Cohen, Joel E. 2003. 'Human Population: The Next Half Century'. *Science*, November. https://doi.org/10.1126/science.1088665.

El-Kholy, Osama. 2012. *The World Environment 1972–1992: Two Decades of Challenge*. Oxford: Springer Science & Business Media.

Falkner, Robert. 2016. 'The Paris Agreement and the New Logic of International Climate Politics'. *International Affairs* 92 (5): 1107–1125.

Flower, Little. 2006. 'Environmental Challenges in the 21st Century'. *AU JT* 9 (4): 248–252.

Fricke, Timothy R., Nina Tahhan, Serge Resnikoff, Eric Papas, Anthea Burnett, Suit May Ho, Thomas Naduvilath, and Kovin S. Naidoo. 2018. 'Global Prevalence of Presbyopia and Vision Impairment from Uncorrected Presbyopia: Systematic Review, Meta-Analysis, and Modelling'. *Ophthalmology* 125 (10): 1492–1499. https://doi.org/10.1016/j.ophtha.2018.04.013.

Hasnat, G.N. Tanjina, Md Alamgir Kabir, and Md Akhter Hossain. 2018. 'Major Environmental Issues and Problems of South Asia, Particularly Bangladesh'. In *Handbook of Environmental Materials Management*, 1–40. Cham, Switzerland: Springer Nature.

Heris, Mostapha Kalami, and Shahryar Rahnamayan. 2020. 'Multi-Objective Optimal Control of Dynamic Integrated Model of Climate and Economy: Evolution in Action'. In *2020 IEEE Congress on Evolutionary Computation (CEC)*, 1–8. https://doi.org/10.1109/CEC48606.2020.9185688.

Hickmann, Thomas, Oscar Widerberg, Markus Lederer, and Philipp Pattberg. 2021. 'The United Nations Framework Convention on Climate Change Secretariat as an Orchestrator in Global Climate Policymaking'. *International Review of Administrative Sciences* 87 (1): 21–38.

Hierlmeier, Jodie. 2001. 'UNEP: Retrospect and Prospect—Options for Reforming the Global Environmental Governance Regime 2002 Student Notes Issue: Note'. *Georgetown International Environmental Law Review* 14 (4): 767–806.

Hockstad, L., and L. Hanel. 2018. 'Inventory of U.S. Greenhouse Gas Emissions and Sinks'. *Cardiac: EPA-EMISSIONS*. Environmental System Science Data Infrastructure for a Virtual Ecosystem (ESS-DIVE) (United States). https://doi.org/10.15485/1464240.

Horowitz, Cara A. 2016. 'Paris Agreement'. *International Legal Materials* 55 (4): 740–755.

Howe, Joshua P. 2014. *Behind the Curve: Science and the Politics of Global Warming*. Seattle: University of Washington Press.

Hunter, David B. 2010. 'Implications of the Copenhagen Accord for Global Climate Governance'. *Sustainable Development Law & Policy* 10 (2): 5.

Kellogg, William W. 1979. 'The World Climate Conference: A Conference of Experts on Climate and Mankind, Held in Geneva, Switzerland, during 12–23 February 1979'. *Environmental Conservation* 6 (2): 162.

Kuyper, Jonathan, Heike Schroeder, and Björn-Ola Linnér. 2018. 'The Evolution of the UNFCCC'. *Annual Review of Environment and Resources* 43 (1): 343–368. https://doi.org/10.1146/annurev-environ-102017-030119.

McEntire, David A. 2021. *Disaster Response and Recovery: Strategies and Tactics for Resilience*. Hoboken, NJ: John Wiley & Sons.

Nordhaus, William D. 1994. *Managing the Global Commons: The Economics of Climate Change*. Vol. 31. Cambridge, MA: MIT Press.

Olivier, Jos G. J., and Jeroen A. H. W. Peters. 2017. *Trends in Global CO2 and Total Green-house Gas Emissions: 2017 Report*. The Hague: PBL Netherlands Environmental Assessment Agency.

Oppenheimer, Michael, and Annie Petsonk. 2005. 'Article 2 of the UNFCCC: Historical Origins, Recent Interpretations'. *Climatic Change* 73 (3): 195–226. https://doi.org/10.1007/s10584-005-0434-8.

Ottinger, Richard L. 2009. 'Copenhagen Climate Conference-Success or Failure'. *Pace Environmental Law Review* 27: 411.

Özkan, Arda. 2016. 'Güvenlik Paradigmasında Sınıraşan Bir Çevre Sorunsalı: "Nükleer Zarar"'. *Alternatif Politika* 8 (1): 128–159.

Özmen, M. Tamer. 2009. 'Sera Gazı-Küresel Isınma ve Kyoto Protokolü'. *İMO Dergisi* 453 (1): 42–46.

Pearce, Laurie. 2003. 'Disaster Management and Community Planning, and Public Participation: How to Achieve Sustainable Hazard Mitigation'. *Natural Hazards* 28 (2): 211–228. https://doi.org/10.1023/A:1022917721797.

Petsonk, Carol Annette. 1989. 'Role of the United Nations Environment Programme (UNEP) in the Development of International Environmental Law Recent Developments in International Organisations'. *American University Journal of International Law and Policy* 5 (2): 351–392.

Rogelj, Joeri, Julia Nabel, Claudine Chen, William Hare, Kathleen Markmann, Malte Meinshausen, Michiel Schaeffer, Kirsten Macey, and Niklas Höhne. 2010. 'Copenhagen Accord Pledges Are Paltry'. *Nature* 464 (7292): 1126–1128.

Shahadin, Maizatul Syafinaz, Nurul Syakima Ab. Mutalib, Mohd Talib Latif, Catherine M. Greene, and Tidi Hassan. 2018. 'Challenges and Future Direction of Molecular Research in Air Pollution-Related Lung Cancers'. *Lung Cancer* 118 (April): 69–75. https://doi.org/10.1016/j.lungcan.2018.01.016.

Streck, Charlotte, Paul Keenlyside, and Moritz Von Unger. 2016. 'The Paris Agreement: A New Beginning'. *Journal for European Environmental & Planning Law* 13 (1): 3–29.

Theodore, Louis. 2008. *Air Pollution Control Equipment Calculations*. Hoboken, NJ: John Wiley & Sons.

UN Copenhagen Accord. 2009. 'Draft Decision-/CP. 15'. In *Conference of the Parties to the L NFCC, Fifteenth Session, Copenhagen*, 7:18. https://www.c2es.org/content/cop-15-copenhagen/

UNGA. 1972. 'Institutional and Financial Arrangements for International Environmental Cooperation'. *Resolution* 2997: 15. https://daccess-ods.un.org/tmp/9449297.18971252.html

UNGA. 2013. 'Resolution Adopted by the General Assembly on 20 December 2013'. *Agenda* 21: 7. https://research.un.org/en/docs/ga/quick/regular/68

United Nations. 1992. 'United Nations Framework Convention on Climate Change. FCCC/INFORMAL/84. GE. 05–62220 (E) 200705'. https://treaties.un.org/doc/treaties/1994/03/19940321%2004-56%20am/ch_xxvii_07p.pdf

United Nations. 1997. 'Report of the Conference of the Parties on Its Third Session, Held at Kyoto From 1 to 11 December 1997'. https://unfccc.int/process-and-meetings/conferences/past-conferences/kyoto-climate-change-conference-december-1997/cop-3/cop-3-reports

United Nations. 2011. 'Report of the Conference of the Parties on Its Sixteenth Session, Held in Cancun from 29 November to 10 December 2010'. In *United Nations Framework Convention on Climate Change*. https://digitallibrary.un.org/record/708138

United Nations. 2015. 'Paris Agreement'. In *Report of the Conference of the Parties to the United Nations Framework Convention on Climate Change (21st Session, 2015: Paris)*. https://unfccc.int/resource/docs/2015/cop21/eng/10.pdf

United Nations. 2021. 'The Climate Crisis—A Race We Can Win'. *United Nations.* www.un.org/en/un75/climate-crisis-race-we-can-win.

Wenzel, Friedemann, Fouad Bendimerad, and Ravi Sinha. 2007. 'Megacities—Megarisks'. *Natural Hazards* 42 (3): 481–491. https://doi.org/10.1007/s11069-006-9073-2.

14 Integrated Energy and Environmental Policy Framework

*Gisleine Cunha-Zeri, Evandro Albiach Branco,
André Gonçalves, Sérgio Pulice, Marcelo Zeri*

14.1 INTEGRATED ENERGY: AN INTRODUCTION

Renewable sources have been used to provide energy since the most ancient times, being the only option available to past generations and still for many with limited energy access (Smil 2017). Examples are the use of muscle power (human and animal workforce), wind power (windmills), hydro power (water clocks and waterwheels), and solar power (solar furnaces), as well as traditional biomass burning (e.g., wood, peat, straw, etc.) (Sørensen 1991).

However, the energy systems have changed dramatically since the Industrial Revolution: the advent of fossil fuels has played and continues to play a crucial role in global energy systems by bringing technological, economic, and social development (Smil 2017). In contrast, fossil energy (coal, oil, and gas) also come with many negative impacts on the environment and human health, for instance, carbon dioxide emissions, which are the greatest driver of global environmental change, and local air pollution, causing respiratory diseases and other health issues (Armaroli and Balzani 2011). Besides, high dependence on fossil fuels could threaten efforts to reduce global temperature, tackle climate change, and achieve sustainability (IPCC 2018).

After a century of unparalleled production growth and consumption expansion, the world is now tackling energy shortages, changing demand patterns, and facing unequal resource distribution, geopolitical struggles over energy supply, and dramatic environmental limitations (Smil 2003); therefore, the importance of integrated energy is emerging as a key approach for the much-needed worldwide energy transition (Wang et al. 2020). Integrated energy connects the energy-related sectors of electricity, heat, cooling, and transport in a way that a greater proportion of renewable energy (solar, wind, wave, biomass) can be efficiently used, aiming at sustainable energy supply and the alleviation of environmental impacts (Xiang et al. 2020).

The initial step towards planning an integrated energy system would be tracking the amount of energy produced and consumed over the years from a diverse range of sources (fossil fuels, nuclear, and renewables) and adjusting it to scale (regional, sub-regional, or at a higher level). For instance, regarding the global primary energy consumption from 1800 to 2019, until the mid-19th century the dominant source of

DOI: 10.1201/9781032715438-14

energy was traditional biomass burning (i.e., wood, crop waste, charcoal); with the Industrial Revolution (1760–1840) came the advance of coal, oil, and gas; hydropower emerged by the turn of the 20th century; in the 1960s came nuclear energy; finally, solar and wind energies appeared in the 1980s (Ritchie and Roser 2020). Concerning global production, in 2019 most of the energy came from fossil fuels [oil (33.1%), coal (27%), gas (24.3%), a total of 84.3%]; despite producing more energy from renewables each year, only 11.4% came from hydropower (6.5%), wind (2.2%), solar (1.1%), biofuels (0.7%), and other renewables (0.9%) (Ritchie and Roser 2020). It is important to highlight that emissions from low-carbon sources (nuclear power and renewable energy) are not zero, since they may produce some carbon, but very small compared to fossil fuels (Schlömer et al. 2014).

In this sense, other benefits of integrated energy can be perceived in (i) improving people's health due to a lower amount of dangerous pollutants in the air; (ii) decreasing electricity prices insofar as solar and wind powers are becoming more cost effective; (iii) decreasing dependence on climate conditions, as hydropower station drawbacks in periods of droughts, and (iv) replacing fossil fuels by various sources of renewable energy, just to cite a few. Figure 14.1 presents a summary of the conceptualization of integrated energy, including energy sources, benefits, and constraints.

While the traditional energy development is focused on the supply side, that is, both public and private companies receive projections of demand and aspire to provide these needs, the integrated energy approach views energy as a dynamic input to promote economic development and energy security, especially in geographic areas usually excluded under traditional energy planning (OAS 1988). In fact, integrated energy systems take into consideration the demand and supply sides, as well as environmental concerns (Dhakouani, Znouda, and Bouden 2021). Demand control mechanisms already exist in current energy systems but are mostly focused on energy

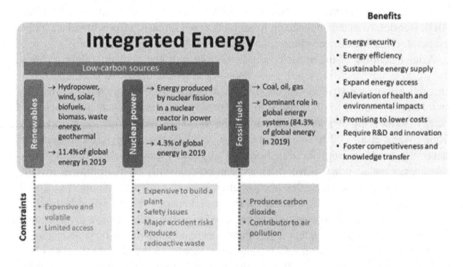

FIGURE 14.1   Conceptualization of integrated energy: energy sources, benefits, and constraints.

market solutions (price as a response from the offer/demand balance). Integrated energy systems should go beyond, acting through regulation (policy frameworks) to tailor energy demand (substituting sources, fostering new technologies) that could deliver general benefits (health, lower emissions, reliability, affordability, etc.).

The concept of integrated energy systems (IESs) has arisen under the current pressure on energy security, resources shortages, and concerns about environmental protection. IESs are basically the integration of heterogeneous energy sources and technologies (renewable and non-renewable) into an energy system, aiming to efficiently provide multiple and reliable sources of energy and, at the same time, address environmental and health issues (Zhang et al. 2021). Integrated energy management is a concept closely connected to sustainability, since it requires linkages among energy production, energy consumption, human wellbeing, economic benefits, and environmental quality (Yildiz 2021), and whose implementation can support the transition to a sustainable energy system (Boer and Zuidema 2015). Actually, IESs encompass the demands of multiple driving forces, such as environment, economy, society, technology, and public policies (Han et al. 2021).

The challenges in designing IESs include (i) the introduction of flexible technologies capable of embodying renewable energy sources (and their fluctuating production patterns) in established energy systems, in order to achieve sustainable development and energy security (H. Lund 2007); (ii) the complexity of the IES requires the development of synergistic and complementary devices, in addition to a well-planned allocation scheme for energy supply, considering the provision of energy to users from a variety of heterogeneous sources (Wang et al. 2020); (iii) the intricacy of formulating energy policies to simultaneously provide socioeconomic and environmental benefits (Dubey 2020); and (iv) designing flexible regulations and policy strategies to stimulate the implementation of IES (H. Lund and Münster 2006).

14.2 THE CASE OF BRAZIL

14.2.1 How Energy Is Produced and Used in Brazil

Brazil may stand as a typical example of country where integrated energy systems should deal with a great diversity of factors all together. The country holds a large population, moderate industrialization, and a diversified economy, imposing high energy demands. On the other hand, the large territory extending from tropics to subtropics provides a variety of energy resources, ranging from oil and gas reserves to biofuels, biomass, wind, solar, and hydro, among other sources. Even though multiple solutions may exist, there are still many challenges concerning energy affordability that impact the access to electricity and clean cooking for a high share of economically vulnerable populations (Coelho and Goldemberg 2013).

In the last 50 years, Brazil experienced a deep transition on energy production and consumption. Figure 14.2 shows a snapshot of the relative contribution by sources for energy consumption from 1970 to 2020 (ONS 2021). Industrialization and urbanization reduced the woody biomass demand (typically for cooking) from a 46% share of annual energy demand to less than 7% in the period, while oil demand rose above 50%. In the meantime, a variety of sources began to play an important role in the

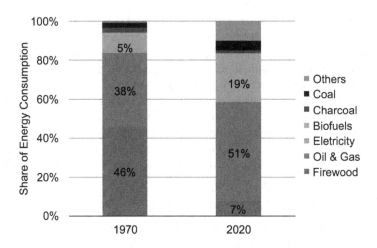

FIGURE 14.2　Evolution of relative energy consumption by source in Brazil from 1970 to 2020.

energy matrix, like biofuels (sugarcane), natural gas, and renewables (hydro, wind, and solar). On the consumption side, electricity faced the highest expansion, reaching a share of 19%, but oil demand stands high, pushed mostly by transportation, impacting energy efficiency. As a consequence, greenhouse gas (GHG) emissions from the energy sector have increased their share in recent decades, reaching around 20% (445.4 GtCO2e) of the country's total emissions, typically dominated by land-use change and forestry sector (EPE 2022; SEEG-Brazil 2022).

The Brazilian electrical matrix can be classified as a hydrothermal system benefiting from a geographically interconnected grid. Hydropower plants are at the heart of Brazilian energy development, which holds more than 109 GW of installed capacity that accounts roughly for two thirds of total electricity production (ANEEL 2021). Despite that, in recent years most of the hydropower projects have been built over small catchments, since a large part of the hydro power potential remains on the Amazon basin, subject to extensive discussion due social and environment impacts (Moretto et al. 2012; Stickler et al. 2013). Flooding of indigenous land, relocation of local communities, deforestation, methane emissions, and impacts on hydrological cycle and on ecological balance are some concerns (Cavalcante and Santos 2012; Fearnside 2016). Potential local benefits associated with economic development are controversial, even the improvement of public services and average income to families promoted by high investments and royalties (Almeida-Prado et al. 2016). Moreover, the addition of hydropower perpetuates the dependence on a single source, making the electrical supply more vulnerable to droughts, thus affecting the energy security of the country. From an integrated energy systems perspective, that situation is not desirable, since diversification and synergies are key factors for building resilient energy systems.

14.2.2　Opportunities for Integrated Energy Planning in Brazil

Energy planning is related not only to physical security of coupling supply and demand sides, but it also affects environmental and social policies at a country level. Integrated energy planning should consider these interactions, as discussed in the previous section.

Nonetheless, the complex nature of this problem demands a gradual inclusion of broader methods for system analysis and planning. Spatially and temporally resolved computational models are evolving in this direction and proving a mandatory tool for integrated energy planning. In any case, those models rely typically on scenario-based simulations, exploring regulatory and normative futures, but they still lack the consideration of political dimension behind the possible technical solutions. This means that a conceptual model including different stakeholders and their values and beliefs, and considering how they can affect the decision-making process, is desirable to assess the chances of implementing a specific solution. In the following sub-sections, we assess some technologies that could build up an integrated energy and environmental system in Brazil.

14.2.2.1 Dealing With Power Intermittency

Variable renewable energies (VREs), like solar, wind, and run-of-the-river hydropower, for instance, introduce a weather-dependent signal that causes strong variations on power supply to the grid. This intermittency is probably the major concern on increasing the share of VREs in the energy system and thus cutting GHG emissions (Kroposki et al. 2017; Tong et al. 2021). Even though technical solutions exist, the dispatchable power sector lobby defies the evidence in favor of conventional plants, like large hydropower and fossil-based thermal plants. Some strategies to overcome this reluctance include spatial planning of power sources, hybrid power systems, and energy storage technologies.

14.2.2.1.1 Spatial Planning

Climate-dependent resources like wind, solar, and hydro are part of a continuous fluid system (atmosphere), where energy and water vertical fluxes are roughly balanced worldwide. This means that patterns arise, allowing a complementarity between regions and sources, or simply due to a spatial smoothing effect, as shown in Figure 14.3, where

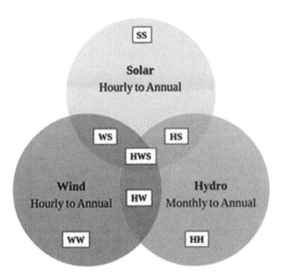

FIGURE 14.3 Energy sources dependent on climate and possible interfaces and timescales of integration focusing on synergies and complementarities.

abbreviations indicate the type of variability in each sector (W-wind, S-solar, H-hydro). For example, "SS" accounts for "Solar-Solar" spatial complementarity in the territory, while "WS" accounts for "Wind-Solar" spatial and temporal complementarity.

In Brazil there is a well-known seasonal complementarity between hydro and wind resources in the inner northeast region, which is important to keep hydropower reservoirs at higher levels, allowing water withdrawals for population consumption and agricultural irrigation (Silva et al. 2016; Cantão et al. 2017). Thus, intermittency is inversely proportional to spatial extent, in a way that a geographically distributed portfolio of VREs smooths the power variability in the electricity grid (Hoff and Perez 2012). It can also be achieved by grid interconnections at national and international levels, building continental-scale electricity grids (Kempton et al. 2010). This evidence suggests that energy system planning should integrate solutions at multiple spatial scales.

14.2.2.1.2 Hybrid Power Plants

One of the most promising solutions for reducing intermittency of electricity supply is the hybridization of two or more power sources in a single generation plant. Hybridization will also help to sustain VRE cost reduction due to synergistic combination of generation profiles, which improves the plant average capacity factor (as exemplified in Figure 14.4), diminishing infrastructure idleness, thus diluting costs. These features, in addition to the grid infrastructure sharing, make hybrid systems economically competitive for providing energy to large-scale grids (Carvalho, Guardia, and Lima 2019; Campos, Nascimento, and Rüther 2020; Peterseim et al. 2014; Jakub Jurasz et al. 2018). Among several options, some of them have shown

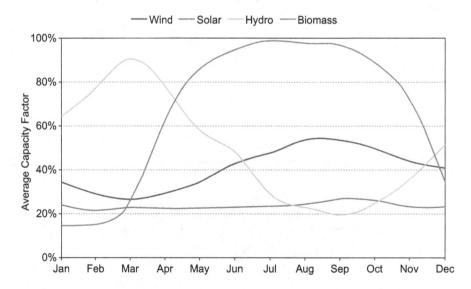

FIGURE 14.4 Average capacity factor along the year for wind, solar, hydro, and biomass power plants in Brazil (ONS 2021).

prominence in Brazil: wind-hydro and solar-hydro hybrid systems take advantage of hydropower modulation and storage capacity to stabilize generation profile (Denault, Dupuis, and Couture-Cardinal 2009) and, in some cases, the flooded area may improve wind and solar potential (Pimenta and Assireu 2015; Gonçalves et al. 2020).

Wind-solar plants may also play an important role in the energy system integration in Brazil, due to the overlap between windiest and sunniest regions, combined with the complementary pattern of daily profiles, once wind speeds up during late night in most Brazilian wind parks (Campos, Nascimento, and Rüther 2020; Tong et al. 2021). Other hybrid solutions transcending power, heat, and fuel sectors may provide even more intricate energy integration. Some examples are the biomass-thermal power cogeneration plants that produces ethanol and electricity from sugar-cane. To overcome off-season operation, it can also be integrated with a concentrated solar-thermal collector, providing heat year-round and improving power capacity (Soria et al. 2016; Peterseim et al. 2014).

In a general approach, the use of solar-thermal collectors for providing heat for agro-industrial processes (such as fruit sanitization, slaughterhouse refrigeration, textile dyeing, and plaster mining, among others) reduces not only electricity and fuel demand but also native firewood extraction, contributing to the environmental targets of an integrated energy system. The concepts of hybridization stand out due to the great potential of better exploiting available network capacity when active network management is involved (J. Jurasz et al. 2020; Sun and Harrison 2019).

14.2.2.1.3   Energy Storage

The straightforward solution for power intermittency is energy storage, where several technologies are available. Traditional lead-acid batteries, commonly used in isolated power systems, are gradually being replaced by modern lithium batteries, providing a fast and modular solution for intermittency. Costs are reducing deeply, but maximum storage capacity and short life cycle are the main drawbacks. On the other hand, upstream water reservoirs are probably the oldest form of storing energy and are still the backbone of seasonally stabilized electric grids in hydropower dependent countries, like Brazil. Hydropower plants are dispatchable at near real time, absorbing variability of VREs. The inclusion of reversible and pumped-storage hydropower plants amplifies its modulation capacity, allowing higher VRE shares in the system. The saved water may provide other services to population like agricultural irrigation, water supply, and flood protection, being part of an integrated energy system planning. Other storage technologies, like hydrogen (P. D. Lund et al. 2015) or plug-in vehicles (Soares M.C. Borba, Szklo, and Schaeffer 2012), for instance, introduce an inter-sectoral impact, affecting industrial heat, feedstock, and transportation demands.

14.2.2.2   Integrating Demand Planning

Demand response is given by the capacity of a customer to shift or reduce their energy usage from peak time to off-peak time in response to time-based price rates (spot prices) or incentive-based programs (Xiang et al. 2020). The continuous technological development introduced smart metering systems, responsible for bidirectional metering and communication between network operators and end users, which allows demand response in near-real time. This is especially important for systems experiencing high levels of wind and solar power, which imposes high intra-day variability in the energy offering. Furthermore, distributed energy storage systems may add extra complexity to the grid management, since behind-the-meter batteries or electric vehicles can automatically modulate the load profile to take advantage of hourly lower energy prices. Besides, the energy market mechanism is the basis for traditional demand response integration, working adequately in well-developed markets, like oil and gas. However, highly variable energy systems depend upon detailed integration planning in order to guarantee a timely energy supply for periods when demand response may not be enough to absorb variability (low price elasticity of energy demand). Regulatory actions can also play an important role in demand-side planning, providing incentives for large consumers to adopt co-generation or load-shift measures (demand relocation). These are important measures to allow a higher penetration of weather-modulated low carbon energy sources in the system.

14.2.2.3 Imposing Socio-Economic and Environmental Constraints

To summarize the information in this section, the main processes that convert energy sources for final use are presented in a conceptual diagram (Figure 14.5).

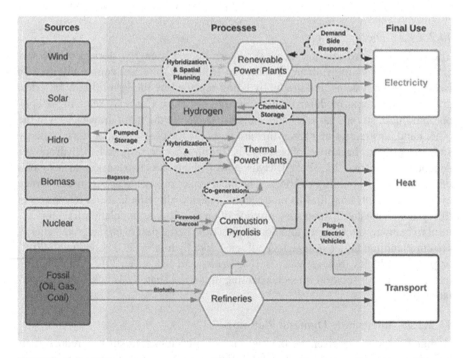

FIGURE 14.5 Main processes that convert energy sources for final use. Opportunities for integrated planning are highlighted in dotted circles, representing potential gains in process efficiency and lower environmental impacts (i.e., GHG emissions and air quality).

Opportunities for efficiency gains and environmental impact reduction (i.e., GHG emissions and air quality) are also highlighted in Figure 14.5. However, it is important to emphasize that the integration of energy systems does not rely only on innovation and technology to deliver a reliable, efficient, and low-carbon network. When it comes to public goods or assets, like health, environment, and social justice, technical solutions may not meet people's expectations adequately. In this scenario, regulatory measures are essential to define constraints for energy system planning. Certainly, it must be assumed that policymakers are concerned and willing to propose the most appropriate public policies. However, lobbies and different chains of beliefs may disturb the decision-making processes. The most paradoxical example in Brazil is the construction of large hydroelectric dams in the Amazon region. In additional to several studies showing deep environmental impacts on the river-forest ecology (Fearnside 2016; Santos et al. 2018; Timpe and Kaplan 2017), there are also long-term side effects to riverside populations, such as deforestation and increased social vulnerability and violence (Hess and Fenrich 2017; Soito and Freitas 2011; de Faria and Jaramillo 2017). On the other hand, hydropower is acknowledged as the backbone of the Brazilian electrical system, providing flexibility due to dispatchability and storage capacity.

Another prominent dispute regarding the water-food-energy nexus is that some hydropower reservoirs in Brazil sustain multiple functions, other than energy storage, like cropland irrigation, public water supply, navigation, and ecosystem services (Ferraz de Campos et al. 2021). During periods of drought, the trade-off between these various functions is complex, and rules are often overloaded. An integrated energy system should account for these events and provide contingency plans to minimize social, economic, and environmental costs.

The ubiquitous example for which technical and economical optimization may fail to meet people's concerns is the use of fossil fuel for transportation and power generation. High GHG emissions that accelerate global warming, in addition to other pollutants that affect human health in cities, are issues that require an appropriate regulatory framework. In sum, an integrated energy and environmental policy framework must address the best practices for integrating the physical system (synergies) within a reasonable regulatory framework that promotes social prosperity and environmental sustainability.

14.3 CHALLENGES IN PLANNING AN ENERGY POLICY FRAMEWORK

The idea of complementarity and integration of energy systems discussed in the previous section requires the prediction and feasibility of political structures to leverage and sustain policies for innovation projects. Such ideas are frequently discussed in the scientific literature and the bureaucratic space of the state, but the paths to design and implement them in terms of public policies remain a challenge.

We argue in this section that leverage mechanisms, such as continuous spaces to promote diversity of actors, must be observed in the planning and implementation of energy policy in Brazil to promote paths to more sustainable energy systems. We also discuss that the specificities and uncertainties of the Brazilian power sector, contrasting with European examples, add more challenges and obstacles to the promotion of learning processes based on the diversity of actors involved in political and decision-making processes.

14.3.1 SPECIFICITIES FROM THE CASE OF THE BRAZILIAN POWER SECTOR

The specificities of global south countries can bring contributions to the research agenda on transition to sustainability, especially for energy issues, not only because of the technological characteristics or natural availability but also because of their history of continuous exploitation and colonization and consequent characterization as less consolidated democracies (Relva et al. 2021).

We may define the governance structure of the Brazilian power sector to be hybrid, with actors from the private sector, regulatory agencies, government, and planning bodies (de Oliveira 2007; J. Dutra and Menezes 2022). The central figure in the governance is the Ministry of Mines and Energy (MME) (Branco 2020; Motta 2006; Hochberg and Poudineh 2021). This central organization, with broad competence, coordinates the planning and formulation of sector policies. There are other secretariats, departments, regulatory agencies, and promoters of the system's operation down in the MME hierarchy. Externally at the governmental level of command and control, the private sector explores resources for the generation and distribution of energy through authorizations and concessions.

After the sector's reform, driven by a serious energy crisis in 2001 (J. Dutra and Menezes 2022), the governance system had its priority defined as keeping the energy supply at the lowest cost and secure, which reduces systemic risks and is predictable in terms of profitability for companies (Goldemberg and Moreira 2005; Sauer 2015; de Oliveira 2007). Schaeffer et al. (2015) refer to the new model of the Brazilian power sector as a "growth-based" and conservative model, which is historically based on hydraulic sources and technology.

Strongly centralized energy planning systems based on cost reduction, as in the Brazilian case, can lead to the emergence of institutional lock-in mechanisms (Unruh 2002; Lazaro et al. 2022). Such lock-in mechanisms, for example, can exacerbate the lack of debates on climate change both in the political strategies level of planning, as in the operational level, and of the parameters and models that underlie the sector (Hochstetler 2021; Lucon, Romeiro, and Fransen 2015; Schaeffer et al. 2015).

The Energy Research Office (EPE, in the Portuguese acronym) is a key actor in the planning processes. It was created in 2004, aiming to "provide services to the Ministry of Mines and Energy (MME) in the area of studies and research aimed at subsidizing the planning of the energy sector, covering electricity, oil and natural gas and their derivatives and biofuels" (EPE 2022). The creation of a company dedicated to energy research was considered one of the great advances in the reform of the power sector, as it brought in specialists from different areas to guide the planning and development of the Brazilian power sector. Although EPE's work is widely recognized, considering the centralizing aspect of its current institutional structure, it is important to question what degree of permeability the planning process has with innovative proposals or even if there is space to propose and anchor innovative solutions, such as those based on assumptions of integrated energy.

The implementation of the policy to guarantee the supply for the National Integrated System relies mainly on an auction system. The free market system (focused on large consumers) seeks to promote the free bilateral initiative of economic agents by directly linking the consumer to the supplier (Tolmasquim et al. 2021). This instrumental structure has been successful in promoting the expansion of the installed capacity of alternative renewable sources and, at the same time, has faced academic and technical controversies for consolidated sources (Diógenes, Claro, and Rodrigues 2019).

As previously presented, the main source of electricity supply in Brazil is hydraulic, a mature technology, historically explored and improved. However, in recent decades its inherent controversies and impacts have pressured the political and institutional agenda of the sector to rethink such a "hydrocentric" position of the energy system. On the one hand, water availability and relative technological stability make it essential for a secure supply to the energy system. On the other hand, its vulnerability to more frequent climatic variations (Pereira de Lucena et al. 2010; Schaeffer et al. 2012; Lima et al. 2020; Ruffato-Ferreira et al. 2017) and socio-environmental impacts (Moran et al. 2018; Mayer et al. 2021) put its hegemonic viability under discussion.

Conversely, wind technology has accelerated and currently accounts for 11.89% of the electricity matrix (ANEEL 2022; Lucena and Lucena 2019). Its acceleration phase, still under consolidation, had integrated tariff processes (R. M. Dutra and Szklo 2008), specific auctions (Bayer 2018), and macroeconomic determinants (Costa et al. 2022). It is important to highlight that the success of the acceleration of wind technology was due to the development of the governance in the power sector and the continuous improvement of the auction system.

Regarding the transparency of the system design decision, the power sector usually acts ambiguously. On the one hand, draft legislation and programs are frequently proposed and discussed with stakeholders and the Congress. For instance, the PROINFA (Programme of Incentives for Alternative Electricity Sources) and the PROCEL (National Energy Conservation Programme), among others, are widely articulated programs that generate important results for the sector (Costa et al. 2022). At the same time, many important decisions are taken through exclusive normative acts (provisional measures and decrees) or suspension acts, which suspend any ongoing judicial decisions that question the executive branch to "avoid serious harm to order, to health, security, and the public economy" (Brazil 1985, 1992; Scabin, Pedroso Jr, and Cruz 2015; Sampaio and McEvoy 2016; Villas-Boas et al. 2015; Nascimento 2017; Fearnside 2018, 2019). This centralizing and asymmetrical aspect of executive power dates back to the history of the great expansion of hydraulics in the 60s and 80s, a period in which the country lived under an authoritarian military regime (Souza Braga 2020).

This brief overview of the sector illustrates that any process of leveraging new ideas, technologies, or initiatives may face a highly complex, centralized structure, backed by a history of authoritarian remnants, operated by coalitions and consolidated interests that want to maintain its hegemony (Soares, Gava, and Puppim de Oliveira 2021; Bradshaw and de Martino Jannuzzi 2019; Souza Braga 2020).

14.3.2   Uncertainties in the Brazilian Power Sector and the Bases for Political Innovation

The question that emerges at this moment is whether there are sufficient institutional bases and arrangements in Brazil to receive and support innovative and heterodox ideas, such as the integrated energy system. In the following, we will bring some examples of elements of the planning and implementation of the electric sector policy that will help us to elucidate this issue.

The main planning documents prepared by the Energy Research Office are the PNE (National Energy Plan) and the PDE (The Ten-Year Energy Expansion Plan). The PNE 2050, the most recent strategic document, aims to:

> outline the energy sector's long-term expansion strategy, from the planner's point of view. Built from the main relevant issues on the horizon, the PNE 2050 explores, through scenarios, the various aspects of the sector's evolution in a perspective of various changes in the production and use of energy, commonly grouped in the so-called energy transition.
>
> (MME/EPE 2020, p. 4)

As a complement to the objective, the Minister of Mines and Energy, in the prologue of the document, declares:

> The strategy is guided by four major objectives—energy security, adequate return on investments, availability of access to the population, and socio-environmental criteria—where scenarios are used to discuss how to maximize the benefits of the current transition to Brazil and prevent regrets.
>
> (MME/EPE 2020, p. 9)

When analyzing the ten principles included in the plan, the importance of economy and profitability over other beliefs becomes stronger. For example, when it situates sustainability as a principle, economic beliefs and the exploitation of resources are the basic concepts to sustain the principle rather than natural conservation, biodiversity loss issues, or even climate change mitigation and adaptation, as mentioned in the plan:

> The energy sector must be aligned with the promotion of sustainable development, based on the best international practices and economic efficiency, seeking to take advantage of the comparative advantages of national natural resources or through public policies that value their environmental attributes.
>
> (MME/EPE 2020, p. 15)

This overvaluation of economic beliefs is materialized in other points of the PNE 2050 that refers to the socio-environmental dimension. Consolidated constitutional rights may be threatened by the necessity to guarantee energy supply by expanding new hydro projects on protected areas—legally recognized, such as Indigenous Lands, Quilombola Territories and Conservation Units. Even if such a possibility can be considered relevant from a technical point of view, its mere proposition indicates the supremacy of the energy security dimension over environmental and human rights.

The inclusion of climate change projections and modelling in the PNE 2050 also reveals a certain ambiguity. Although the subject is recognized for its relevance and transversality, its adoption does not advance because "more complex evaluations are necessary, as they involve many uncertainties and enormous challenges, whose understanding must be deepened" (MME/EPE 2020, p. 40). Such an argument is not convincing, since several studies have attributed climate risks to both hydraulic and wind power sources (Pereira de Lucena et al. 2010; Schaeffer et al. 2012; Herrera, Cosenz, and Dyner 2019; Lima et al. 2020; Ruffato-Ferreira et al. 2017). Despite the technical discussion, the conflicting values of energy planning evidence the sector's skepticism, even in the face of emerging advances in climate science, as demonstrated by recent reports from the Intergovernmental Panel on Climate Change (IPCC).

From these excerpts, important and contemporary scientific challenges are distinguished, such as scenario building, uncertainties, and energy transition, as well as the integration of different dimensions in energy planning. Additionally, energy security and profit continue to have priority over access to energy by the population, as well as socio-environmental issues. From a strictly discursive perspective, it is possible to identify the contradictions between the views and beliefs proposed in the PNE 2050.

Aside from the design of strategies and the implementation of public policies, the literature has shown that even though relevant initiatives have been proposed, debated, and implemented, the process of integration between the power sector and other sectors is still quite fragile, especially at the interface with the complex Brazilian environmental policy. For instance, the auction mechanism, central to inducing energy supply, has faced problems of legal uncertainty and contractual guarantees that impact investor confidence. One of the main complaints is the supporting infrastructure and environmental licensing processes (Tolmasquim et al. 2021; Anatolitis, Azanbayev, and Fleck 2022). The auction mechanism does not seem appropriate to address this issue, given the need for strategic institutional integration between other government departments to properly link procedural steps and flows and address effective solutions (Herrera, Cosenz, and Dyner 2019; Kwon 2018).

Other tax and financial incentive policies also surpass the scope of the MME, covering both the state and the private sector, or even the discussion of industrial policies that cover several economic sectors at the same time. Hochstetler (2021) discusses the complexity of these connections between political arenas, demonstrating that it would be ideal for the planner to engage in a "green spiral process", involving several areas at multiple levels. Something complex, but not complicated.

These examples illustrate that a centralized state structure—as in the Brazilian case—reinforces the bias and ambiguities that are reflected in the planning and execution of policies for the power sector. This is not just a superficial criticism of the system, but rather it puts into perspective the idea that the political and instrumental beliefs of the central actors end up conditioning the course of the sector as a whole, reinforcing an institutional lock-in, even that other sustainable values are also present in the sector. In this sense, it seems important to advance in the understanding of the political processes necessary for the elaboration of new agendas, specifically in the context of the transition to sustainability.

14.3.3 AGENDA SETTING AND TRANSITIONAL ARENAS AS LEVERAGE POINTS

Now, we are back to the political science and policy process research agenda, which for some decades has studied the formation of public problems and its support by advocacy coalitions as important elements of political agenda formation, the recognition of public problems, and the leverage of heterodox or innovative initiatives (Hill 2005; Birkland 2019; Gerston 2010).

The Agenda Vetting and Agenda Setting processes have as premises the formation of discussion processes, whether they are the content discussed, the consolidation of spaces for debate, or the role of those who define spaces and agendas (Carpenter 2014). The proposition of alternatives—an inherent part of the Agenda Setting process—has a relationship with the diversity of participants in the political arena (Kingdon 2003).

Tosun, Biesenbender, and Schulze (2015) reinforce the importance of understanding the process of incorporating alternatives from a broader perspective, typically based on *Agenda Shaping* processes, which include (i) innovations in the political agenda (Agenda Setting), (ii) the processes of positioning and valuing a given alternative (Structuring Agenda), (iii) even placing barriers—explicit or hidden—to a given alternative (Exclusion Agenda). Such processes determine the planning results and the direction of public policy to a vision of the central institutions or institutions that hold the power to define the structures and spaces for discussion. These themes are supported by practical examples, such as the transition arenas and recent research on the decentralization of the energy transition in Europe. Although they are attempts to incorporate the concepts described previously, the results achieved still demonstrate uncertainties and improvement points, characteristic of frontier themes.

In essence, the diversity of actors and confluent or divergent positions, declared or not, make up the fundamental ideology for agenda formation and visualization of public problems. We argue that the creation of spaces that allow the emergence of this diversity of information can be understood as a leverage point that allows the emergence and development of new paths, such as the requirements for the feasibility of projects based on integrated energy systems. At this point, we take the case of Dutch transitional arenas as an example to illustrate the possibility of the emergence of new information.

The idea of transitional arenas is located within the technical framework of transitional management and is understood as a study perspective within the field of the energy transition (Loorbach and Rotmans 2010). This perspective aims to investigate the influence of structural changes in sociotechnical systems, considering aspects of innovation or orientation of public policies (Kern and Smith 2008; van den Bergh and Bruinsma 2008).

The management of transitions combines bottom-up and top-down processes to modulate the definition of objectives and strategies together, as it assumes complexity and uncertainty as contextual elements present in the analysis (Kemp, Loorbach, and Rotmans 2007). For this reason, it aims to foster learning processes that are associated with long-term coordinated and participatory strategies within sociotechnical systems (Rotmans and Loorbach 2008).

Transition arenas are environments in which different actors are placed face to face in a continuous process of engagement and learning (Kanger and Schot 2019). Processes of this nature tend to be more accepted by the engaged community, since they do not promote ruptures but the gradual incorporation of innovations based on previously designed and negotiated bases (Kanger and Schot 2019). Furthermore, it is considered essential to face the issue of paradigm shifts when confronting internal values consolidated in the systems. In other words, transition arenas are:

> an informal network of frontrunners within which a group process unfolds, often in an unplanned and unforeseen way. This puts high demands on the group composition. In terms of group dynamics, a group is much more than the sum of the individuals. In general, it takes a few iterations before a stable, diverse and representative constellation has been formed for a transition arena. Some frontrunners leave, new ones enter, which gives some dynamics that might be fruitful for the group process. In this sense an arena process is an evolutionary process with continuous mutations.
>
> (Loorbach and Rotmans 2010)

One of the most cited experiences in the transition management area is the Dutch energy transition policy, which not only used the model to assess its situation but also to design new management policies and strategies (Kemp 2010; Rotmans, Kemp, and Van Asselt 2001; Kemp, Loorbach, and Rotmans 2007). In 2001, the country adopted the transition management approach as a method to chart innovation paths for the energy system. The policy model adopted aimed to tackle persistent structural problems by establishing six debate platforms, bringing together individuals from the public and private sectors to develop thematic proposals and development trajectories and suggest experiments on energy (Dietz, Brouwer, and Wterings 2008; Kemp 2010; Kern and Smith 2008).

Mediated by new institutions created over the years, this Dutch experience resulted in changes in the implemented public policy strategies, mainly in the Dutch energy research strategy. In addition, another relevant impact highlighted is the treatment of support strategies for renewable energy experiments, such as the capture and storage of carbon dioxide (CO_2). The main criticism of this case is the lack of satisfactory treatment of deep issues, such as long-term political commitments versus short-term results, uncertainties for investments, balancing of incentives, and control policies (Kern and Smith 2008).

This European experience shows us that, regardless of the governance model, the feasibility of incorporating any paths aimed at transitions to sustainability necessarily involves reflexive approaches to the core beliefs that underlie the sector, its way of operating, discussing, and planning. In other words, it is essential to enable spaces, structures, and reflexive processes to reposition planning strategies and directions (Hendriks and Grin 2007; Smith, Voß, and Grin 2010; Voß, Smith, and Grin 2009).

This task of repositioning core beliefs inexorably demands the input of new information, different perspectives of the world, dialogue, and examples, regardless of the governance model. We consider the guarantee of consolidated spaces for discussion a basic pillar for sustaining innovative ideas and also that this theoretical proposition does not limit or exclude other forms of thought and beliefs already established.

Conversely, establishing diversified spaces and systematic discussions will likely aggregate, improve, and update the electric sector governance, jointly guaranteeing the expansion of supply in a safe, profitable way and adapted to the imperative climatic and socio-environmental conditions of our time.

Finally, we do not intend here to discuss the problem as a mere need for sector reform or to import experiences that are radically distant from the political and institutional reality of Brazil. The central point is to highlight how different institutional arrangements and environments can allow or restrict the advancement of certain agendas, as well as that engaging in reflexive processes (Hendriks and Grin 2007; Grin 2020; Voß, Smith, and Grin 2009) and policy learning processes (Weible and Ingold 2018; Sabatier and Jenkins-Smith 1993; von Malmborg 2021) may provide essential inputs to the sector. Those arguments are, therefore, ideas to be introduced and discussed within the power sector.

14.4 CONCLUDING REMARKS

Integrated energy is a key approach to pave the path for the urgent worldwide demand for a cleaner and more efficient energy supply. The integration of heterogeneous energy sources and technologies (renewable and non-renewable) aims not only to efficiently provide energy but also to address environmental and health concerns. It is important to emphasize that a regulatory and policy framework is essential to properly plan and implement an integrated energy system, in addition to technological and innovative advances. Some elements are vital in the complex transition to an integrated energy policy framework, regardless of country governance models, such as the formation of agenda setting to consolidate the policy formulation process, the inclusion of different actors and stakeholders, with the aim of increasing representativeness and dialogue in discussions on policy strategies, and decentralizing decision-making. These approaches can provide important inputs for the implementation of technologies and innovations in an integrated energy policy framework, bringing potential solutions to address current global energy and environmental challenges.

REFERENCES

Almeida-Prado, Fernando, Simone Athayde, Joann Mossa, Stephanie Bohlman, Flavia Leite, and Anthony Oliver-Smith. 2016. "How Much Is Enough? An Integrated Examination of Energy Security, Economic Growth and Climate Change Related to Hydropower Expansion in Brazil." *Renewable and Sustainable Energy Reviews* 53: 1132–1136. https://doi.org/10.1016/j.rser.2015.09.050.

Anatolitis, Vasilios, Alina Azanbayev, and Ann-Katrin Fleck. 2022. "How to Design Efficient Renewable Energy Auctions? Empirical Insights from Europe." *Energy Policy* 166 (March): 112982. https://doi.org/10.1016/j.enpol.2022.112982.

ANEEL. 2021. "List of Electric Power Plants—Generation Information System of the National Electric Energy Agency (ANEEL) (SIGA)." www.aneel.gov.br/documents/655808/0/BD+SIGA+01102021/c11baa85-463a-44e0-dd7c-4a3e4cc60cce?version=1.1&download=true.

ANEEL. 2022. "SIGA—ANEEL Generation Information System (in Portuguese)." The Brazilian Electricity Regulatory Agency (ANEEL). https://dados.gov.br/dataset/siga-sistema-de-informacoes-de-geracao-da-aneel.

Armaroli, Nicola, and Vincenzo Balzani. 2011. "Fossil Legacy." In *Energy for a Sustainable World: From the Oil Age to a Sun-Powered Future*, edited by Nicola Armaroli and Vincenzo Balzani, 97–122. Weinheim, Germany: Wiley-VCH.

Bayer, Benjamin. 2018. "Experience with Auctions for Wind Power in Brazil." *Renewable and Sustainable Energy Reviews* 81 (November 2016): 2644–2658. https://doi.org/10.1016/j.rser.2017.06.070.

Bergh, J. van den, and F. Bruinsma. 2008. "The Transition to Renewable Energy: Background and Summary." In *Managing the Transition to Renewable Energy. Theory and Practice from Local, Regional and Macro Perspectives*, edited by J. van den Bergh and F. Bruinsma, 1–15. Cheltenham and Northampton: Edward Elgar.

Birkland, Thomas A. 2019. *An Introduction to the Policy Process: Theories, Concepts, and Models of Public Policy Making*. 5th ed. New York, NY: Routledge.

Boer, Jessica de, and Christian Zuidema. 2015. "Towards an Integrated Energy Landscape." *Proceedings of the Institution of Civil Engineers—Urban Design and Planning* 168 (5): 231–240. https://doi.org/10.1680/udap.14.00041.

Bradshaw, Amanda, and Gilberto de Martino Jannuzzi. 2019. "Governing Energy Transitions and Regional Economic Development: Evidence from Three Brazilian States." *Energy Policy* 126 (September 2017): 1–11. https://doi.org/10.1016/j.enpol.2018.05.025.

Branco, Evandro Albiach. 2020. *Land Use and Cover Dynamics and Large Hydropower Plants in Amazon (in Portuguese)*. São Paulo: Universidade de São Paulo. https://doi.org/10.11606/T.106.2020.tde-17112021-121110.

Brazil. 1985. *Law Nº 7347/1985—It Regulates the Public Civil Action of Liability for Damages Caused to the Environment, to the Consumer, to Goods and Rights of Artistic, Aesthetic, Historical, Tourist, and Scenic Value (Vetoed) and Makes Other Provisions (in Portuguese)*. Official Diary of the Union. Brazil: Presidency of the Republic, Civil House, Sub-office for Legal Affairs.

Brazil. 1992. *Law Nº 8437/1992—Provides for the Granting of Precautionary Measures against Acts of the Government and Other Measures (in Portuguese)*. Official Diary of the Union. Brazil: Presidency of the Republic, Civil House, Sub-office for Legal Affairs.

Campos, Rafael, Rafael L. Nascimento, and Ricardo Rüther. 2020. "The Complementary Nature between Wind and Photovoltaic Generation in Brazil and the Role of Energy Storage in Utility-Scale Hybrid Power Plants." *Energy Conversion and Management* 221 (October): 113160. https://doi.org/10.1016/j.enconman.2020.113160.

Cantão, Mauricio P., Marcelo R. Bessa, Renê Bettega, Daniel H.M. Detzel, and João M. Lima. 2017. "Evaluation of Hydro-Wind Complementarity in the Brazilian Territory by Means of Correlation Maps." *Renewable Energy* 101: 1215–1225. https://doi.org/10.1016/j.renene.2016.10.012.

Carpenter, Charli. 2014. *"Lost" Causes. Agenda Vetting in Global Issue Networks and the Shaping of Human Security*. Ithaca and London: Cornell University Press.

Carvalho, Diego B., Eduardo C. Guardia, and José W. M. Lima. 2019. "Technical-Economic Analysis of the Insertion of PV Power into a Wind-Solar Hybrid System." *Solar Energy* 191: 530–539. https://doi.org/10.1016/j.solener.2019.06.070.

Cavalcante, Maria Madalena de Aguiar, and Leonardo José Cordeiro Santos. 2012. "Hydroelectric at the Madeira River: Tensions about the Territory Using and the Natural Resources in the Amazon." *Confins*, no. 15 (June). https://doi.org/10.4000/confins.7758.

Coelho, Suani T., and José Goldemberg. 2013. "Energy Access: Lessons Learned in Brazil and Perspectives for Replication in Other Developing Countries." *Energy Policy* 61: 1088–1096. https://doi.org/10.1016/j.enpol.2013.05.062.

Costa, Evaldo, Ana Carolina Rodrigues Teixeira, Suellen Caroline Silva Costa, and Flavia L. Consoni. 2022. "Influence of Public Policies on the Diffusion of Wind and Solar PV Sources in Brazil and the Possible Effects of COVID-19." *Renewable and*

Sustainable Energy Reviews 162 (December 2021): 112449. https://doi.org/10.1016/j.rser.2022.112449.

Denault, Michel, Debbie Dupuis, and Sébastien Couture-Cardinal. 2009. "Complementarity of Hydro and Wind Power: Improving the Risk Profile of Energy Inflows." *Energy Policy* 37 (12): 5376–5384. https://doi.org/10.1016/j.enpol.2009.07.064.

Dhakouani, Asma, Essia Znouda, and Chiheb Bouden. 2021. "The Role of Social Discount Rate in Energy Modelling." In *Energy and Environmental Security in Developing Countries*, edited by Muhammad Asif, 475–500. Cham, Switzerland: Springer Nature Switzerland. https://doi.org/10.1007/978-3-030-63654-8_19.

Dietz, Frank, Hugo Brouwer, and Rob Wterings. 2008. "Energy Transition Experiments in the Netherlands." In *Managing the Transition to Renewable Energy—Theory and Practice from Local, Regional and Macro Perspectives*, edited by J. van den Bergh and F. Bruinsma, 385. Cheltenham and Northamptom: Edward Elgar.

Diógenes, Jamil Ramsi Farkat, João Claro, and José Coelho Rodrigues. 2019. "Barriers to Onshore Wind Farm Implementation in Brazil." *Energy Policy* 128 (September 2018): 253–266. https://doi.org/10.1016/j.enpol.2018.12.062.

Dubey, Kankana. 2020. "Energy Policy and Climate Change." In *Dynamics of Energy, Environment and Economy*, edited by Hassan Qudrat-Ullah and Muhammad Asif, 209–224. Cham, Switzerland: Springer Nature Switzerland. https://doi.org/10.1007/978-3-030-43578-3_10.

Dutra, Joisa, and Flavio Menezes. 2022. "Energy Transition in the Brazilian Electric Power System." *Competition and Regulation in Network Industries*: 178359172210887. https://doi.org/10.1177/17835917221088765.

Dutra, Ricardo Marques, and Alexandre Salem Szklo. 2008. "Incentive Policies for Promoting Wind Power Production in Brazil: Scenarios for the Alternative Energy Sources Incentive Program (PROINFA) under the New Brazilian Electric Power Sector Regulation." *Renewable Energy* 33 (1): 65–76. https://doi.org/10.1016/j.renene.2007.01.013.

EPE. 2022. "National Energy Balance—Base Year 2021 (in Portuguese)." Brasília, Brazil. www.epe.gov.br/sites-pt/publicacoes-dados-abertos/publicacoes/PublicacoesArquivos/publicacao-675/topico-631/BEN_S%C3%ADntese_2022_PT.pdf.

Faria, Felipe A. M. de, and Paulina Jaramillo. 2017. "The Future of Power Generation in Brazil: An Analysis of Alternatives to Amazonian Hydropower Development." *Energy for Sustainable Development* 41: 24–35. https://doi.org/10.1016/j.esd.2017.08.001.

Fearnside, Philip M. 2016. "Environmental and Social Impacts of Hydroelectric Dams in Brazilian Amazonia: Implications for the Aluminum Industry." *World Development* 77 (January): 48–65. https://doi.org/10.1016/j.worlddev.2015.08.015.

Fearnside, Philip M. 2018. "Belo Monte: Actors and Arguments in the Struggle Over Brazil's Most Controversial Amazonian Dam." *Revista Nera*, no. 42 (March): 162–185. https://doi.org/10.47946/rnera.v0i42.5691.

Fearnside, Philip M. 2019. "Environmental Justice and Amazonian Dams: 11—Safety Suspensions (in Portuguese)." *Amazônia Real*, August 28, 2019. https://amazoniareal.com.br/justica-ambiental-e-barragens-amazonicas-11-suspensoes-de-seguranca/.

Ferraz de Campos, Érica, Enio Bueno Pereira, Pieter van Oel, Fernando Ramos Martins, André Rodrigues Gonçalves, and Rodrigo Santos Costa. 2021. "Hybrid Power Generation for Increasing Water and Energy Securities During Drought: Exploring Local and Regional Effects in a Semi-Arid Basin." *Journal of Environmental Management* 294 (September): 112989. https://doi.org/10.1016/j.jenvman.2021.112989.

Gerston, Larry N. 2010. *Public Policy Making: Process and Principles*. 3rd ed. New York, NY: Routledge.

Goldemberg, José, and José Roberto Moreira. 2005. "Energy Policy in Brazil (in Portuguese)." *Estudos Avançados* 19 (55): 215–228. https://doi.org/10.1590/S0103-40142005000300015.

Gonçalves, André R., Arcilan T. Assireu, Fernando R. Martins, Madeleine S. G. Casagrande, Enrique V. Mattos, Rodrigo S. Costa, Robson B. Passos, et al. 2020. "Enhancement of

Cloudless Skies Frequency Over a Large Tropical Reservoir in Brazil." *Remote Sensing* 12 (17): 1–26. https://doi.org/10.3390/rs12172793.

Grin, John. 2020. "'Doing' System Innovations from Within the Heart of the Regime." *Journal of Environmental Policy & Planning* 22 (5): 682–694. https://doi.org/10.1080/1523908X.2020.1776099.

Han, Xie, Li Qifen, Yang Yongwen, Song Lifei, Ning Ning, and Jiang Xiumei. 2021. "Research on Flexibility Analysis of Integrated Energy System." Edited by M.S. Nazir and H.A. Aziz. *E3S Web of Conferences* 257 (May): 02005. https://doi.org/10.1051/e3sconf/202125702005.

Hendriks, Carolyn M., and John Grin. 2007. "Contextualizing Reflexive Governance: The Politics of Dutch Transitions to Sustainability." *Journal of Environmental Policy and Planning* 9 (3–4): 333–350. https://doi.org/10.1080/15239080701622790.

Herrera, Milton M., Federico Cosenz, and Isaac Dyner. 2019. "How to Support Energy Policy Coordination? Findings from the Brazilian Wind Industry." *Electricity Journal* 32 (8): 106636. https://doi.org/10.1016/j.tej.2019.106636.

Hess, Christoph Ernst Emil, and Eva Fenrich. 2017. "Socio-Environmental Conflicts on Hydropower: The São Luiz Do Tapajós Project in Brazil." *Environmental Science and Policy* 73: 20–28. https://doi.org/10.1016/j.envsci.2017.03.005.

Hill, Michael James. 2005. *The Public Policy Process*. 4th ed. Harlow, England: Pearson Longman.

Hochberg, Michael, and Rahmatallah Poudineh. 2021. "The Brazilian Electricity Market Architecture: An Analysis of Instruments and Misalignments." *Utilities Policy* 72 (January 2020): 101267. https://doi.org/10.1016/j.jup.2021.101267.

Hochstetler, Kathryn. 2021. *Political Economies of Energy Transition: Wind and Solar in Brazil and South Africa*. Cambridge, New York, Melbourne, New Delhi, and Singapore: Cambridge University Press.

Hoff, Thomas E., and Richard Perez. 2012. "Modeling PV Fleet Output Variability." *Solar Energy* 86 (8): 2177–2189. https://doi.org/10.1016/j.solener.2011.11.005.

IPCC. 2018. *Global Warming of 1.5°C. An IPCC Special Report on the Impacts of Global Warming of 1.5°C above Pre-Industrial Levels and Related Global Greenhouse Gas Emission Pathways, in the Context of Strengthening the Global Response to the Threat of Climate Change*. Edited by Valérie Masson-Delmotte, Panmao Zhai, Hans-Otto Pörtner, Debra Roberts, Jim Skea, Priyadarshi R. Shukla, Anna Pirani, et al. www.ipcc.ch/site/assets/uploads/sites/2/2019/06/SR15_Full_Report_Low_Res.pdf.

Jurasz, Jakub, F.A. Canales, A. Kies, M. Guezgouz, and A. Beluco. 2020. "A Review on the Complementarity of Renewable Energy Sources: Concept, Metrics, Application and Future Research Directions." *Solar Energy* 195 (January): 703–724. https://doi.org/10.1016/j.solener.2019.11.087.

Jurasz, Jakub, Paweł B. Dąbek, Bartosz Kaźmierczak, Alexander Kies, and Marcin Wdowikowski. 2018. "Large Scale Complementary Solar and Wind Energy Sources Coupled with Pumped-Storage Hydroelectricity for Lower Silesia (Poland)." *Energy* 161 (October): 183–192. https://doi.org/10.1016/j.energy.2018.07.085.

Kanger, Laur, and J. Schot. 2019. "Deep Transitions: Theorizing the Long-Term Patterns of Socio-Technical Change." *Environmental Innovation and Societal Transitions* 32 (September 2017): 7–21. https://doi.org/10.1016/j.eist.2018.07.006.

Kemp, René. 2010. "The Dutch Energy Transition Approach." *International Economics and Economic Policy* 7 (2): 291–316. https://doi.org/10.1007/s10368-010-0163-y.

Kemp, René, Derk Loorbach, and Jan Rotmans. 2007. "Transition Management as a Model for Managing Processes of Co-Evolution towards Sustainable Development." *International Journal of Sustainable Development and World Ecology* 14 (1): 78–91. https://doi.org/10.1080/13504500709469709.

Kempton, Willett, Felipe M. Pimenta, Dana E. Veron, and Brian A. Colle. 2010. "Electric Power from Offshore Wind via Synoptic-Scale Interconnection." *Proceedings of the*

National Academy of Sciences of the United States of America 107 (16): 7240–7245. https://doi.org/10.1073/pnas.0909075107.

Kern, Florian, and Adrian Smith. 2008. "Restructuring Energy Systems for Sustainability? Energy Transition Policy in the Netherlands." *Energy Policy* 36 (11): 4093–4103. https://doi.org/10.1016/j.enpol.2008.06.018.

Kingdon, J. W. 2003. *Agendas, Alternatives, and Public Policies.* Longman Classics Series. New York, NY: Longman.

Kroposki, Benjamin, Brian Johnson, Yingchen Zhang, Vahan Gevorgian, Paul Denholm, Bri Mathias Hodge, and Bryan Hannegan. 2017. "Achieving a 100% Renewable Grid: Operating Electric Power Systems with Extremely High Levels of Variable Renewable Energy." *IEEE Power and Energy Magazine* 15 (2): 61–73. https://doi.org/10.1109/MPE.2016.2637122.

Kwon, Tae Hyeong. 2018. "Policy Synergy or Conflict for Renewable Energy Support: Case of RPS and Auction in South Korea." *Energy Policy* 123 (September): 443–449. https://doi.org/10.1016/j.enpol.2018.09.016.

Lazaro, L. L.B., R. S. Soares, C. Bermann, F. M.A. Collaço, L. L. Giatti, and S. Abram. 2022. "Energy Transition in Brazil: Is There a Role for Multilevel Governance in a Centralized Energy Regime?" *Energy Research and Social Science* 85 (November 2021): 102404. https://doi.org/10.1016/j.erss.2021.102404.

Lima, M. A., L. F.R. Mendes, G. A. Mothé, F. G. Linhares, M. P.P. de Castro, M. G. da Silva, and M. S. Sthel. 2020. "Renewable Energy in Reducing Greenhouse Gas Emissions: Reaching the Goals of the Paris Agreement in Brazil." *Environmental Development* 33 (September 2016): 100504. https://doi.org/10.1016/j.envdev.2020.100504.

Loorbach, Derk, and Jan Rotmans. 2010. "The Practice of Transition Management: Examples and Lessons from Four Distinct Cases." *Futures* 42 (3): 237–246. https://doi.org/10.1016/j.futures.2009.11.009.

Lucena, Juliana de Almeida Yanaguizawa, and Klayton Ângelo Azevedo Lucena. 2019. "Wind Energy in Brazil: An Overview and Perspectives under the Triple Bottom Line." *Clean Energy* 3 (2): 69–84. https://doi.org/10.1093/ce/zkz001.

Lucon, Oswaldo, Viviane Romeiro, and Taryn Fransen. 2015. *Bridging the Gap between Energy and Climate Policies in Brazil.* São Paulo, Brazil: World Resources Institute.

Lund, Henrik. 2007. "Renewable Energy Strategies for Sustainable Development." *Energy* 32 (6): 912–919. https://doi.org/10.1016/j.energy.2006.10.017.

Lund, Henrik, and Ebbe Münster. 2006. "Integrated Energy Systems and Local Energy Markets." *Energy Policy* 34 (10): 1152–1160. https://doi.org/10.1016/j.enpol.2004.10.004.

Lund, Peter D., Juuso Lindgren, Jani Mikkola, and Jyri Salpakari. 2015. "Review of Energy System Flexibility Measures to Enable High Levels of Variable Renewable Electricity." *Renewable and Sustainable Energy Reviews.* https://doi.org/10.1016/j.rser.2015.01.057.

Malmborg, Fredrik von. 2021. "Exploring Advocacy Coalitions for Energy Efficiency: Policy Change through Internal Shock and Learning in the European Union." *Energy Research & Social Science* 80 (August): 102248. https://doi.org/10.1016/j.erss.2021.102248.

Mayer, Adam, Laura Castro-Diaz, Maria Claudia Lopez, Guillaume Leturcq, and Emilio F. Moran. 2021. "Is Hydropower Worth It? Exploring Amazonian Resettlement, Human Development and Environmental Costs with the Belo Monte Project in Brazil." *Energy Research and Social Science* 78 (January): 102129. https://doi.org/10.1016/j.erss.2021.102129.

MME/EPE. 2020. "National Energy Plan 2050 (in Portuguese)." Brasilia, Brazil: Ministry of Mines and Energy (MME), Energy Research Company (EPE). http://antigo.mme.gov.br/documents/36208/468569/Relatório+Final+do+PNE+2050/77ed8e9a-17ab-e373-41b4-b871fed588bb.

Moran, Emilio F., Maria Claudia Lopez, Nathan Moore, Norbert Müller, and David W. Hyndman. 2018. "Sustainable Hydropower in the 21st Century." *Proceedings of the National Academy of Sciences*, 201809426. https://doi.org/10.1073/pnas.1809426115.

Moretto, Evandro Mateus, Carina Sernaglia Gomes, Daniel Rondinelli Roquetti, and Carolina de Oliveira Jordão. 2012. "History, Trends and Perspectives in the Spatial Planning of Brazilian Hydroelectric Plants: The Former and Current Amazon Frontier (in Portuguese)." *Ambiente & Sociedade* 15 (3): 141–164. https://doi.org/10.1590/S1414-753X2012000300009.

Motta, C. M. 2006. "The Brazilian Neoliberal Model and the Electricity Sector: Restructuring and Crises (1995–2005) (in Portuguese)." Pontifícia Universidade Católica de São Paulo (PUC/SP).

Nascimento, Sabrina Mesquita do. 2017. "Violence and State of Exception in the Brazilian Amazon: A Study on the Implementation of the Belo Monte Hydroelectric Plant on the Xingu River (PA) (in Portuguese)." Doctorate degree, Belem, PA, Brazil: Universidade Federal do Pará. http://repositorio.ufpa.br/jspui/handle/2011/10428.

OAS. 1988. "Integrated Energy Development—Experiences of the Organization of American States." Washington, DC: OAS.

Oliveira, Adilson de. 2007. "Political Economy of the Brazilian Power Industry Reform." In *The Political Economy of Power Sector Reform*, edited by David G. Victor and Thomas C. Heller, 31–75. Cambridge: Cambridge University Press. https://doi.org/10.1017/CBO9780511493287.003.

ONS. 2021. "Power Generation History—Brazilian Transmission System Operator." www.ons.org.br/Paginas/resultados-da-operacao/historico-da-operacao/geracao_energia.aspx.

Pereira de Lucena, André Frossard, Alexandre Salem Szklo, Roberto Schaeffer, and Ricardo Marques Dutra. 2010. "The Vulnerability of Wind Power to Climate Change in Brazil." *Renewable Energy* 35 (5): 904–912. https://doi.org/10.1016/j.renene.2009.10.022.

Peterseim, Juergen H., Udo Hellwig, Amir Tadros, and Stuart White. 2014. "Hybridisation Optimization of Concentrating Solar Thermal and Biomass Power Generation Facilities." *Solar Energy* 99 (January): 203–214. https://doi.org/10.1016/j.solener.2013.10.041.

Pimenta, Felipe M., and Arcilan T. Assireu. 2015. "Simulating Reservoir Storage for a Wind-Hydro Hybrid System." *Renewable Energy* 76: 757–767. https://doi.org/10.1016/j.renene.2014.11.047.

Relva, Stefania Gomes, Vinícius Oliveira da Silva, André Luiz Veiga Gimenes, Miguel Edgar Morales Udaeta, Peta Ashworth, and Drielli Peyerl. 2021. "Enhancing Developing Countries' Transition to a Low-Carbon Electricity Sector." *Energy* 220. https://doi.org/10.1016/j.energy.2020.119659.

Ritchie, Hannah, and Max Roser. 2020. "Energy." Published Online at Our World in Data. 2020. https://ourworldindata.org/energy.

Rotmans, Jan, René Kemp, and Marjolein Van Asselt. 2001. "More Evolution than Revolution: Transition Management in Public Policy." *Foresight* 3 (1): 15–31. https://doi.org/10.1108/14636680110803003.

Rotmans, Jan, and Derk Loorbach. 2008. "Transition Management: Reflexive Governance of Societal Complexity through Searching, Learning and Experimenting." In *Managing the Transition to Renewable Energy—Theory and Practice from Local, Regional and Macro Perspectives*, edited by J. van den Bergh and F. Bruinsma, 385. Cheltenham and Northamptom: Edward Elgar.

Ruffato-Ferreira, Vera, Renata da Costa Barreto, Antonio Oscar Júnior, Wanderson Luiz Silva, Daniel de Berrêdo Viana, José Antonio Sena do Nascimento, and Marcos Aurélio Vasconcelos de Freitas. 2017. "A Foundation for the Strategic Long-Term Planning of the Renewable Energy Sector in Brazil: Hydroelectricity and Wind Energy in the

Face of Climate Change Scenarios." *Renewable and Sustainable Energy Reviews* 72 (October 2015): 1124–1137. https://doi.org/10.1016/j.rser.2016.10.020.

Sabatier, Paul A., and Hank C. Jenkins-Smith. 1993. "The Policy-Oriented Learning." In *Policy Change and Learning: And Advocacy Coalition Approach*, edited by Paul A. Sabatier and Hank C. Jenkins-Smith, 41–56. Nashville, TN: Westview Press.

Sampaio, Alexandre Andrade, and Matthew McEvoy. 2016. "The Dynamics of Energy Policy Securitization in Brazil and the Consequences for Tribal Peoples." *Homa Publica— Revista Internacional de Derechos Humanos y Empresas* 1 (1): 1–19. https://periodicos.ufjf.br/index.php/HOMA/article/view/30459/20495.

Santos, Rangel E., Ricardo M. Pinto-Coelho, Rogério Fonseca, Nadson R. Simões, and Fabrício B. Zanchi. 2018. "The Decline of Fisheries on the Madeira River, Brazil: The High Cost of the Hydroelectric Dams in the Amazon Basin." *Fisheries Management and Ecology* 25 (5): 380–391. https://doi.org/10.1111/fme.12305.

Sauer, Ildo Luís. 2015. "The Genesis and Permanence of the Electric Sector Crisis in Brazil (in Portuguese)." *Revista USP*, no. 104: 145. https://doi.org/10.11606/issn.2316-9036.v0i104p145-174.

Scabin, Flávia Silva, Nelson Novaes Pedroso Jr, and Julia Cortez da Cunha Cruz. 2015. "Judicialization of Large Enterprises in Brazil: A View on the Impacts of the Installation of Hydroelectric Plants on Local Populations in the Amazon (in Portuguese)." *Revista Pós Ciências Sociais* 11 (22): 129–150. http://periodicoseletronicos.ufma.br/index.php/rpcsoc/article/view/3418.

Schaeffer, Roberto, André Frossard Pereira de Lucena, Régis Rathmann, Alexandre Szklo, Rafael Soria, and Rodriguez Chavez. 2015. "Who Drives Climate-Relevant Policies in Brazil?" Rio de Janeiro: Universidade Federal do Rio de Janeiro, Instituto Alberto Luiz Coimbra de Pós-Graduação e Pesquisa de Engenharia, Energy Planning Program–PPE/COPPE/UFRJ.

Schaeffer, Roberto, Alexandre Salem Szklo, André Frossard Pereira de Lucena, Bruno Soares Moreira Cesar Borba, Larissa Pinheiro Pupo Nogueira, Fernanda Pereira Fleming, Alberto Troccoli, Mike Harrison, and Mohammed Sadeck Boulahya. 2012. "Energy Sector Vulnerability to Climate Change: A Review." *Energy* 38 (1): 1–12. https://doi.org/10.1016/j.energy.2011.11.056.

Schlömer, Steffen, Thomas Bruckner, Lew Fulton, Edgar Hertwich, Alan McKinnon, Daniel Perczyk, Joyashree Roy, et al. 2014. "Annex III: Technology-Specific Cost and Performance Parameters." In *Climate Change 2014: Mitigation of Climate Change. Contribution of Working Group III to the Fifth Assessment Report of the Intergovernmental Panel on Climate Change*, edited by O. Edenhofer, R. Pichs-Madruga, Y. Sokona, E. Farahani, S. Kadner, K. Seyboth, A. Adler, et al., 28. Cambridge: Cambridge University Press.

SEEG-Brazil. 2022. "The Greenhouse Gas Emission and Removal Estimating System (SEEG)." http://seeg.eco.br/en/o-que-e-o-seeg.

Silva, Allan Rodrigues, Felipe Mendonça Pimenta, Arcilan Trevenzoli Assireu, and Maria Helena Constantino Spyrides. 2016. "Complementarity of Brazil's Hydro and Offshore Wind Power." *Renewable and Sustainable Energy Reviews* 56: 413–427. https://doi.org/10.1016/j.rser.2015.11.045.

Smil, Vaclav. 2003. *Energy at the Crossroads: Global Perspectives and Uncertainties*. Cambridge, MA: The MIT Press.

Smil, Vaclav. 2017. *Energy and Civilization: A History*. Cambridge, MA: The MIT Press. https://mitpress.mit.edu/books/energy-and-civilization.

Smith, Adrian, Jan Peter Voß, and John Grin. 2010. "Innovation Studies and Sustainability Transitions: The Allure of the Multi-Level Perspective and Its Challenges." *Research Policy* 39 (4): 435–448. https://doi.org/10.1016/j.respol.2010.01.023.

Soares, Ítalo Nogueira, Rodrigo Gava, and José Antônio Puppim de Oliveira. 2021. "Political Strategies in Energy Transitions: Exploring Power Dynamics, Repertories of Interest

Groups and Wind Energy Pathways in Brazil." *Energy Research and Social Science* 76 (October 2020). https://doi.org/10.1016/j.erss.2021.102076.

Soares, M.C. Borba, Bruno, Alexandre Szklo, and Roberto Schaeffer. 2012. "Plug-in Hybrid Electric Vehicles as a Way to Maximize the Integration of Variable Renewable Energy in Power Systems: The Case of Wind Generation in Northeastern Brazil." *Energy* 37 (1): 469–481. https://doi.org/10.1016/j.energy.2011.11.008.

Soito, João Leonardo da Silva, and Marcos Aurélio Vasconcelos Freitas. 2011. "Amazon and the Expansion of Hydropower in Brazil: Vulnerability, Impacts and Possibilities for Adaptation to Global Climate Change." *Renewable and Sustainable Energy Reviews* 15 (6): 3165–3177. https://doi.org/10.1016/j.rser.2011.04.006.

Sørensen, Bent. 1991. "A History of Renewable Energy Technology." *Energy Policy* 19 (1): 8–12. https://doi.org/10.1016/0301-4215(91)90072-V.

Soria, Rafael, André F.P. Lucena, Jan Tomaschek, Tobias Fichter, Thomas Haasz, Alexandre Szklo, Roberto Schaeffer, Pedro Rochedo, Ulrich Fahl, and Jürgen Kern. 2016. "Modelling Concentrated Solar Power (CSP) in the Brazilian Energy System: A Soft-Linked Model Coupling Approach." *Energy* 116: 265–280. https://doi.org/10.1016/j.energy.2016.09.080.

Souza Braga, F. de. 2020. "A Ditadura Militar e a Governança Da Água No Brasil : Ideologia, Poderes Político-Econômico e Sociedade Civil Na Construção Das Hidrelétricas de Grande Porte." In *A Ditadura Militar e a Governança Da Água No Brasil*. Universiteit Leiden. https://doi.org/10.1201/9781003047896.

Stickler, Claudia M., Michael T. Coe, Marcos H. Costa, Daniel C. Nepstad, David G. McGrath, Livia C. P. Dias, Hermann O. Rodrigues, and Britaldo S. Soares-Filho. 2013. "Dependence of Hydropower Energy Generation on Forests in the Amazon Basin at Local and Regional Scales." *Proceedings of the National Academy of Sciences* 110 (23): 9601–9606. https://doi.org/10.1073/pnas.1215331110.

Sun, Wei, and Gareth P. Harrison. 2019. "Wind-Solar Complementarity and Effective Use of Distribution Network Capacity." *Applied Energy* 247 (April): 89–101. https://doi.org/10.1016/j.apenergy.2019.04.042.

Timpe, Kelsie, and David Kaplan. 2017. "The Changing Hydrology of a Dammed Amazon." *Science Advances* 3 (11): 1–14. https://doi.org/10.1126/sciadv.1700611.

Tolmasquim, Maurício T., Tiago de Barros Correia, Natália Addas Porto, and Wikus Kruger. 2021. "Electricity Market Design and Renewable Energy Auctions: The Case of Brazil." *Energy Policy* 158. https://doi.org/10.1016/j.enpol.2021.112558.

Tong, Dan, David J. Farnham, Lei Duan, Qiang Zhang, Nathan S. Lewis, Ken Caldeira, and Steven J. Davis. 2021. "Geophysical Constraints on the Reliability of Solar and Wind Power Worldwide." *Nature Communications* 12 (1): 6146. https://doi.org/10.1038/s41467-021-26355-z.

Tosun, Jale, Sophie Biesenbender, and Kai Schulze. 2015. "Building the EU's Energy Policy Agenda: An Introduction." In *Energy Policy Making in the EU—Building the Agenda*, edited by Jale Tosun, Sophie Biesenbender, and Kai Schulze, 1–17. London: Springer. https://doi.org/10.1007/978-1-4471-6645-0_1.

Unruh, Gregory C. 2002. "Escaping Carbon Lock-In." *Energy Policy* 30: 317–325.

Villas-Boas, Andre, Biviany Rojas Garzon, Carolina Reis, Leonardo Amorim, and Leticia Leite. 2015. *Belo Monte—There Are No Conditions for the Operating License*. São Paulo, Brazil: Instituto Socioambiental (ISA).

Voß, Jan Peter, Adrian Smith, and John Grin. 2009. "Designing Long-Term Policy: Rethinking Transition Management." *Policy Sciences* 42 (4): 275–302. https://doi.org/10.1007/s11077-009-9103-5.

Wang, Qingwei, Yongli Wang, Yuze Ma, Yang Ma, Huanran Dong, Chengyuan Qi, Fuhao Song, and Feifei Huang. 2020. "Planning Optimization of Integrated Energy

System Considering Economy and Integrated Energy Efficiency." *IOP Conference Series: Earth and Environmental Science* 546 (August): 022035. https://doi.org/10.1088/1755-1315/546/2/022035.

Weible, Christopher M., and Karin Ingold. 2018. "Why Advocacy Coalitions Matter and Practical Insights about Them." *Policy and Politics* 46 (2): 325–343. https://doi.org/10.1332/030557318X15230061739399.

Xiang, Yue, Hanhu Cai, Chenghong Gu, and Xiaodong Shen. 2020. "Cost-Benefit Analysis of Integrated Energy System Planning Considering Demand Response." *Energy* 192 (February): 116632. https://doi.org/10.1016/j.energy.2019.116632.

Yildiz, I. 2021. *Greenhouse Engineering: Integrated Energy Management*. Boca Raton, FL: CRC Press—Taylor & Francis Group.

Zhang, Haijing, Weishuai Wang, Lingkai Zhu, Zhuoxin Sun, Decao Ma, and Wenjiao Chen. 2021. "Prospect of Typical Application of Integrated Energy System." *IOP Conference Series: Earth and Environmental Science* 651 (February): 022030. https://doi.org/10.1088/1755-1315/651/2/022030.

15 Decision-Making in Energy and Environmental Systems Based on Water-Energy-Food Security Nexus Principles

Abdolvahhab Fetanat, Mohsen Tayebi,
Hossein Mofid

15.1 INTRODUCTION

A quick increase in global population growth, urbanization, growing prosperity, and climate change are putting unsustainable pressures on resources. The need for water, food, and energy is predicted to increase by 30–50% in the next two decades, while economic inequalities incentivize short-term reactions in the production and consumption of resources and impair long-term sustainability. Lack of these three resources can lead to political and social instability, geopolitical incompatibility, and irreparable environmental damages. Any policy that focuses on sectors of the water-energy-food (WEF) security nexus by considering its interconnections and principles has satisfying consequences. The fears of global resource shortage that followed the first wave of sustainable development (SD) in the 20th century are still quite evident in the second decade of the 21st century (Bizikova et al. 2013). WEF systems are interconnected. These interconnections become more complicated as the demand for resources rises with growing population and altering consumption patterns. In the meantime, major issues—especially climate change and land-utilization patterns—limit the affordability and reliability of the existing system for meeting the growing demand. These dynamics pose considerable risks to sustainable development and resource security in many businesses, communities, and governments (Rabia et al. 2015). Recently, more attention has been given to advancing frameworks that can describe and assess the relevance of the elements of the WEF security nexus. These frameworks also help researchers in administering case studies and, finally, recognizing effective actions and policies. The approach of the WEF nexus is crucial and pivotal to giving an investment framework in land usage to increase productivity

DOI: 10.1201/9781032715438-15

while improving natural, social, and economic capital as protection against long-term risks of WEF security. With the identification of the relationships among the principles of the WEF nexus, there are challenges that should be considered in all three portions when analyzing consequences and planning for decisions, policies, actions, and investments. Such challenges are as follows:

- A robust push to utilize biofuels, which could affect the availability of water and land for other targets, specifically to produce food (Bazilian et al. 2011; Hellegers et al. 2008),
- The increase of requirements for drinking water, while water demand is quickly developing for production of food, for activities of energy processing, and for urban needs (Hellegers et al. 2008),
- Regions of irrigable land utilizing water that is significant to produce food and hydropower (McCornick, Awulachew, and Abebe 2008), and
- The rising dependence on energy-intensive water desalination as a source of drinking water and irrigation, specifically in fast-growing regions (Bazilian et al. 2011).

The WEF security nexus is of great importance in various sectors such as industries, agriculture, environment, and health. In this way, the principles of water security (access, safety, and affordability), energy security (availability, accessibility, affordability, acceptability, applicability, and adaptability), and food security (availability, access, utilization, and stability) can be used as the assessment criteria for the decision-making in these sectors. In the field of the management of energy and environmental (E&E) systems, which has many aspects with considerable uncertainty, multi-criteria decision-making (MCDM) methods have many applications to select sites and technologies (Fetanat et al. 2019; Fetanat and Khorasaninejad 2015).

The technology of hydroelectric power has been applied in the world for many years. It is one of the largest and most popular sources in sector of renewable energy (Cai, Ye, and Gholinia 2020; Fei, Xuejun, and Razmjooy 2019). The approach of utilizing hydroelectric energy is mostly associated with stored energy behind dams. However, lakes constructed behind dams often originate climate change. Moreover, it can sometimes result in social, environmental, and health impacts for the residents of the adjoining regions (Tian et al. 2020; Popa et al. 2020). There is another approach for utilizing this energy. It is the utilization of small hydropower (SHP) stations, which is used for generating energy by river water flow (Mayeda and Boyd 2020; Cai, Ye, and Gholinia 2020). SHP stations have major potential for general electricity production with few social and environmental issues (Khodaei et al. 2018; Mayeda and Boyd 2020; Tian et al. 2020). These stations are placed in the path of rivers that stream from mountainous heights to the exit of the watershed and produce electricity by controlling the water stream of the rivers (Ghorbani, Makian, and Breyer 2019). SHP stations supply electricity from natural sources and do not cause any damage or pollution to the environment (Ghorbani, Makian, and Breyer 2019; Craig et al. 2019). Hydropower has received an extensive amount of interest at various scales (Bhandari, Saptalena, and Kusch 2018). The location of an SHP plant that can be recognized as an E&E system can be assessed on the basis of sustainability concepts by applying various decision-making tools (Tamm and Tamm 2020; Alp, Akyüz, and Kucukali 2020; Tian et al. 2020). As a result, in the current chapter, the site selection is analyzed for an E&E system called an SHP plant based on the principles of the WEF security nexus using an MCDM method to improve the agriculture situation of Behbahan city in the south of Iran. Applying these principles in decisions associated with E&E systems can help in the implementation of circular economics strategies and policies and, as a suitable tool, navigate local societies towards the Sustainable Development Goals (SDGs) (Fetanat, Tayebi, and Mofid 2021). The WEF nexus security and SHP are interconnected. The relationship of the main factors with this plant is illustrated in Figure 15.1. The four elements that affect this relationship are population growth, urbanization, growing prosperity, and climate change.

Currently, the WEF nexus has been taken into consideration as a better approach for understanding the complicated interactions among systems related to these sources. The security of these three linked sources is urgent to survive humanity. The approach to the WEF nexus requires analyzing the criteria affecting the livelihood of humans and the contribution of institutions for the improvement of managing resources. The efficacies of the management issues of water, energy, and food are obviously perceived in the E&E systems operations. The energy generated from the water flow can be used as a section of the required energy for society. The produced energy is clean energy. The circulation of water by water turbines used in these plants increases the dissolved oxygen in water sources, which improves the quality

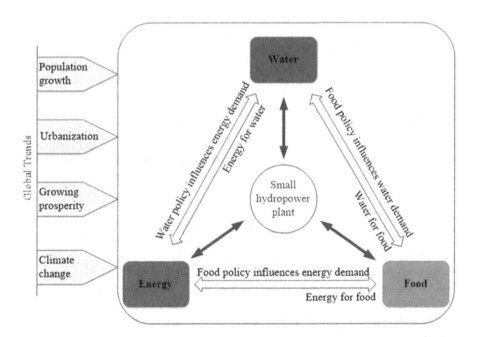

FIGURE 15.1   The interfaces among water, energy, food, and a small hydropower plant.

Adapted from Rasul and Sharma 2016; Fetanat, Tayebi, and Mofid (2021).

of water. Also, energy and water are applied for agricultural purposes, and agriculture is a food producer. In this regard, proper site selection for these plants considering WEF security nexus principles can raise the nexus security of three vital sources of water, energy, and food and mitigate the challenges of resource management.

15.1.1 LITERATURE REVIEW

The close relationship among the three vital sources of water, energy, and food has caused the WEF nexus approach to be taken into consideration as an issue of global concern. This approach was first presented during a conference on WEF security in 2011 in Bonn, Germany (Hoff 2011). The relevance of protecting these three sources was one of the findings of the conference. In 2013, the United Nations Economic and Social Council (ECOSOC) released a report on the status of water, energy, and food for the Asia-Pacific area. In 2014, the Food and Agriculture Organization of the United Nations (FAO) called for policy-makers in the area of sustainable agricultural development to focus on the safety of agriculture and food as a reference. In recent years, the significance of the WEF security nexus is perceptible in many sections of study (for deep reviews, consider the following references: Amer, Adeel, and Saleh 2017; Gondhalekar and Ramsauer 2017; Yuling et al. 2016; Howarth and Monasterolo 2016; Kurian 2017; Owen, Scott, and Barrett 2018; Parkinson et al. 2018; Rasul and Sharma 2016).

Various industries and sectors are moving towards SD during the period of transition. Thus, various management issues that were formerly taken into consideration only technically and economically now should be assessed from other aspects, including social, ethical, and environmental issues. In recent years, owing to the complex and multidimensional nature of such management issues and the capabilities of the MCDM techniques, many studies have focused on employing proper MCDM techniques (see Nie et al. 2018; Ren 2018a, 2018b; Ren and Dong 2018; Zhang et al. 2017; Fetanat and Khorasaninejad 2015)). Site and technology selections are samples of these multidimensional problems.

Various studies have been carried out to analyze the potential of SHP from rivers water flow, and most of them concentrate on traditional techniques and contained field examination. Usually, high-potential locations for SHP stations are located in distant mountain regions and impassable paths. Thus, the assessment of these locations via traditional techniques requires a lot of time and money because accessing these areas is a difficult and complex process (Aghajani and Ghadimi 2018; Tian et al. 2020; Alp, Akyüz, and Kucukali 2020; Tamm and Tamm 2020; Bhandari, Saptalena, and Kusch 2018). There are regions near rivers that have more potential than other parts of the watershed, which can expand the capacity of generating electricity through electricity project development. Finding locations with high potential for SHP plants is needed, but it is hard to do through traditional approaches. Because some locations with higher potential may be ignored and the error rate in detecting and selecting sustainable and appropriate locations may increase (Tian et al. 2020; Tamm and Tamm 2020).

Selecting an appropriate site should be done with attention and care to various principles affecting the issue. Regional priorities and different environmental situations have great significance for using an SHP plant in the E&E sector. The multiplicity of influential variables and the existence of conflicts among these variables in selecting the most proper site for producing clean energy from water flow have led to the use of MCDM techniques.

A precise assessment of the mentioned studies shows that the effort to give appropriate innovative structure for decision-making principles, especially from the resource security policy-making perspective, has been less realized in the problem literature. Most articles have concentrated upon the management factors of the decision-making structure of site selection for an SHP plant and not enough attention has been paid to the assessment principles. Selecting an exhaustive framework of principles is significant, as much as the selection of the most suitable decision-making techniques. In most studies, the assessment principles have been utilized on the basis of the general aspects of sustainability concepts, without regard to the way that these aspects affect the different sectors of E&E system management. From the point of view of policymaking, a proper evaluation framework that conducts society in the direction of sustainable growth is more important than the approach to decision-making (Fetanat et al. 2019).

The great potential of the SHP plant as an environmentally friendly option and its contribution to secure water, energy, and food emphasize the idea that the principles of WEF nexus security can be an appropriate assessment framework for decision-making in this sector. The WEF security nexus is a novel field with significant specifications that can assist researchers by providing them a useful framework of suitable and reliable principles. The international community is currently attempting a transition to SD. Applying the principles of the WEF security nexus as an evaluation framework, which consists of the sustainability aspects needed to assess the site selection of an SHP plant, can be constructive and valuable. Moreover, choosing a framework of WEF security nexus principles can be a useful tool to advance social acceptance. In addition, the WEF security nexus inclusion in the actions of E&E system management can bring in managers' opinions for constructing an SHP plant that helps to reduce the utilization of fossil fuels and mitigates greenhouse gas emissions (especially carbon). Also, it can create job opportunities and new business. Installing and establishing these types of plants affects sustainability and adaptation to the climate change issues of the region.

This work is intended to be a basis for a discussion among experts working on the practical applications and theoretical approaches of the WEF security nexus. The final goal of the research is to consider and place the principles of the WEF security nexus as an effective and useful framework for E&E system management and decision-making.

15.1.2 Rationale for the Study

The unique situation of an SHP plant as an influential option (in the circular economy-supply chain of the three vital sources of water, energy, and food) requires the utilization of the WEF security nexus principles for sustainable site management of the plant. This decision includes a broad range of economic, social, human, and environmental elements that can significantly influence the sustainable site selection of the SHP plant. To achieve this, first, an evaluation framework must be determined based upon the principles of the WEF security nexus and then enforced with the assistance of suitable decision-making techniques as a decision support system (DSS). This work contributes to present studies by presenting the WEF security nexus as a robust assessment framework for the first time. In the novel presented framework, the principles of water security, energy security, and food security have been assessed. This framework has been enforced by the use of a MCDM technique on the basis of simultaneous evaluation of criteria and alternatives (SECA) under a fuzzy environment. The SECA technique focuses on obtaining the weights of the principles and ranking the alternatives according to the WEF security nexus concepts. The assessment framework and MCDM technique have been employed to select the most sustainable site for constructing an SHP plant in one of the southern cities of Iran, Behbahan. The region of study has high degrees of installed capacity; the potential of SHP plant remains untapped and is often hindered by a lack of proper managerial practices. Four sites on the Maroun River are selected as the alternatives. They are Beheshtak (A1), Karya (A2), Kharestan (A3), and Lasbid (A4). These four candidates are the most appropriate regions to construct the SHP plant in Behbahan. The principles associated with the nexus of WEF security have been used as the criteria framework. Assessment of the criteria to select the most sustainable site has been fulfilled based upon the opinions of five experts (decision-makers) from multiple related organizations. Therefore, the suggested DSS and findings of this work can be a guide for policymakers and managers. Also, it can give them novel insights to assess and select sustainable sites for SHP plants.

The other parts of the research are as follows. First, the decision tree of the research is given in Section 15.2.1. Second, the principles of the WEF security nexus as the assessment framework are described in Section 15.2.2. Third, the alternatives studied are introduced in Section 15.2.3. Fourth, the MCDM technique is presented in Section 15.2.4. In Section 15.3, results and discussion are shown. Managerial insights and conclusions are explained in Sections 15.3.3 and 15.4, respectively.

15.2   METHODS

This research gives a robust DSS based upon the fuzzy SECA technique and the principles of the WEF security nexus for the sustainable site selection of an SHP plant as an environmentally friendly option in E&E system management. A schematic of the problem flowchart is indicated in Figure 15.2.

15.2.1   Decision Tree of the Research

The decision tree has three levels. The first, second, and third levels are the target, the WEF security nexus principles, and the proposed alternatives, respectively. The target was to select the most sustainable site to construct the SHP plant. Then, to define the assessment framework, the principles of the WEF security nexus are presented. The introduced framework contains three sectors: a) water security, b) energy security, and c) food security. Different principles are introduced for each sector.

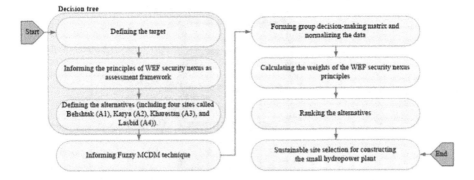

FIGURE 15.2 The flowchart of the research.

- **In the water security sector:**
  1. Access
  2. Safety
  3. Affordability

- **In the energy security sector:**
  1. Availability
  2. Accessibility
  3. Affordability
  4. Acceptability
  5. Applicability
  6. Adaptability

- **In the food security sector:**
  1. Availability
  2. Access
  3. Utilization
  4. Stability

The studied alternatives of the research are four proposed areas, Beheshtak (A1), Karya (A2), Kharestan (A3), and Lasbid (A4). The three levels of the decision tree of the study are shown in Figure 15.3.

15.2.2 Principles of WEF Security Nexus as the Assessment Framework

The concepts of the WEF security nexus attempt to merge the economic, environmental, health, and human aspects into planning and decision-making in various sectors of the WEF system as an exhaustive assessment framework. The use of this novel united framework can mitigate the major challenges in storage. It is also can be helpful to reduce poverty and increase the security of the three vital sources of water, energy, and food. It will assist us in controlling the very important issues of climate change via reconciliation and adaptation of the trends of conflict in various

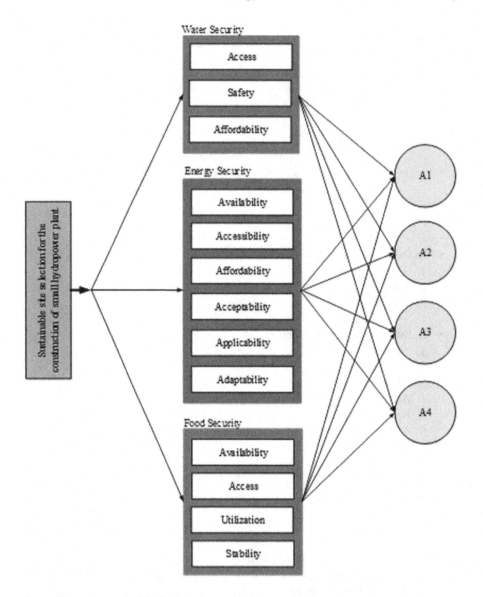

FIGURE 15.3 The decision tree of the research.

sectors of producing, distributing, and consuming energy. Hence, the principles of the WEF security nexus have been chosen as a valuable and useful framework for assessing various scenarios of implementing projects related to E&E systems such as an SHP plant as a significant renewable energy solution to adequately respond to the challenge of rural areas. Also, it can help in advancing SD in these regions. Pursuant to the nexus of water-energy-food security, three sets of principle systems (water, energy, and food security principles) have been taken into consideration. Figure 15.4 shows the principle framework, the relevant principles, and the details of each one.

Principles framework	Principles	Principles details
Water security	Access:	The geographical location of water resources is significant so that people can utilize it properly.
	Safety:	The health of people is crucial to a society, so utilizing clean water resources should be of the utmost significant.
	Affordability:	It is significant that people utilize water resources at the least cost.
Energy security	Availability:	One of the needs of human life is energy utilization. It must be made available to the public in a variety of ways.
	Accessibility:	The measure of government support for the people is comprised in this principle.
	Affordability:	All people, including the poor, should not pay more than 10 percent of their income for energy services. This indicates that people's financial power is important in the use of energy resources.
	Acceptability:	The projects of energy sources should be enforced in a way that addresses regional priorities and environmental issues.
	Applicability:	Using new systems and technologies in the utilization of energy sources is significant.
	Adaptability:	Energy resource projects must be capable of adaptation to new environmental and technological conditions.
Food security	Availability:	People deserve high-quality food. So food should be fairly distributed to the public.
	Access:	It is significant to have suitable access to food.
	Utilization:	The way of food should be consumed is significant.
	Stability:	Keeping food safe from any contamination is a concern.

FIGURE 15.4 The principles of the WEF security nexus and their details (Bizikova et al. 2013; Ren, Andreasen, and Sovacool 2014; Sovacool et al. 2017).

Thirteen general principles are specified on the basis of the WEF security nexus. They are explained as follows (Bizikova et al. 2013; Ren, Andreasen, and Sovacool 2014; Sovacool et al. 2017).

The principles of water security include:

- **Access (P_1):** This principle considers the utilization of all people in a fair and equitable manner by reference to the geographical location of the water source.
- **Safety (P_2):** It measures issues such as monitoring people's health in decision-making.
- **Affordability (P_3):** The issues of economy are covered in this principle. It is significant that water resources are given at a reasonable cost and made available to people.

The principles of energy security are:

- **Availability (P_4):** This principle addresses issues such as the diversification of energy sources and assistance in saving energy in the region.
- **Accessibility (P_5):** The principle covers geopolitical factors. This is a principle to measure government support.
- **Affordability (P_6):** It includes assessing the programs of economy. This covers subjects such as creating job opportunities and income for the region's people. All people, including the poor, should not pay more than 10% of their income for energy services.
- **Acceptability (P_7):** The social acceptance of new technologies and E&E systems such as an SHP plant are evaluated in this principle.
- **Applicability (P_8):** This principle addresses technological issues and shows the maturity of using and employing clean energy in the region.
- **Adaptability (P_9):** In this principle, the interchangeable features are analyzed. The target is to utilize clean energy sources and new technologies instead of old technologies and fossil fuel sources in energy production.

The principles of food security include:

- **Availability (P_{10}):** It considers people's utilization of quality food fairly. People deserve high-quality food.
- **Access (P_{11}):** It is significant for making food at a proper level of quality, sufficiently, and at a low cost.
- **Utilization (P_{12}):** The utilization and consumption of food are of great significance.
- **Stability (P_{13}):** Preserving food from any contamination is considered in this principle.

The significance of the principles as well as the priority of the scenarios based on the principles are computed from the experts' perspective.

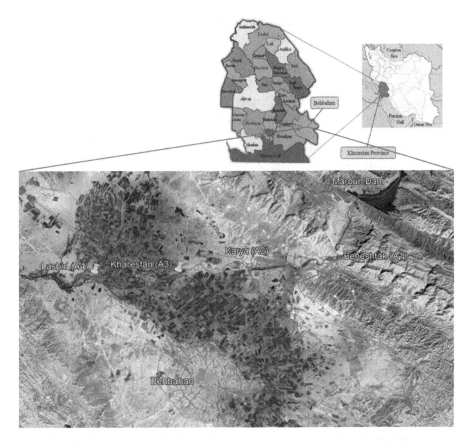

FIGURE 15.5   The four selected alternatives (locations) on Maroun River of Behbahan city and the position of Behbahan and Maroun Dam.

15.2.3 ALTERNATIVES

Behbahan city, with an area of 3715 km², is located in the south-west of Khuzestan province in Iran. This city is between longitude 50 degrees and 91 minutes to 50 degrees and east 25 minutes latitude 30 degrees and 45 minutes to 30 degrees and 32 minutes north. Its highest altitude is 180.93 and its lowest altitude 14.267 meters from sea level (see Figure 15.5). The maximum temperature in July and August reaches 50° C, and the minimum temperature in February and March is 0° C. Behbahan city, owing to natural features and being located in a hot and semi-arid climate, has hot summers for five to seven months and cold short-term winters with a low rate of precipitation. The mean of the minimum annual temperature is 18.1° C, and the mean of the maximum annual temperature is 32.37° C. The maximum precipitation occurs in this region during 30 days, between the months of August to December (Fetanat and Tayebi 2021).

- **Maroun River:** The main river in Behbahan is the Maroun River, which is located in the north of Behbahan city. The Maroun River is one of the rivers in southwestern Iran. It originates from the Zagros Mountains in Kohgiluyeh and Boyer-Ahmad Provinces and, passing through Behbahan and Aghajari, finally flows into the Shadegan Wetland in Khuzestan Province and in the rainy season flows into the Persian Gulf. The Maroun River plays a key role in shaping the life of rural communities on its outskirts. The approximate area of the Maroun River Basin is 4,600 square kilometers, and the length of the Maroun River is about 280 kilometers. The main source of the Maroun River is from the Zagros Mountains. The flow of the Maroun River at the end of the dam is between 5 and 150 m³/s. In estuarine areas, the mean flow is 30 m³/s. The water temperature of this river in different seasons of the year varies between 10 and 30° C, and the water depth of the river is between 1 and 15 meters.
- **Maroun Reservoir Dam:** It is built on the Maroun River, northeast of Behbahan, with a height of 165 meters, a length of 345 meters, a width of 15 meters, and a total reservoir of 1,200 million m³. This dam, 340 meters long and 175 meters high, was built in 1998. This earthen dam is located 19 km north of Behbahan. It is of a sandy gravel type with a clay core. The construction of the dam on the lake began in 1989 and ended in 1998.

The SHP plant belongs to the renewable energy technology group and is an environmentally attractive form of electricity generation due to its potential in small rivers. Applying an SHP plant as a source of renewable energy in order to mitigate existing environmental effects in electricity generation and has the maximum utilization of water, a renewable resource (Okot 2013). There are multiple sites available in the region of study for exploiting the potential of an SHP plant. Alternatives (sites) are usually chosen on the basis of geographic location, climate conditions, access to land, and skilled workforce in the region (Nizami et al. 2016). Therefore, the four sites, 1) Beheshtak, 2) Karya, 3) Kharestan, and 4) Lasbid, are considered scenarios to implement and use an SHP in Behbahan (on the Maroun River of Behbahan). These areas are located along the Maroun River. Descriptions of these sites are as follows:

- **Beheshtak (A1):** This region is located in about 13.5 km from Behbahan city. There is a small bridge over the river.
- **Karya (A2):** This region is located in 11 km from Behbahan city. It has been welcomed by many people as a suitable recreational place for different ages. There is a small dam near this site.
- **Kharestan (A3):** Kharestan is located 3 km from Behbahan city and next to the Maroun River, with an area of 25 hectares in the shape of an oval. This place is a very good place to spend holidays and leisure. There is a big play pool as well as other facilities for children and recreational places such as beautiful parks with lush trees and spaces for playing volleyball, football, and other sports such as swimming in this area. These features have made it one of the most attractive places on weekends.

- **Lasbid (A4):** Lasbid is one of the suburban rural areas of the central part of Behbahan city, and the farmers exploiting these lands are residents of Lasbid village, residents of Chahar Asyab village, and farmers living in Behbahan city. This area is a plain region with rugged hills and is located along the Maroun River.

15.2.4 MCDM Technique under a Fuzzy Environment

In this work, a DSS is employed as an analytical and systematic approach in decision-making based on the SECA technique under a fuzzy environment for installation and establishment of an SHP in Behbahan.

15.2.4.1 Fuzzy Set Theory

Because of the unavailable, unquantifiable, and incomplete information, many decisions in the real environment cannot be clearly made. The theory of the fuzzy set as a mathematical method was developed for solving these types of problems (Zadeh 1965). This theory is more beneficial than traditional sets for decision-making when dealing with vague terms. Experts make decisions on the basis of their past knowledge and experiences. Hence, estimations of them are often a function of ambiguity and linguistic terminology. To merge the opinions, ideas, and experiences of experts, it is better to translate the linguistic estimate to a fuzzy number. Therefore, the requirements for fuzzy logic in problems of decision-making in the actual environment are presented. In summary, the necessary definitions for fuzzy logic and linguistic expressions are in explained in Fetanat, Tayebi, and Shafipour (2021). The linguistic-variable method is usually utilized by experts for expressing their evaluations, which is very advantageous in dealing with uncertain, incomplete, and unspecific conditions in traditional quantitative terms. Linguistic values can be indicated with fuzzy numbers. In particular, triangular fuzzy numbers (TFNs) are commonly employed (Lin and Wu 2008; Fetanat et al. 2019). The stages of fuzzy assessment computation are explained as follows (Fetanat and Khorasaninejad 2015):

Stage I. The effects of principles are assessed with each other by a team of experts with several members. This is performed by pairwise comparisons by asking "How much effect does a principle have compared to another principle regarding our preferences or interests?" The value of relative effect is specified by employing TFNs, as given in Table 15.1, to show no effect to excellent effect.

Stage II. The replies of experts (N experts) are integrated by utilizing the arithmetic average to compute TFNs. For instance, the triangular fuzzy number $(l_{F_{i,j}}, m_{F_{i,j}}, u_{F_{i,j}})$ for the relative effect between principle i and principle j is computed as follows:

$$l_{F_{i,j}} = \frac{1}{N}\sum_{y=1}^{N} B_{F_{ijy}} \ , \ m_{F_{i,j}} = \frac{1}{N}\sum_{y=1}^{N} B_{F_{ijy}} \ , \ u_{F_{i,j}} = \frac{1}{N}\sum_{y=1}^{N} B_{F_{ijy}} \tag{15.1}$$

TABLE 15.1
The Fuzzy Linguistic Scale for the SECA Method

Linguistic Term	Triangular Fuzzy Numbers
None	(0, 0, 0.1)
Very low	(0, 0.1, 0.2)
Low	(0.1, 0.2, 0.3)
Fairly low	(0.2, 0.3, 0.4)
More or less low	(0.3, 0.4, 0.5)
Medium	(0.4, 0.5, 0.6)
More or less good	(0.5, 0.6, 0.7)
Fairly good	(0.6, 0.7, 0.8)
Good	(0.7, 0.8, 0.9)
Very good	(0.8, 0.9, 1)
Excellent	(0.9, 1, 1)

where $B_{F_{ijy}}$ is the relative effect value between principle i and principle j specified by expert y.

Stage III. Each TFN is defuzzified into a crisp number by applying the Yager prioritization method according to Eq. (15.2):

$$C_{i,j} = \int_0^1 \frac{1}{2}\left(\left(\tilde{C}_{i,j}\right)^l_\alpha + \left(\tilde{C}_{i,j}\right)^u_\alpha \right) d\alpha = \frac{l_{F_{i,j}} + 2m_{F_{i,j}} + u_{F_{i,j}}}{4} \qquad (15.2)$$

where $\tilde{C}$ is the TFN.

15.2.4.2 SECA Technique

In the present part, a novel technique is suggested to handle the problems of MCDM. The purpose of this technique is to specify the weights of principles and the overall performance scores of options simultaneously. A multi-target non-linear mathematical approach is systematized in this part to achieve this purpose. To systematize the mathematical technique, two kinds of reference points are explained for the weights of principles. The first kind is on the basis of within-principle variation information determined by the standard deviation. The second is associated with between-principle variation information specified according to the correlation measure. The multi-target approach attempts to maximize the overall performance of each option and minimize the deviation of principle weights from the reference points. To maximize the overall performance of each option, a weighted sum method is utilized as a target. Also, we take the benefit of the sum of squared deviations from the reference points to specify the other targets of the method (Keshavarz-Ghorabaee et al. 2018).

Presume that we have a problem of MCDM with m alternatives and n principles, and the weight of each principle $(w_j, j \in \{1, 2, ..., n\})$ is undecided. The matrix of decision related to this problem can be defined as follows:

$$D = \begin{bmatrix} d_{11} & d_{12} & \cdots & d_{1j} & \cdots & d_{1n} \\ d_{21} & d_{22} & \cdots & d_{2j} & \cdots & d_{2n} \\ \vdots & \vdots & \ddots & \vdots & \ddots & \vdots \\ d_{i1} & d_{i2} & \cdots & d_{ij} & \cdots & d_{in} \\ \vdots & \vdots & \ddots & \vdots & \ddots & \vdots \\ d_{m1} & d_{m2} & \cdots & d_{mj} & \cdots & d_{mn} \end{bmatrix} \tag{15.3}$$

In this matrix, d_{ij} indicates the value of performance related to the ith ($i \in \{1,2,...,m\}$) alternative on ($j \in \{1,2,...,n\}$) principle and $r_{ij} > 0$. According to this matrix, the normalized decision matrix can be constructed utilizing Eqs. (15.4) and (15.5):

$$D^N = \begin{bmatrix} d_{11}^N & d_{12}^N & \cdots & d_{1j}^N & \cdots & d_{1n}^N \\ d_{21}^N & d_{22}^N & \cdots & d_{2j}^N & \cdots & d_{2n}^N \\ \vdots & \vdots & \ddots & \vdots & \ddots & \vdots \\ d_{i1}^N & d_{i2}^N & \cdots & d_{ij}^N & \cdots & d_{in}^N \\ \vdots & \vdots & \ddots & \vdots & \ddots & \vdots \\ d_{m1}^N & d_{m2}^N & \cdots & d_{mj}^N & \cdots & d_{mn}^N \end{bmatrix} \tag{15.4}$$

where

$$d_{ij}^N = \begin{cases} \dfrac{d_{ij}}{\max_i d_{ij}} & \text{if } j \in AP \\[3ex] \dfrac{\min_i d_{ij}}{d_{ij}} & \text{if } j \in NP \end{cases} \tag{15.5}$$

In Eq. (15.5), AP and NP are the sets of advantageous and non-advantageous principles, respectively.

$V_j = [d_{ij}^N]_{m \times 1}$ shows the vector of the jth ($j \in \{1,2,...,n\}$) principle. The standard deviation of the parameters of each vector (σ_j) can get the within-principle variation information. For capturing the between-principle variation information from the matrix of decision, we need to compute the correlation between each two vectors of principles. Allow us to show by r_{jb} the correlation between the jth and b^{th} vectors (j and $b \in \{1,2,...,n\}$).

Next, the following summation ($\neq_j$) can indicate the degree of conflict between the jth principle and the other principles (Diakoulaki, Mavrotas, and Papayannakis 1995):

$$\pi_j = \sum_{b=1}^n (1 - r_{jb}) \tag{15.6}$$

A rise in the variation inside the vector of a principle (σ_j), as well as a rise in the degree of conflict between a principle and the other principles ($\neq_j$), intensifies the target significance of that principle. Thus, it defines the values normalized of σ_j and $\neq_j$ as the reference points for the principle weights. These values can be computed by Eqs. (15.7) and (15.8):

$$\sigma_j^N = \frac{\sigma_j}{\sum_{b=1}^{n} \sigma_b} \tag{15.7}$$

$$\pi_j^N = \frac{\pi_j}{\sum_{b=1}^{n} \pi_b} \tag{15.8}$$

According to this explanation, a model of multi-target non-linear programming is systematized as follows:

Model I:

$$\max \ S_i = \sum_{j=1}^{n} w_j d_{ij}^N, \quad \forall i \in \{1, 2, ..., m\}, \tag{15.9}$$

$$\min \ \lambda_g = \sum_{j=1}^{n} (w_j - \sigma_j^N)^2, \tag{15.10}$$

$$\min \ \lambda_h = \sum_{j=1}^{n} (w_j - \pi_j^N)^2, \tag{15.11}$$

$$\text{s.t.} \quad \sum_{j=1}^{n} w_j = 1, \tag{15.12}$$

$$w_j \le 1, \quad \forall j \in \{1, 2, ..., n\}, \tag{15.13}$$

$$w_j \ge \varepsilon, \quad \forall j \in \{1, 2, ..., n\}. \tag{15.14}$$

where Eq. (15.9) maximizes the overall performance of each option. Eqs. (15.10) and (15.11) minimize the deviation of weights of principles from the reference points for each principle. Eq. (15.12) guarantees that the sum of weights is equal to 1. Eqs. (15.13) and (15.14) set the weights of principles to some values in the period $[\varepsilon, 1]$. It should be noted that ε is defined as a small positive parameter taken into consideration as a lower limit for principles weights. In the present work, it is set to 10^{-3}.

The optimization of model I utilizes some methods of the multi-target optimization and converts model I to model II as follows:

Model II:

$$\max \ U = \lambda_k - \beta(\lambda_g + \lambda_h), \tag{15.15}$$

$$\text{s.t.} \quad \lambda_k \le S_i, \quad \forall i \in \{1, 2, ..., m\}, \tag{15.16}$$

$$S_i = \sum_{j=1}^{n} w_j d_{ij}^N, \quad \forall i \in \{1, 2, ..., m\}, \tag{15.17}$$

$$\lambda_g = \sum_{j=1}^{n} (w_j - \sigma_j^N)^2, \tag{15.18}$$

$$\lambda_h = \sum_{j=1}^{n} (w_j - \pi_j^N)^2, \tag{15.19}$$

$$\sum_{j=1}^{n} w_j = 1, \tag{15.20}$$

$$w_j \leq 1, \quad \forall j \in \{1, 2, ..., n\}, \tag{15.21}$$

$$w_j \geq \varepsilon, \quad \forall j \in \{1, 2, ..., n\}. \tag{15.22}$$

Based on the target function of model II, the minimum overall performance score of the options (λ_k) is maximized. Because the deviations from the reference points must be minimized. They are deducted from the function of the target with a coefficient β ($\beta \geq 0$). The coefficient affects the importance of reaching the reference points related to the weights of the principles. The overall performance score of the options (S_i) and target weights of the principles (w_j) are specified by resolving model II.

15.3 RESULTS AND DISCUSSION

In the present part, the results of the proposed technique are given. The results are presented in two sub-sections. First, we analyze the target weights obtained from the proposed technique. Second, the overall scores of alternative performances and the prioritization of alternatives are assessed. It should be noted that the model is resolved by utilizing the LINGO version 17.0 software.

The opinions of five experts (decision-makers) from university and water, electricity, agriculture, and environment organizations have been collected, and, according to the stages in Sections 15.2.4.1 and 15.2.4.2, the members of the decision matrix (matrix D in Eq. (15.3)) have been calculated.

The decision matrix of the problem along with the kind of each principle is indicated in Table 15.2. The normalized decision matrix, which is produced by utilizing Eqs. (15.4) and (15.5), is shown in Table 15.3. In these two tables, there are four alternatives (A1 to A4), which need to be investigated regarding 13 principles of the WEF security nexus (P1 to P13).

TABLE 15.2

The Decision Matrix of the Problem

	P1	P2	P3	P4	P5	P6	P7	P8	P9	P10	P11	P12	P13
A1	0.6800	0.5400	0.4600	0.4000	0.3800	0.5200	0.5400	0.6200	0.4000	0.3800	0.4600	0.5200	0.7200
A2	0.8000	0.7800	0.8200	0.6200	0.6800	0.7200	0.7400	0.8000	0.8200	0.6800	0.7800	0.7600	0.9400
A3	0.7600	0.6800	0.7200	0.5200	0.4600	0.5800	0.6800	0.6800	0.7600	0.5500	0.6200	0.6800	0.8000
A4	0.5200	0.4600	0.4000	0.3800	0.2600	0.4000	0.4600	0.5800	0.4600	0.2600	0.2200	0.4000	0.6800

TABLE 15.3

The Normalized Decision Matrix of the Problem

	P1	P2	P3	P4	P5	P6	P7	P8	P9	P10	P11	P12	P13
A1	0.8500	0.6923	0.5610	0.6452	0.5588	0.7222	0.7297	0.7750	0.4878	0.5588	0.5897	0.6842	0.7660
A2	1.0000	1.0000	1.0000	1.0000	1.0000	1.0000	1.0000	1.0000	1.0000	1.0000	1.0000	1.0000	1.0000
A3	0.9500	0.8718	0.8780	0.8387	0.6765	0.8056	0.9189	0.8500	0.9268	0.8088	0.7949	0.8947	0.8511
A4	0.6500	0.5897	0.4878	0.6129	0.3824	0.5556	0.6216	0.7250	0.5610	0.3824	0.2821	0.5263	0.7234

TABLE 15.4

The Sets of Principles Weights Specified by the Change of β

	β									
	0.1	0.2	0.3	0.4	0.5	1	2	3	4	5
w1	0.1864	0.1729	0.1607	0.1516	0.1458	0.1248	0.1118	0.1075	0.1053	0.1040
w2	0.0100	0.5200	0.6578	0.6889	0.7095	0.6500	0.5959	0.5778	0.5682	0.5633
w3	0.0100	0.0100	0.0422	0.2872	0.4340	0.6291	0.7021	0.7264	0.7385	0.7458
w4	0.0603	0.9253	0.9642	0.9477	0.9388	0.8212	0.7378	0.7101	0.6964	0.6879
w5	0.0100	0.0100	0.0100	0.0100	0.0100	0.4727	0.6772	0.7455	0.7797	0.8001
w6	0.0100	0.2544	0.5312	0.6344	0.6968	0.7225	0.7108	0.7069	0.7043	0.7037
w7	0.0751	0.9631	0.9672	0.9321	0.9126	0.7734	0.6792	0.6479	0.6318	0.6227
w8	0.3368	0.2289	0.1861	0.1611	0.1462	0.1064	0.8410	0.7665	0.7290	0.7069
w9	0.0194	0.1005	0.1246	0.1343	0.1395	0.1406	0.1387	0.1381	0.1378	0.1376
w10	0.0100	0.0100	0.0100	0.0100	0.0100	0.3000	0.5049	0.5730	0.6073	0.6276
w11	0.0100	0.0100	0.0100	0.0100	0.0100	0.2445	0.5744	0.6844	0.7399	0.7725
w12	0.0100	0.0100	0.2495	0.4135	0.5128	0.6117	0.6366	0.6449	0.6488	0.6516
w13	0.3324	0.2265	0.1844	0.1597	0.1450	0.1056	0.8350	0.7612	0.7246	0.7021

15.3.1 ASSESSMENT OF PRINCIPLES WEIGHTS

In the present sub-section, we solve model II using the provided data in Table 15.3 and various values for the β parameter ($\beta = 0.1, 0.2, 0.3, 0.4, 0.5, 1, 2, 3, 4$, and 5). By applying model II via LINGO software, ten sets of principles weights are specified. The principles weights associated with varying values of β are given in Table 15.4. The weight variation is indicated in Figure 15.6. Based on Figure 15.6, it can be seen that the principles weights are more stable when the values of the β parameter are greater than 3 ($\beta \geq 3$).

15.3.2 ASSESSMENT OF THE ALTERNATIVES PREFERENCES

In the present sub-section, we find the overall performance score of each alternative (S_i) by resolving model II according to the provided data in Table 15.3 and the same values of β as those used for the assessment of principles weights. The overall performance scores of the alternatives are given in Table 15.5, and the corresponding

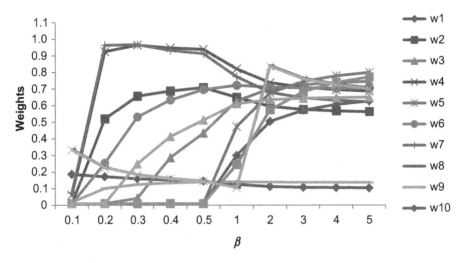

FIGURE 15.6   The variation of the principles weights associated with β.

TABLE 15.5

Overall Performance Scores of the Alternatives in Various Values of β

	β									
	0.1	0.2	0.3	0.4	0.5	1	2	3	4	5
S1	0.7731	0.7341	0.7207	0.7103	0.7041	0.6794	0.6645	0.6595	0.6570	0.6555
S2	1	1	1	1	1	1	1	1	1	1
S3	0.8734	0.8807	0.8815	0.8821	0.8824	0.8697	0.8617	0.8591	0.8577	0.8569
S4	0.6938	0.6620	0.6464	0.6349	0.6280	0.5898	0.5631	0.5542	0.5497	0.5471

TABLE 15.6

Prioritizations of Alternatives in Various Values of β

		β									
		0.1	0.2	0.3	0.4	0.5	1	2	3	4	5
Priority	A1	3	3	3	3	3	3	3	3	3	3
	A2	1	1	1	1	1	1	1	1	1	1
	A3	2	2	2	2	2	2	2	2	2	2
	A4	4	4	4	4	4	4	4	4	4	4

prioritizations of the reviewed alternatives are illustrated in Table 15.6. Also, Figure 15.7 shows a graphical view of performance scores. As can be seen in Figure 15.7 and Table 15.5, these scores for the alternatives are more distinguishable and stable when the values of β are greater than $3(\beta \geq 3)$. In Table 15.6, it has an entirely stable prioritization for all values of β.

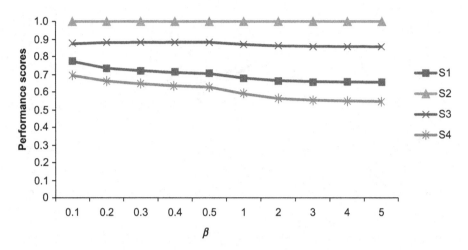

FIGURE 15.7 Variation of the overall performance scores of the alternatives associated with β.

The results of Table 15.6 and Figure 15.7 indicate that alternative A2 is the best site to construct the SHP plant on the Maroun River of Behbahan city. Then, alternatives A3 and A1 are the best sites, and ultimately, alternative A4 is the worst site (in other words, A2 > A3 > A1 > A4). If the rankings of the principle weights are altered, the best site of the SHP plant may change as a result.

According to the results, we can deduce that the findings of the suggested technique are valid, and SECA can be taken into consideration as an efficient technique to deal with MCDM problems. To utilize the proposed technique, the proposed value of β is $\beta = 3$. Nevertheless, it can be set based on the preferences of the decision-makers and specifications of the decision-making issue. The proper value of β can be estimated by doing a sensitivity analysis on the basis of the decision matrix of the study.

It should be noted that the factors (members) of the matrix of decision should be greater than zero ($d_{ij} > 0$) to utilize SECA. If there are some factors with negative and/or zero values, they should be transformed.

15.3.3 Managerial Insights

The present work intended to use a decision-making system on the basis of the principles of the WEF security nexus for sustainable site selection for constructing an SHP plant in the study region. It is introduced as a low-carbon energy resource that can bring environmental, economic, and social benefits and help to navigate the region towards SD goals. The system of decision-making was carried out by the use of an MCDM technique based on fuzzy set theory that gives permission to multiple stakeholders to contribute to the procedure. An explanatory instance with four sites for extending clean energy resources was reviewed in the Behbahan, Iran. Multiple advantages of the decision-making system proposed in the present research for policymakers, stakeholders, and the community are given as follows:

- The reduction of environmental impacts by implementing and using clean energy and the improvement of economic and social issues by creating job opportunities and recreational regions for the resident population, as well as ameliorating the social acceptance of applying an SHP plant.
- Assisting in making knowledgeable decisions with a more comprehensive vision of selecting a sustainable site for the SHP plant.
- The contribution to planning and improving the supply chains of three vital sources of water, energy, and food via the implementing circular economy strategies in the area.
- The development of the area and local community towards SD targets in sector of sustainability management issues.

15.4 CONCLUSIONS

The WEF security nexus is a significant and broad issue. Approximately all human function needs water, energy, and food. With populations increasing and appetites rising for these three limited resources, the concept of the WEF security nexus has been taken into consideration by many policymakers, organizations, and communities. In addition, discrete institutions (e.g., different levels of administration, different ministries) that often lack coordination make most of the decisions of policy with potential influences on the principles of the WEF security nexus. The challenges raised by this nexus are partly the effect of such fragmented policy-making employed for resources that are interdependent and increasingly scant. Against this background, a systematic method to planning resources in line with the principles of the WEF security nexus would be profitable as an assessment framework for friendly policy-making in this nexus. Especially, and from a policy-making perspective in energy and environmental systems, this would entail understanding the relationships of water, food, and energy security principles in decisions. MCDM techniques could be very useful and effective for this target. The WEF security nexus principles are at the heart of developing three fields of sustainability, economy, and environment. In this way, the principles of water security (access, safety, and affordability), energy security (availability, accessibility, affordability, acceptability, applicability, and adaptability), and food security (availability, access, utilization, and stability) have been taken into consideration. The site selection for constructing an SHP plant as a clean energy producer is assessed based on the principles of the WEF security nexus to help in the improvement of sustainable development targets in Behbahan city, Iran. Four candidates are specified on the Maroun River of Behbahan. They contain the regions Beheshtak, Karya, Kharestan, and Lasbid. A systematic methodology based on the MCDM technique called the SECA technique under a fuzzy environment was selected to support the integrated decision to choose the most sustainable site to construct an SHP plant in Behbahan. SHP plant as an appropriate case can improve people's quality of life by creating jobs, boosting the local economy, and promoting environmental and health policies in the region.

According to the WEF security nexus principles and the findings, the Karya region is the best site. Then, the regions of Kharestan and Beheshtak are the best sites, and finally, the Lasbid region is the worst site (in other words $A2 > A3 > A1 > A4$).

In conclusion, we hope the decision-making system based on the WEF security nexus principles and MCDM method will assist future innovation improvements to be more efficient and practical in sustainable site selection for E&E systems. This work assists planners in better making a decision. Applying the methodology for all countries and utilization of the assessment principles for other E&E systems in Behbahan city, Iran, and the world are suggested for continuing this work.

REFERENCES

Aghajani, Gholamreza, and Noradin Ghadimi. 2018. "Multi-Objective Energy Management in a Micro-Grid." *Energy Reports* 4: 218–225. https://doi.org/10.1016/j.egyr.2017.10.002.

Alp, Ahmet, Adil Akyüz, and Serhat Kucukali. 2020. "Ecological Impact Scorecard of Small Hydropower Plants in Operation: An Integrated Approach." *Renewable Energy* 162: 1605–1617. https://doi.org/10.1016/j.renene.2020.09.127.

Amer, Kamel, Zafar Adeel, and Walid Saleh. 2017. *The Water, Energy, and Food Security Nexus in the Arab Region*. 1st ed. Springer International Publishing. Cham, Switzerland. https://doi.org/10.1007/978-3-319-48408-2.

Bazilian, Morgan, Holger Rogner, Mark Howells, Sebastian Hermann, Douglas Arent, Dolf Gielen, Pasquale Steduto, et al. 2011. "Considering the Energy, Water and Food Nexus: Towards an Integrated Modelling Approach." *Energy Policy* 39 (12): 7896–7906. https://doi.org/10.1016/j.enpol.2011.09.039.

Bhandari, Ramchandra, Lena Ganda Saptalena, and Wolfgang Kusch. 2018. "Sustainability Assessment of a Micro Hydropower Plant in Nepal." *Energy, Sustainability and Society* 8 (3). https://doi.org/10.1186/s13705-018-0147-2.

Bizikova, Livia, Dimple Roy, Darren Swanson, David Henry Venema, and Matthew McCandless. 2013. *The Water–Energy–Food Security Nexus: Towards a Practical Planning and Decision-Support Framework for Landscape Investment and Risk Management*. 6th ed. International Institute for Sustainable Development. Winnipeg, Canada.

Cai, Xiaowen, Feng Ye, and Fatemeh Gholinia. 2020. "Application of Artificial Neural Network and Soil and Water Assessment Tools in Evaluating Power Generation of Small Hydropower Stations." *Energy Reports* 6: 2106–2118. https://doi.org/10.1016/j.egyr.2020.08.010.

Craig, Michael, Jin Zhao, Gia Schneider, Abe Schneider, Sterling Watson, and Greg Stark. 2019. "Net Revenue and Downstream Flow Impact Trade-Offs for a Network of Small-Scale Hydropower Facilities in California." *Environmental Research Communications* 1 (1): 11001. https://doi.org/10.1088/2515-7620/aafd62.

Diakoulaki, D., G. Mavrotas, and L. Papayannakis. 1995. "Determining Objective Weights in Multiple Criteria Problems: The Critic Method." *Computers and Operations Research* 22 (7): 763–770. https://doi.org/10.1016/0305-0548(94)00059-H.

Fei, Xi, Ruan Xuejun, and Navid Razmjooy. 2019. "Optimal Configuration and Energy Management for Combined Solar Chimney, Solid Oxide Electrolysis, and Fuel Cell: A Case Study in Iran." *Energy Sources, Part A: Recovery, Utilization and Environmental Effects*, 1–21. https://doi.org/10.1080/15567036.2019.1680770.

Fetanat, Abdolvahhab, and Ehsan Khorasaninejad. 2015. "A Novel Hybrid MCDM Approach for Offshore Wind Farm Site Selection: A Case Study of Iran." *Ocean and Coastal Management* 109. Elsevier Ltd: 17–28. https://doi.org/10.1016/j.ocecoaman.2015.02.005.

Fetanat, Abdolvahhab, Hossein Mofid, Mojtaba Mehrannia, and Gholamreza Shafipour. 2019. "Informing Energy Justice Based Decision-Making Framework for Waste-to-Energy Technologies Selection in Sustainable Waste Management: A Case of Iran."

Journal of Cleaner Production 228. Elsevier B.V.: 1377–1390. https://doi.org/10.1016/j. jclepro.2019.04.215.

Fetanat, Abdolvahhab, and Mohsen Tayebi. 2021. "Sustainable Design of the Household Water Treatment Systems Using a Novel Integrated Fuzzy QFD and LINMAP Approach: A Case Study of Iran." *Environment, Development and Sustainability.* https://doi. org/10.1007/s10668-021-01284-5.

Fetanat, Abdolvahhab, Mohsen Tayebi, and Hossein Mofid. 2021. "Water-Energy-Food Security Nexus Based Selection of Energy Recovery from Wastewater Treatment Technologies: An Extended Decision Making Framework under Intuitionistic Fuzzy Environment." *Sustainable Energy Technologies and Assessments* 43: 100937. https:// doi.org/10.1016/j.seta.2020.100937.

Fetanat, Abdolvahhab, Mohsen Tayebi, and Gholamreza Shafipour. 2021. "Management of Waste Electrical and Electronic Equipment Based on Circular Economy Strategies: Navigating a Sustainability Transition Toward Waste Management Sector." *Clean Technologies and Environmental Policy* 23: 343–369. https://doi.org/10.1007/s10098-020-02006-7.

Ghorbani, Narges, Hamed Makian, and Christian Breyer. 2019. "A GIS-Based Method to Identify Potential Sites for Pumped Hydro Energy Storage—Case of Iran." *Energy* 169: 854–867.https://doi.org/10.1016/j.energy.2018.12.073.

Gondhalekar, Daphne, and Thomas Ramsauer. 2017. "Nexus City: Operationalizing the Urban Water-Energy-Food Nexus for Climate Change Adaptation in Munich, Germany." *Urban Climate* 19. Elsevier: 28–40.

Hellegers, Petra, David Zilberman, Pasquale Steduto, and Peter McCornick. 2008. "Interactions between Water, Energy, Food and Environment: Evolving Perspectives and Policy Issues." *Water Policy* 10 (March): 1–10. https://doi.org/10.2166/wp.2008.048.

Hoff, Holger. 2011. "Understanding the Nexus. Background Paper for the Bonn2011 Nexus Conference, Germany. Stockholm." *Stockholm Environment Institute*, no. November: 1–52.

Howarth, Candice, and Irene Monasterolo. 2016. "Understanding Barriers to Decision Making in the UK Energy-Food-Water Nexus: The Added Value of Interdisciplinary Approaches." *Environmental Science & Policy* 61. Elsevier: 53–60.

Keshavarz-Ghorabaee, Mehdi, Maghsoud Amiri, Edmundas Kazimieras Zavadskas, Zenonas Turskis, and Jurgita Antucheviciene. 2018. "Simultaneous Evaluation of Criteria and Alternatives (SECA) for Multi-Criteria Decision-Making." *Informatica (Netherlands)* 29 (2): 265–280. https://doi.org/10.15388/Informatica.2018.167.

Khodaei, Hossein, Mahdi Hajiali, Ayda Darvishan, Mohammad Sepehr, and Noradin Ghadimi. 2018. "Fuzzy-Based Heat and Power Hub Models for Cost-Emission Operation of an Industrial Consumer Using Compromise Programming." *Applied Thermal Engineering* 137: 395–405. https://doi.org/10.1016/j.applthermaleng.2018.04.008.

Kurian, Mathew. 2017. "The Water-Energy-Food Nexus: Trade-Offs, Thresholds and Transdisciplinary Approaches to Sustainable Development." *Environmental Science & Policy* 68. Elsevier: 97–106.

Lin, Chi-Jen, and Wei-Wen Wu. 2008. "A Causal Analytical Method for Group Decision-Making under Fuzzy Environment." *Expert Systems with Applications* 34: 205–213. https://doi.org/10.1016/j.eswa.2006.08.012.

Mayeda, A. M., and A. D. Boyd. 2020. "Factors Influencing Public Perceptions of Hydropower Projects: A Systematic Literature Review." *Renewable and Sustainable Energy Reviews* 121: 109713. https://doi.org/10.1016/j.rser.2020.109713.

McCornick, Peter G., Seleshi B. Awulachew, and Michael Abebe. 2008. "Water–food–energy–environment Synergies and Tradeoffs: Major Issues and Case Studies." *Water Policy* 10 (S1): 23–36. https://doi.org/10.2166/wp.2008.050.

Nie, Ru xin, Zhang peng Tian, Jian qiang Wang, Hong yu Zhang, and Tie li Wang. 2018. "Water Security Sustainability Evaluation: Applying a Multistage Decision Support Framework in Industrial Region." *Journal of Cleaner Production* 196: 1681–1704. https://doi.org/10.1016/j.jclepro.2018.06.144.

Nizami, A. S., O. K. M. Ouda, M. Rehan, A. M. O. El-maghraby, J. Gardy, A. Hassanpour, S. Kumar, and I. M. I. Ismail. 2016. "The Potential of Saudi Arabian Natural Zeolites in Energy Recovery Technologies." *Energy* 108. Elsevier Ltd: 162–171. https://doi.org/10.1016/j.energy.2015.07.030.

Okot, David Kilama. 2013. "Review of Small Hydropower Technology." *Renewable and Sustainable Energy Reviews* 26: 515–520. https://doi.org/10.1016/j.rser.2013.05.006.

Owen, Anne, Kate Scott, and John Barrett. 2018. "Identifying Critical Supply Chains and Final Products: An Input-Output Approach to Exploring the Energy-Water-Food Nexus." *Applied Energy* 210. Elsevier: 632–642.

Parkinson, Simon C., Marek Makowski, Volker Krey, Khaled Sedraoui, Abdulrahman H. Almasoud, and Ned Djilali. 2018. "A Multi-Criteria Model Analysis Framework for Assessing Integrated Water-Energy System Transformation Pathways." *Applied Energy* 210. Elsevier: 477–486.

Popa, F., G. E. Dumitran, L. I. Vuta, E. I. Tica, B. Popa, and A. Neagoe. 2020. "Impact of the Ecological Flow of Some Small Hydropower Plants on Their Energy Production in Romania." *Journal of Physics: Conference Series* 1426. https://doi.org/10.1088/1742-6596/1426/1/012043.

Rabia, Ferroukhi, Nagpal Divyam, Lopez-Peña Alvaro, Hodges Troy, H. Mohtar Rabi, Daher Bassel, Mohtar Samia, and Keulertz Martin. 2015. *Renewable Energy in the Water, Energy & Food Nexus*. International Renewable Energy Agency. Abu Dhabi, UAE.

Rasul, Golam, and Bikash Sharma. 2016. "The Nexus Approach to Water–energy–food Security: An Option for Adaptation to Climate Change." *Climate Policy* 16 (6). Taylor & Francis: 682–702.

Ren, Jingzheng. 2018a. "Sustainability Prioritization of Energy Storage Technologies for Promoting the Development of Renewable Energy: A Novel Intuitionistic Fuzzy Combinative Distance-Based Assessment Approach." *Renewable Energy* 121. Elsevier Ltd: 666–676. https://doi.org/10.1016/j.renene.2018.01.087.

Ren, Jingzheng. 2018b. "Multi-Criteria Decision Making for the Prioritization of Energy Systems under Uncertainties after Life Cycle Sustainability Assessment." *Sustainable Production and Consumption* 16. Elsevier B.V.: 45–57. https://doi.org/10.1016/j.spc.2018.06.005.

Ren, Jingzheng, K.P Andreasen, and Benjamin K. Sovacool. 2014. "Viability of Hydrogen Pathways That Enhance Energy Security: A Comparison of China and Denmark." *International Journal of Hydrogen Energy* 39 (28). Elsevier Ltd: 15320–15329. https://doi.org/10.1016/j.ijhydene.2014.07.150.

Ren, Jingzheng, and Liang Dong. 2018. "Evaluation of Electricity Supply Sustainability and Security: Multi-Criteria Decision Analysis Approach." *Journal of Cleaner Production* 172. Elsevier Ltd: 438–453. https://doi.org/10.1016/j.jclepro.2017.10.167.

Sovacool, Benjamin K., Matthew Burke, Lucy Baker, and Chaitanya Kumar Kotikalapudi. 2017. "New Frontiers and Conceptual Frameworks for Energy Justice." *Energy Policy* 105. Elsevier Ltd: 677–691. https://doi.org/10.1016/j.enpol.2017.03.005.

Tamm, Ottar, and Toomas Tamm. 2020. "Verification of a Robust Method for Sizing and Siting the Small Hydropower Run-of-River Plant Potential by Using GIS." *Renewable Energy* 155: 153–159. https://doi.org/10.1016/j.renene.2020.03.062.

Tian, Yizhi, Feng Zhang, Zhi Yuan, Zihang Che, and Nicholas Zafetti. 2020. "Assessment Power Generation Potential of Small Hydropower Plants Using GIS Software." *Energy Reports* 6: 1393–1404. https://doi.org/10.1016/j.egyr.2020.05.023.

Yuling, Melissa, Leung Pah, Elias Martinez-hernandez, Matthew Leach, and Aidong Yang. 2016. "Designing Integrated Local Production Systems: A Study on the Food-Energy-Water Nexus." *Journal of Cleaner Production* 135. Elsevier Ltd: 1065–1084. https://doi.org/10.1016/j.jclepro.2016.06.194.

Zadeh, L. A. 1965. "Fuzzy Sets." *Information and Control* 8: 338–353.

Zhang, Hong-yu, Hong-gang Peng, Jing Wang, and Jian-qiang Wang. 2017. "An Extended Outranking Approach for Multi-Criteria Decision-Making Problems with Linguistic Intuitionistic Fuzzy Numbers." *Applied Soft Computing Journal* 59. Elsevier B.V.: 462–474. https://doi.org/10.1016/j.asoc.2017.06.013.

16 Policy Dynamics for Energy, Environment and Sustainable Development in the Year 2060

Seeme Mallick

16.1 INTRODUCTION

A policy plan is needed to reach an economically efficient and environmentally sustainable future. The polices set in motion during the 2020s for efficiency in the energy sector could result in a green Industrial Revolution by the middle of the century for Pakistan. Environmentally and ecologically sustainable development by 2060 would result due to the success of an efficient policy plan, a policy plan that would change the direction for the economy from economic growth to environmentally sustainable development. Following the example of China's 14th Five-Year Plan for Modern Energy System, this plan has been prepared (Chinese Government June 1, 2022).

One of the most important topics under environmentally sustainable development is the dynamics of energy transition towards environmentally sustainable energy. The sustainable energy transition is moving away from fossil fuels and increasing consumption of renewable energy (Demirtas 2013). One of the main reasons for investing in renewable energy infrastructure, globally, is the de-carbonization of the energy sector. The future role of fossil fuels would reduce as the energy transition unfolds. Environmental impact assessments (EIAs) of each renewable energy project, construction, and operation would highlight the contribution to the environmental betterment of the area compared to fossil fuels. The main focus in the future would be on transport technologies. The role of electric mobility based on electric vehicles (EVs) is the future of transport. The upcoming technology of hydrogen and fuel cells has future prospects of emission-free transport, globally and nationally.

The main concern in many countries globally is increased dependence on nuclear power technologies. The challenges and prospects of large-scale nuclear fission technology and as yet experimental nuclear fusion technology are numerous. The challenges are the safety of the infrastructure built around these technologies in each country. The prospects are a positive outlook for the energy supply to be cost effective in the long-run and in abundance. This brings hope to the developing countries like Pakistan. The goal of this chapter is to initiate the process of changing the

DOI: 10.1201/9781032715438-16

energy mix and bringing forward renewable energy resources to the national main grid. It has the following main objectives:

1. To design a policy dynamic that interlinks energy with the environment
2. To achieve environmentally sustainable development
3. To design policies that change the energy mix

In this chapter, the policy plan focuses on the following policy discussion:

Energy: For the development of the energy sector, finance and investment policies would set the right direction for the expansion of economic activities. The infrastructure policies would bring flexibility to the energy sector for both expansion and technological innovation. The policies for research and development (R&D) for technology development would select the right mix of energy resource technologies that would keep Pakistan on a sustained path towards development. No policy set is complete without policies for human resource (HR) development, education, and training, in this case for the energy sector.

Environment: The policy for restoration, conservation, and management of the environment would be central to environmental planning in Pakistan. Finance and investment for environmental management planning would need to be incorporated into mainstream economic planning. The R&D policy for technology development for the environment would focus on both restoration and conservation of natural resources. Most importantly, HR development, education, and training for the environment would prepare a workforce that is environmentally sensitive.

Sustainable Development: Transition from fossil fuels to renewable energy is an integral part of the path leading from economic growth to environmentally sustainable development. The policy discussion would mainly revolve around the impact of the energy and environment policies on the transition from economic growth to environmentally sustainable development.

Finance and investment for renewable energy would play an important role in setting the path for sustainable development in Pakistan. The policy for restoration, conservation, and management of the environment would play a central role in sustainable development. R&D for technology development for the low-emission or zero-emission fuels differentiates economic growth from sustainable development. HR development, education, and training for the environment in Pakistan would prepare a workforce that would work alongside the mainstream workforce in converting the economic growth process into environmentally sustainable development. These policy plans would focus on connectivity of energy and environment policies with the sustainable development policies. An integrated energy and environmental policy framework would result in sustainable development. The policy discussion would then include a selective set of the United Nations (UN) Sustainable Development Goals (SDGs).

The chapter is structured as follows: after the introduction, the second section presents a literature review. The third section presents "Policies for Energy Transition for Meeting Environmental and Sustainable Development Standards". The fourth section presents a discussion on the topic "From Economic Growth to Sustainable Development Pathway". The fifth section presents a discussion on the topic "The Envisioned Future and Technology for Energy Innovation". The sixth section presents "Policy Dynamic to Reach the Desired Future". The chapter ends with a short conclusion section.

16.2   LITERATURE REVIEW

The environment or natural assets provide essential ecosystem services. Like any other capital asset, environmental services diminish as the ecosystem depreciates due to anthropogenic activities. Sustained flow of these ecosystem services depend on two activities: restoration of the environment and maintaining the environment in its pristine condition. The cost of environmental management (CEM) is a fund allocated for these two activities (Mallick, Sinden, and Thampapillai 2000).

To highlight sustainability or for sustainable development in Pakistan, the criteria for the Public Sector Development Project (PSDP) would change. Green investment would stimulate the greening of some of the projects under the PSDP. This would result in a "clean and green resource efficient economy" or environmentally sustainable development in Pakistan (Mallick May 2021).

To meet the energy needs in the year 2047, it is assumed that the Pakistan energy sector would consume energy from local and imported resources like local hydropower; local nuclear; local coal gasification; local oil, gas, and coal; local solar, wind, and wave energy; imported hydro-power; imported oil, gas, and coal; and imported nuclear (Mallick 2018).

On a global scale, an attempt is made at identifying a sustainable circular economy design in the year 2050 based on renewable energy consumption for water and food security (Mallick September 2021). The tasks that would identify the sustainable circular economy design are:

- Environmental engineering for hydrology and water supply technology in the arid conditions for the Great Green Wall
- Civil engineering and infrastructure development for renewable energy production in arid areas
- Seawater desalination technology and hydroponic farming technology
- Environmental engineering technology for artificial aquifer design
- Biosphere technology for designing habitable artificial ecosystems

A case study is presented where the environmental costs of emission are estimated for coal-powered plants (Oyewo, Olarewaju, Cloete, and Adenuga 2021). This builds the case for change in the energy mix in the future where renewable energy has a much greater part.

In a policy paper, the task is to design a combination of nine renewable energy resources that could become significant in the future. In addition to this, taking nine industries that would follow the circular economy design. For the implementation of this idea by the year 2060, the author takes a few countries from each continent to achieve this combination of renewable energy and circular economy (Mallick 2021*).

16.3 POLICIES FOR ENERGY TRANSITION FOR MEETING ENVIRONMENTAL AND SUSTAINABLE DEVELOPMENT STANDARDS

A central topic for this chapter is the energy transition from fossil fuel consumption to renewable energy technologies. All the policies presented for discussion focus on this energy transition. The proportion of fossil fuels is low compared to the renewable energy in this future scenario for the year 2060. Table 16.1 presents a possible combination for future energy infrastructure installation. Out of the total of 400,000 MW electricity, only 32.5% is thermal energy: oil, gas, coal, and coal gasification, and the remaining 67.5% are renewable energy: hydropower, nuclear fission and fusion, solar and wind energy, and ocean wave and tide energy.

Table 16.2 presents five sectors including multiple industries that consume 400,000 MW electricity by the year 2060. This consists of all four types of electricity

TABLE 16.1
The Pakistan Energy and Electricity Plan in the Year 2060

	Energy Supply Source:		Electricity Estimated Installed Capacity—MW	Electricity Estimated Installed Capacity—%
A	Hydropower		120,000	30
B	Thermal energy: oil, gas, coal, and coal gasification		130,000	32.5
C	Nuclear fusion and fission			25
C1		Nuclear fission: uranium and thorium	80,000	
C2		Nuclear fusion: ITER Tokamak, artificial sun	20,000	
D	Renewable energy			12.5
D1		Solar energy systems	20,000	
D2		Wind energy systems	20,000	
D3		Ocean wave energy and tide energy	10,000	
			400,000	**100**

(*Source:* Assumptions made by the author)

TABLE 16.2
Five Sectors Electricity Allocation by the Year 2060

Sector	Energy Demand Sector	Electricity Allocation—Thermal Energy MW	Electricity Allocation—Hydropower, Nuclear, and Renewable MW	Total Electricity Allocation MW
Urban centers	Energy demand in residential and commercial urban areas Architecture design—buildings with natural light and climate control Building construction industry—material innovation	20,000 (25%)	60,000 (75%)	80,000
Agriculture	Agricultural inputs and farming Crop processing industry: food, beverage, and dairy industry Textile and furniture industry	25,000 (31.25%)	55,000 (68.75%)	80,000
Industry	Machinery and consumer product and appliances manufacturing industry Transport and automotive vehicle manufacturing industry Mobile phone and electronics: internet and gaming industry	25,000 (31.25%)	55,000 (68.75%)	80,000
Transport	Construction of roads, railway lines, airports, and shipping ports Operations and energy demand for transport hubs	30,000 (37.5%)	50,000 (62.5%)	80,000
Aerospace technologies	Aviation (airplane design) and space travel (rocket design) Aerospace (aeronautics and astronautics) for launch into space Satellite, rocket, and spaceship: design, manufacturing and operations in the orbital space and beyond	30,000 (37.5%)	50,000 (62.5%)	80,000
	Totals (Average)	130,000 (32.5%)	270,000 (67.5%)	400,000

(*Source:* Assumptions made by the author)

resources: hydropower, thermal energy, nuclear, and renewable. Two separate columns present electricity sources as thermal energy in one column and hydropower, nuclear, and renewable in another column. The five sectors are:

- Urban centers
- Agriculture
- Industry
- Transport
- Aerospace technologies

Table 16.2 presents these five sectors and the percentages for energy allocation in each of them. The assumption made here is that for simplification of estimation, an equal amount of energy could be allocated to each of the five sectors; each sector is allocated 20% of electricity, 80,000 MW. As there would be three different combinations of fossil fuel and renewable energy resources, there would be a difference in value among the five sectors. The main reason for keeping aviation with aerospace and satellite technology is the experimentation in the fuel technology.

As presented in Table 16.2, the assumption is that from 2024 to 2060, each of the five energy demand sectors and industries would increase its energy demand to 80,000 MW each. The energy infrastructure expansion would be based on this assumption. The residential and stationary energy demand and the architecture design and building construction industry are part of the urban centers sector. The electricity allocation for this sector from thermal energy would be 20,000 MW and from hydropower, nuclear, and renewable energy would be 60,000 MW.

The agriculture sector includes industries like agricultural inputs and farming; crop processing industry: food, beverage, and dairy industry; and textile and furniture industry. This sector would be allocated 25,000 MW from thermal energy resources and 55,000 MW from hydropower, nuclear, and renewable energy resources for production and exports. The manufacturing industry sector includes three types of industrial sectors: machinery and consumer product and appliances manufacturing industry, transport and automotive vehicle manufacturing industry, and mobile phone and electronics: internet and gaming industry. The manufacturing sector would be allocated 25,000 MW from thermal energy resources and 55,000 MW from hydropower, nuclear, and renewable energy resources for production and export.

The transport sector: road, rail line, airport, and shipping port construction and operation energy demand would be allocated 30,000 MW from thermal energy resources and 50,000 MW from hydropower, nuclear, and renewable energy resources. Aerospace technologies, aviation and space travel: airplane, satellite, rocket, and spaceship industry, would be allocated 30,000 MW from thermal energy resources and 50,000 MW from hydropower, nuclear, and renewable energy resources for production and exports.

These five sectors are discussed in more detail in light of the UN SDGs for the years 2024 to 2030.

16.3.1 Sustainable Development Policy

In the present study, there are five sectors and six selected Sustainable Development Goals in combination. Energy and environment are both part of the UN's SDGs and therefore part of the global and national policy landscapes. Each of the 17 SDGs is important for long-term environmentally sustainable development. SDGs cover a time span from 2015 to 2030. The present study starts in 2024 and completes its goals and objectives by 2060. Its focus is on the combination of the energy resources, both fossil fuels and renewables. During the 36 years of this energy plan and policies, the focus would shift from fossil fuels onto hydropower, nuclear fission and fusion, and other renewable energy resources.

Policy discussion: For environmentally sustainable development, it is imperative that the energy sector transform from a carbon emissions sector to a net-zero emissions energy sector. As energy resources move from the category of experimental renewable to mainstream, carbon emissions could reduce markedly. SDG 7—Affordable and Clean Energy, is central to the present study between 2024 and 2030. The process of change in the energy mix would continue over the remaining three decades from 2031 to 2060.

16.3.1.1 SDG 2—Zero Hunger

16.3.1.1.1 Agriculture

The agriculture sector in the present study is divided into three parts:

- Agricultural inputs and farming
- Crop processing industry: food, beverage, and dairy products
- Textile and furniture industries

SDG2—Zero Hunger would focus on the first two parts between 2024 and 2030. The availability and quality of the agricultural inputs would have direct repercussions on agricultural farming in Pakistan. About 40% of the population in the country lives in the cities. Crop productivity for the category of small subsistence farmers would need to improve to the level of self-reliance. The large farms, producing at an industrial scale, would provide for the urban population. With an increase in the proportion of the urban population to about 50% between 2031 and 2060, there would be more buyers of food crops and fewer growers of food crops. Crop productivity would need to continue to improve due to shrinking land resources as urban and transport sectors take up additional lands.

To meet SDG 2—Zero Hunger, the crop processing industries for food, beverage, and dairy products would need to reach the level of no waste of inputs. This would mean investment in new and seasonally appropriate technology and equipment. The supply chain from farm to factory to market would need to ensure product quality for minimizing waste. From 2031 to 2060, the supply chain system would need to increase capacity to reach net exporter levels.

The third part of the agriculture sector is textile and furniture industries. This sector has direct repercussions on SDG 2—Zero Hunger. The opportunity cost of using land for products other than food crops is going to be high when looking at only the local and national markets. But with the foreign exchange earning potential of these two industries continuing to increase, would mean larger area allocation in the long run from 2031 to 2060.

16.3.1.2   SDG 7—Affordable and Clean Energy

16.3.1.2.1   Transport
The transport sector consists of two parts:

- Construction of roads, railway lines, airports, and shipping ports
- Operations and energy demand for transport hubs

SDG 7—Affordable and Clean Energy is central to the discussion in this chapter regarding the policy dynamics for energy, environment, and sustainable development. In this chapter, SDG 7 is part of the transport sector at two levels: 1. During the construction of transport hubs like shipping ports, airports, railway stations, and roads, moving away from fossil fuels and increasing renewables in the energy mix. 2. With the operation of transport hubs, the energy demand is ongoing. The conversion to affordable and clean energy is the only way to reduce carbon emissions while traffic actually increases.

16.3.1.2.2   Aerospace Technologies
Aerospace technologies consist of three portions:

- Aviation (airplane design) and space travel (rocket design)
- Aerospace (aeronautics and astronautics) for launch into space
- Satellite, rocket, and spaceship design, manufacturing, and operations in orbital space and beyond

SDG 7—Affordable and Clean Energy is very important for aerospace technologies. This sector would be built using new and revolutionary energy technologies. The design technologies would focus on two aspects: 1. airplanes and rockets are designed to be lightweight, fuel efficient, and with accelerated speed and increased capacity. Engines would be designed to be fuel efficient and with the capacity to use new and experimental fuel and energy sources. 2. Satellites would be designed in collaboration with other countries, and satellite operations would be part of a larger or global agenda of communications and weather monitoring. HR would pay special attention to training of aeronautical engineers and astronautical engineers. The major part of R&D in the aerospace technologies sector would be on energy resources that are affordable and available and provide clean energy.

16.3.1.3   SDG 8—Decent Work and Economic Growth

16.3.1.3.1   Industry
The industry sector consists of:

- Machinery and product and appliances manufacturing industry
- Electronic equipment, mobile phone, internet connectivity, and gaming industry
- Transport and automotive vehicle manufacturing industry

SDG 8—Decent Work and Economic Growth brings into focus the role that HR and R&D play in economic development. With R&D comes new and improved energy technologies and machinery and product and appliances design and manufacturing. This increases demand for trained humanpower, resulting in expansion of the whole industry. For future communication, internet connectivity, mobile phones, and electronic equipment would be efficiently designed, affordable, and easily available. The gaming industry has large domestic and export potential. With sustained economic growth based on industries with huge export potential and domestic demand expansion, decent work would be available to the labor force, particularly the youth bulge in the population.

16.3.1.4   SDG 9—Industry, Innovation and Infrastructure

16.3.1.4.1   Industry

The industry sector consists of:

- Machinery and product and appliances manufacturing industry
- Electronic equipment, mobile phone, internet connectivity, and gaming industry
- Transport and automotive vehicle manufacturing industry

SDG 9—Industry, Innovation and Infrastructure would give focus to the industry sector in this chapter. From 2024 to 2030, the manufacturing industry for household products and appliances would innovate to bring consumer-suggested designs to the manufacturing process. Machinery production and manufacturing would focus on energy efficiency and innovation in infrastructure construction.

Internet connectivity infrastructure is based on both hardware equipment and software. To build global-standard, high-speed internet connectivity, Pakistan would need to collaborate with the industrial world leaders. To build the electronics and mobile phone industry and maintain an innovative environment, R&D and HR would need to be focused and market oriented.

The third part of the industry sector is design, innovation, and manufacturing of all transport vehicles. The innovation from SDG 9 applies at two levels: innovation in the energy used for the manufacturing process is affordable and clean energy, and all the engines are going through an evolution process where modifications are taking place to improve efficiency.

The industry sector would build the capacity to meet the domestic demand and develop export potential. With well-built momentum due to SDG 9, the industry sector would expand the infrastructure and innovate technology from 2031 to 2060 based on this momentum.

16.3.1.4.2   Aerospace Technologies

Aerospace technologies consist of three portions:

- Aviation (airplane design) and space travel (rocket design)
- Aerospace (aeronautics and astronautics) for launch into space
- Satellite, rocket, and spaceship design, manufacturing, and operations in orbital space and beyond

For the aerospace technologies industry, SDG 9—Industry, Innovation and Infrastructure would provide the foundations to build this industry between 2024 and 2030. From 2031 to 2060, the industry would innovate in technology and expand the infrastructure. Innovation in renewable energy technologies would accelerate growth in the aerospace industries. The technology required for the satellite-based artificial-moon energy source would need to be developed in collaboration with China, Russia, and other interested parties during the next few decades.

16.3.1.5   SDG 11—Sustainable Cities and Communities

16.3.1.5.1   Urban Centers

Urban centers consist of:

• Energy demand in residential and commercial urban areas
• Architecture design—buildings with natural light and climate control
• Building construction industry—material innovation

SDG 11—Sustainable Cities and Communities would provide special support to urban planning in Pakistan during 2024 to 2030. Cities and communities would become sustainable with planning for climate resilience. If a hundred-year flood become a reality, then cities would need to prepare for that. If communities face prolonged droughts and climate-related outmigration, then new support systems would need to be built for these communities. The main focus of this chapter is change in the energy mix. The stationary energy demand from residential and commercial urban areas would increase with the increase in population and urban spread. Changes in architectural design and construction materials would have significant impacts on this energy demand. During 2031 to 2060, architectural design would change to accommodate natural light and reflection and plants and indoor trees for climate control. These modifications would reduce energy consumption significantly. Additional canals, dike walls, levees, and barrages would need to be built around the urban localities that are in flood-prone areas. Water shortages and prolonged droughts would make planning for water conservation and water use efficiency a main topic in urban planning systems. Urban planning in the future would be about natural resource consumption efficiency and conservation. Sustainability of urban services like electricity, water, transport, and hospitals would ensure smooth running of urban systems.

16.3.1.6   SDG 12—Responsible Consumption and Production

16.3.1.6.1   Agriculture

The agriculture sector in the present study is divided into three parts:

• Agricultural inputs and farming
• Crop processing industry: food, beverage, and dairy products
• Textile and furniture industries

SDG 12—Responsible Consumption and Production presents discussion regarding both consumer behavior and production processes for the agriculture sector.

For agricultural farming production processes, inputs are the focus. Here responsible production would mean technology improvement for the production process where land and input efficiency is enhanced. Better technology would mean the same amount of inputs or land would yield an increased amount of products. Here the consumer's responsibility would be to demand products that use inputs that are less harmful to the environment. This means that agricultural products, production processes, and inputs could all be consumer driven.

Food waste, particularly in the food processing industry, has repercussions not only on the cost of the products but also on availability. The focus of this chapter is on increasing the proportion of renewable energy in the energy mix. Production processes would need to improve at two levels: 1. improve food product output ratio to food crop input, making the availability of larger amounts of products possible, and 2. improve the energy input ratio to product output, resulting in energy efficiency.

SDG 12 is most fundamental for the furniture and forest product industries. Responsible consumers control and reduce demand for furniture from endangered forest types. Buying wood and wood products only from plantation forest and from reforested forests that are well kept and with active regeneration. The production process for the textile industry uses cotton that is input-intensive, particularly pesticides. Research is needed to identify environmentally friendly cotton crops with reduced need for pesticides.

16.3.2 Finance and Investment Policies

Economic growth would take place in two stages: initially, additional industrial activities would take place as exports gradually increase; second, this would then increase demand for energy in each of these industries.

An investment plan based on public finance, public private partnership, and corporate strategic investment would build a rupee reserve fund. This fund is needed for construction of the installed capacity infrastructure for electricity. This would bring rapid progress in infrastructure development for the energy sector. The justification for this large amount of investment in the energy infrastructure would be based on the robust global economy that would be able to absorb exports from Pakistan.

Table 16.3 presents total overnight capital costs of new electricity-generating technologies in 2020 and 2060. The assumption for Table 16.3 is a 1% decrease in cost per year due to technological improvement from 2020 to 2060. Any combination of technological improvement incremental rates and related costs could apply, depending on the rate of technological innovation. Based on the present assumption, in 2060, 400 GW would cost $900 billion. In comparison, Turkey is planning 232.54 GW in 2050 (Enerdata 2022).

Putting Tables 16.3 and 16.4 together gives one a sense of a baseline. The actual costs in 2020 are applied to an estimated future installed capacity of 50 GW in 2030. The infrastructure cost for the year 2030 based on 2020 costs comes out to be $152 billion. Using estimated cost for the year 2030—with 1% cost reduction every year due to technological improvements—the total for 50 GW comes to $138 billion. This difference is more than $14 billion.

TABLE 16.3

Generation Capacity Capital Cost for MW Electricity in 2020, 2030, and 2060

Electricity Source	2020 (Actual Costs) $/MW	2030 (Estimated Costs) $/MW	2060 (Estimated Costs) $/MW
Oil	1,000,000	900,000	670,000
Gas	710,000	642,111	475,000
Coal	3,700,000	3,346,000	2,500,000
Solar PV (fixed)	830,000	751,000	555,247
Solar PV (tracking)	860,000	778,000	580,000
Onshore wind	1,600,000	1,447,000	1,070,400
Offshore wind	6,500,000	5,880,000	4,350,000
Wave energy	5,000,000	4,522,000	3,345,000
Nuclear	7,000,000	6,331,000	4,700,000
Dams	2,500,000	2,261,000	1,670,000

(Source: U.S. Energy Information Administration 2022; and USA Department of Energy 2022)

TABLE 16.4

Pakistan Energy Sector's Progressive Increase in Electricity from 2030 to 2060 (MW)

Energy Type	2030	2040	2050	2060
Oil, gas, coal, and coal gasification	20,750	32,500	65,000	130,000
Hydropower	15,000	30,000	60,000	120,000
Nuclear—fission	10,000	20,000	40,000	80,000
Nuclear—fusion	0	5,000	10,000	20,000
Solar energy systems	2,000	5,000	10,000	20,000
Wind energy systems	2,000	5,000	10,000	20,000
Ocean wave and tide energy	250	2,500	5,000	10,000
Total	**50,000**	**100,000**	**200,000**	**400,000**

(Source: Assumptions made by the author)

The finance and investment policies and plans would give new impetus to the growth in the energy sector infrastructure development in the country. This would support the industrial sector to increase productivity and engage in export markets.

16.3.3 INFRASTRUCTURE POLICIES

Expansion in industrial sector production is based on facilities that are provided in the form of infrastructure expansion. The infrastructure consists of buildings, machinery, electricity, water, and transport hubs for moving raw material, labor, and products.

Like each new infrastructure expansion requirement, considerable new investments would be necessary in the energy and electricity sector for infrastructure expansion.

TABLE 16.5
R&D Investment for the Energy Sector

Year	R&D in US Dollars
2030	1.52 billion (2020 based on actual costs)
2030	1.38 billion (2030 estimated costs)
2060	9 billion (2060 estimated costs)

(Source: Assumptions made by the author)

This would be based on the intensification of the industrial production process. The national grid and provincial energy and electricity generation, transmission, and distribution would increase supply for the industries and transport hubs, including ports, and would prepare to upgrade accordingly. Table 16.4 presents planned progress in increased electricity generation capacity in Pakistan from 2030 to 2060.

16.3.4 RESEARCH AND DEVELOPMENT POLICIES

The policies for R&D would ensure that the energy transition takes place as the demand for energy increases with expanding exports markets. For research innovation in the energy sector, 1% of the energy sector's installed capacity value should be allocated to the R&D agenda. In 2060, the energy sector's estimated worth would be $900 billion, and 1% would be $9 billion. As a starting point, in 2030, the total actual worth of the energy sector would be $152 billion. One percent would be $1.52 billion ($1,520,000,000). With technology improvement cost estimates, the estimated worth of the energy sector in 2030 would be $138 billion; 1% would be $1.38 billion ($1,380,000,000). All five sectors would fully contribute to the R&D funding process. The expectation is that foreign direct investment would provide 50% of the R&D expenses, whereas national investment would fund 50% of the remaining R&D. Table 16.5 presents the progress that R&D would go through from 2030 to 2060.

16.3.5 HUMAN RESOURCES DEVELOPMENT, EDUCATION, AND TRAINING POLICIES

The energy sector–specific needs for HR development, education, and training would focus on the technical and engineering fields. The new technically trained young labor force would take the energy sector on the path of rapid growth and transformation. Table 16.6 presents three sets of energy technologies with different time scales for humanpower requirements:

- Sector expansion and technology upgrade with immediate startup from 2024
- Establishing new sectors from 2030 onwards based on the energy technologies that are still under research experimentation
- Establishing new sectors from 2040 onwards based on the energy technologies that are still at different levels of specialized research experimentation

TABLE 16.6
Future Needs for HR Development in the Energy Sector

	Energy Supply Source:	Starting Year	Future Task for HR
A	Hydropower	2024	Sector expansion
B	Thermal energy: oil, gas, coal, and coal gasification	2024	Sector expansion
C	Nuclear fusion and fission:		
C1	Nuclear fission: uranium and thorium	2024	Sector expansion
C2	Nuclear fusion: ITER Tokamak, artificial sun	2040	Establishing sector
D	Renewable energy:		
D1	Solar energy systems	2024	Sector expansion
D2	Wind energy systems	2024	Sector expansion
D3	Ocean wave energy and tide energy	2030	Establishing sector

(Source: Assumptions made by the author)

16.3.6 POLICIES FOR RESTORATION, CONSERVATION, AND MANAGEMENT OF THE ENVIRONMENT

The policy set for the environment would focus on policies for restoration, conservation, and management of the environment. This is the basis for the other four policies for the environment: finance and investment for environmental management; infrastructure development for environmental management; R&D for technology development for the environment; and human resource development, education, and training for the environment in Pakistan. The management of the environment has two goals:

- Conservation of environmental resources and the ecosystem so that it continues to provide environmental services.
- Restoration of degraded environmental resources and the ecosystems to provide efficient environmental services.

Both these responsibilities need environmental quality standard that are measurable and implementable. To ensure these environmental quality standards, a system of monitoring environmental quality, that is, air and water quality, would become imperative. Once environmental quality is measured at specific locations, in each province and nationally, then the task of restoration of the environment and conservation of the environment begins.

16.4 FROM ECONOMIC GROWTH TO SUSTAINABLE DEVELOPMENT PATHWAYS

Pakistan consumes energy and electricity from two basic sources: fossil fuels and hydropower. The economic development in various Five-Year-Plans from 1947 until the current Plan (Thirteenth Plan: 2024–2029, to be issued by the Government of Pakistan) has mainly focused on these two energy sources. Since the Tenth Five-Year-Plan

(2010–15), the inclusion of renewable energy in the context of climate change was made into mainstream development process, included in the Economic Survey. Since the 1980s, Pakistan has been developing solar energy with public funding. Alternative fuels and new energy resources became a public funding priority in the early 2000s. Nuclear technology for electricity was brought to Pakistan in the 1950s.

From 2024 to 2030, Pakistan has specific targets to meet the clean and renewable energy under UN SDG 7—Affordable and Clean Energy. Between 2031 and 2060, Pakistan will build an infrastructure that will be able to provide 130,000 MW from fossil fuels and 270,000 MW from renewable energy resources, as presented in Table 16.1, section III.

To start on this trajectory in 2024 for the desired outcome in 2060, specific sets of policies will need to be designed. A growth rate of 7.25% per year starting from 2024 would take the electricity demand in Pakistan from 30,000 to 400,000 MW by the year 2060. The energy sector in Pakistan would experience two types of transformation. The growth rate would be rather high and constant, and the other transformation would be shrinkage of the demand for fossil fuels and expansion of the demand for renewable energy resources.

The planning and policies for this trajectory and persistently high growth rate would be designed as presented in Table 16.7. The focus would be on innovations in

---

TABLE 16.7

Policies for Innovation in the Energy, Electricity, and Transport Fuel Technologies

A   **Planning for Energy and Electricity:**
- Transmission and distribution of uninterrupted electricity through national and provincial grid infrastructures[a]
- For efficient transportation of the fossil fuels and the new fuels, planning, design and construction of the sea ports, train-stations and roads
- Planning and financial allocations for the renewable energy infrastructure
- Planning and building dams

B   **Oil, Gas, and Coal:**
- Planning self-sufficiency of oil, gas and coal: discoveries, refineries, and supply
- Innovation in technologies for efficient processing and refinery of fossil fuels

C   **Solar and Wind Technologies:**
- Solar energy technology innovation
- Perovskite solar cell technology, transparent solar panels to be used as building windows[b]
- Design and efficiency innovations in wind turbines
- Reverse meters and solar and wind farms at new housing societies

D   **Electric Vehicles (EV) + New Fuels:**
- Science and technology for dual-fuel vehicles and electric vehicles
- Marketing and pricing of new fuel and electric vehicle charging stations
- Fuel and technological innovations for new fuels for land, water, air, and space travel
- Hydrogen energy; experimentation with other solid, liquid, and gaseous elements as new fuels

---

(Continued)

TABLE 16.7

Policies for Innovation in the Energy, Electricity, and Transport Fuel Technologies *(Continue)*

E	**Experimental Energy Technologies:**
	• Efficiency improvement in nuclear fission electricity generation and transmission
	• Artificial sun: beyond nuclear fission, new focus on nuclear fusion: ITER and Tokomak

(Source: ᵃ NTDC May 2021; ᵇ Metalgrass August 8, 2021).

energy-sector technologies. These innovations would be mainly in energy, electricity, and transport fuel technologies. The following five policy options would prepare the energy sector for an annual growth rate of 7.25% from 2024 to 2060. These policies would set in motion a process that will transform the energy sector into a clean energy sector and economic growth into environmentally sustainable economic development.

16.5   THE ENVISIONED FUTURE AND TECHNOLOGY FOR ENERGY INNOVATION

In 2022, the population of Pakistan is about 220,000,000. The electricity production capacity is about 30,000 MW. This means that 7,333 persons are allocated 1 MW of electricity each in Pakistan. There is a need for policy dynamics to reach an energy-intensive future; this envisioned future is in the year 2060. The assumption is that Pakistan will have a population of 400,000,000 (DAWN 2019). To reach the level of industrialized economy, Pakistan would be providing 400,000 MW of electricity to its citizens; that is, 1 MW of electricity would be provided to 1,000 people. Pakistan would raise funds for developing the energy sector to bring it to this level. The energy sector would consist of a combination of fossil fuels and renewable energy resources. Let us assume that policies would focus on five sectors.

The export potential for each of the five sectors is presented in Table 16.8. This is one of the main justifications for rapid development in the energy supply system.

As can be seen from Table 16.8, some policy sectors have export potential. To reach this level of productivity, another set of industrial policies would be designed. These industrial policies will bring technological revolution in the production process of each of these industries. Processes like circular economy (Martinez-Alier 2021) could be introduced later into industrial production innovation.

Agricultural-dependent industries like farming and agricultural inputs; crop processing industry; food, beverage, and dairy industries; textile industry; and furniture industry all would be able to increase productivity and export 25% of the products.

By 2060, the manufacturing industry for heavy machinery and household products and appliances; the manufacturing industry for transport and automotive vehicles; and the mobile phone, electronics, software programming, internet storage, and internet gaming industries would be able to export 50% of designed and manufactured products for foreign exchange.

TABLE 16.8

Export Percentages for the Five Sectors in 2060

Five Sectors	Energy Demand Sectors	Export Percentage
Urban centers		
	Energy demand in residential and commercial urban areas	
	Architecture design—buildings with natural light and climate control	
	Building construction industry—materials innovation	
Agriculture		25%
	Agricultural inputs and farming	
	Crop processing industry: food, beverage, and dairy industry	
	Textile and furniture industry	
Industry		50%
	Machinery and consumer product and appliances manufacturing industry	
	Transport and automotive vehicle manufacturing industry	
	Mobile phone and electronics: internet and gaming industry	
Transport		
	Construction of roads, railway lines, airports, and shipping ports	
	Operations and energy demand for transport hubs	
Aerospace technologies		10%
	Aviation (airplane design) and space travel (rocket design)	
	Aerospace (aeronautics and astronautics) for launch into space	
	Satellite, rocket, and spaceship: design, manufacturing, and operations in orbital space and beyond	

(Source: Assumptions made by the author)

An innovative, yet-to-be-established export earnings industry would consist of aviation and aerospace technologies and satellite and rocket designs for space travel, along with innovative aviation technology and airplane design industry. The exports from this future industry would be 10% by the year 2060.

Technology for the Next Industrial Revolution: Economic development for the next four decades would focus on building the manufacturing industry and other industrial sectors. The transport industry is divided into land, sea, rivers, air, and eventually space travel. The six sections presented in Table 16.9 represent some of the innovation that would build the industrial sectors, particularly the manufacturing industry, to lead Pakistan on a path of economic growth that is also environmentally sustainable.

Development in the electronics industry is essential, as it operates all other industries, including the energy sector. Training is imperative in the new technologies as markets evolve. Technology innovation is needed, based on strong human resource development for software technology and electronic equipment. To speed up technological innovation, the emphasis would be on artificial intelligence (AI), machine learning, and deep learning.

TABLE 16.9
Innovation in Manufacturing and Other Industries

Town Planning and Building Construction Industry
- Cement and other construction input efficiency standards—production and utilization
- Design innovation in construction of roads, rail stations, airports, and seaports
- Perovskite solar cell window panels for the buildings
- Energy-efficient green buildings
- Eco-friendly buildings—sustainability by rooftop vegetation

Equipment and Machinery for Innovation in Food Security Technologies
- Machinery for large-scale mainstream agriculture and irrigation
- Seawater desalinization, water-logged agriculture land desalinization, and urban saline-water desalinization
- Reverse Karez using desalinized seawater[a]
- Artificial aquifers
- Hydroponics

Land and Water Transport Manufacturing Industry
- Technological design improvements for road and off-road land transport vehicles
- Technological innovation for train and railway design and track systems, including magnetic levitation (maglev)
- Design innovation and manufacturing of marine commercial transportation ships for oceans and boats for rivers
- Marine fishery boats and supply-chain cold storage boat design and manufacturing
- Transport fuel and engine design innovations

Electronic Industry as Central to All Other Industries
- Electronic equipment
- Technology software
- Artificial intelligence
- Machine learning and deep learning technology
- Design and participation in video gaming market

Commercial Aircraft Manufacturing Industry
- Development of commercial air transport facilities
- Technological innovation for small airplanes, helicopters, and passenger-carrying flying drones
- Technological innovation in drone designs and operations
- Innovation in air travel fuel efficiency and sustainability

Satellite and Spacecraft Manufacturing Industry, Including Space Stations and Moon Mining Site Machinery, Plus Experimental Design for Space Travel Fuel
- Innovation in rocket fuel for carrying satellites to space
- Innovations in science and engineering of satellite technology
- Development of space travel technology
- Space station and colonies on Earth's moon, other planets, and their satellites[b]

(Source: [a] and [b] Mallick, September 2021)

As presented earlier in the introduction to finance and investment policies, the export quality of most industries would improve gradually. As export quality improves, the ratio of export to production would increase. This would result in rapid development of industrial infrastructure, along with an increase in electricity and energy demand. This would attract investors interested in investing in the energy sector. With increase in environmental awareness and stewardship, energy sector investors would set a path towards clean energy resources that would be considered less harmful to the environment.

16.6   POLICY DYNAMIC TO REACH THE DESIRED FUTURE

The two sets of policies, energy policy and environmental management policy, bring together an agenda for a long-term environmentally sustainable development.

16.6.1   ENERGY POLICY

The five sectors identify economic activities that require electricity and energy consumption. With specific policies for each of these five sectors, a clearly defined path would emerge. The five sectors are:

Urban centers
Agriculture
Industry
Transport
Aerospace technologies

Each of these five policy sectors will be analyzed using the following policies that are considered here:

- Investment policy for the energy sector
- Infrastructure policy for the energy infrastructure development
- Research and development policy for energy sector technology development
- Policy for human resource development for the energy sector

16.6.1.1   Investment Policy for the Energy Sector

As mentioned earlier in Table 16.2, thermal energy would be about 32.5% of the total energy consumption, whereas hydropower, nuclear, and renewables would be about 67.5%. Even with thermal energy resources much cheaper than most of the renewables, consumer choice would shift towards renewables. Both nuclear and renewable energy resources include experimental energy production technologies during the 2020s that could become stable energy producers by 2060.

Table 16.2 presents the percentage difference of investment in thermal and renewable energy technologies. There needs to be a basis for building momentum for renewable energy to provide more than 60% of the economy-wide demand for energy. By 2060, consumers will be very particular about net-zero emissions and the necessary connection with renewable energy resources. This will build a base for higher demand for renewable energy resources.

In Table 16.2, the assumption is that the stationary energy demand at the urban centers is most likely to use a larger proportion of the renewable energy resources. The agriculture and industry sectors use near average amounts of thermal and renewable energy resources. Transport and aerospace technologies use 37.5% electricity made from fossil fuels. Rockets use highly efficient fossil fuels for ignition and then thrust to tackle gravity, and then various other fuel and energy resources to travel in space.

Using Table 16.3 data, the original estimates for 2020 based on the actual costs in dollars per MW gives thermal energy at 18%, or $5.4 million. Table 16.3 also gives hydropower, nuclear, and renewable at 82%, $24.3 million. The same 18% and 82% are assumed in 2030 and 2060, with a 1% cost decrease due to technological improvement. In 2030 the cost would be $4.9 million and $21.97 million. In 2060, the cost would go down to $3.6 million and $16.3 million.

The basic assumption is that from 80% thermal and 20% hydropower, nuclear, and renewables in 2022, the new demand in 2060 would be 32.5% thermal and 67.5% hydropower, nuclear, and renewables, which is the basis for this investment plan.

The focus is on renewables; only technological improvement could change the per MW dollar cost for renewables. When these change, then a more realistic plan could be designed for future energy based on renewables. In its present form, the six lower-cost energy sources are gas, solar PV (fixed), solar PV (tracking), oil, onshore wind, and dams. Both oil and gas are on this list. In reality, the dollar cost per MW for all (or some) renewables will have to fall below that of thermal energy resources for them to become viable investment options. In the list in Table 16.3, nothing is cheaper than gas. Only both solar forms are cheaper than oil. Only onshore wind and dams are cheaper than coal. The three most expensive installation costs are for nuclear, offshore wind, and wave energy; all three are under the category of renewable energy.

Each of the five sectors would attract investors that are interested in both technological improvements for the policy sector and also technological innovation in the energy sector. In 2020s, energy technologies that are new and experimental would be part of the total energy supply by 2060 with significant investment. With net-zero carbon emissions as a main political topic by 2060, Pakistan would be keen to change the energy mix permanently.

16.6.1.2 Infrastructure Policy for Energy Infrastructure Development

The energy infrastructure includes a national grid connection from the site of energy generation for distribution and transmission nationally. Hydropower is mainly produced in the mountain regions, whereas large industrial estates are located in the plain areas near railway stations and seaports. Renewable energy production sites could be planned much closer to consumers. Nuclear fission and later nuclear fusion both would be built a safe distance from urban centers.

The energy sector future estimation must provide accurate numbers for energy consumption change and changes in the energy mix. The investment and infrastructure planning depend on these estimates (DW July 1, 2021). The assumption is that by 2060, the energy consumption would be 400,000 MW per annum in Pakistan. The other assumption is that the consumer would demand low carbon emission energy technologies. Fossil fuels would contract to 32.5% of the total energy mix. The global average on fossil fuels in an energy mix is 80% in 2021 (Corbett June 15, 2021). Intensive change in infrastructure would be needed between 2024 and 2060 to accommodate 67.5% of renewable energy in the energy mix.

16.6.1.3 Research and Development Policy for Energy Sector Technology Development

In the 2020s, the energy production technology of hydropower and thermal energy, oil, gas, coal, and coal gasification are already widely used globally. Nuclear fission and renewable energy, solar energy and wind energy, are also producing electricity for main grids in many countries in almost all of the continents. New experimental energy-producing technologies, hydrogen, fuel cells and batteries for the electric vehicles, and ocean wave and tide energy structures are under research study in many countries. The experimental technologies, nuclear fusion and ITER Tokamak (artificial sun), have a strong theoretical astrophysics base. It is expected that artificial sun will revolutionize the way energy, heat, light, and electricity are produced and supplied globally.

Investment in R&D would improve the technologies for the energy and electricity sector. R&D policy would focus on both efficiency improvement of already existing energy technologies and designing new energy technologies. R&D funding by the public sector would be 10%, public private partnership would be 30%, and corporate funding would be 60% in each of the five sectors.

16.6.1.4 Policy for Human Resource Development for the Energy Sector

Human resource development in the energy sector would mainly focus on engineers for energy technology and structural engineers for energy technology infrastructure. The HR policy would prepare a workforce that is dedicated to each of the energy sectors. As presented in Table 16.10, each of the energy sectors would either go through an expansion or would be established during 2024 to 2060.

Human resource development based on additional market demand is a good justification for redirecting funding for new industrial technology training and technology training for new resources of energy, heat, light, and electricity. As new technologies become mainstream, the technical training demands would change accordingly.

16.6.2 ENVIRONMENTAL MANAGEMENT POLICY

There are two tasks for the environmental management policy:

- Restoring the degraded environment by returning environmental services to their optimal levels
- Maintaining the environment and natural resources in pristine condition and protecting environmental services

The five set of policies that are to be considered for environmental management are:

- Policy for restoration, conservation, and management of the environment
- Investment policy for environmental management
- Infrastructure policy for environmental management

TABLE 16.10

Future Transition of the Energy Technologies

Task for HR	Energy Technology Development and Expansion
1 Sector expansion and technology upgrade with immediate startup from 2024:	
	• Hydropower • Thermal energy: oil, gas, coal, and coal gasification • Nuclear fission: uranium, and thorium • Solar energy systems • Wind energy systems
2 Establishing new sectors from 2030 onwards based on the energy technologies still under research experimentation:	
	• Ocean wave energy and tide energy[a]
3 Establishing new sectors from 2040 onwards based on the energy technologies that are still at different levels of specialized research experimentation:	
	• Artificial sun—nuclear fusion: ITER Tokamak[b]

(Source: [a] Clean Energy Ideas 2019, [b] New Scientist 2009).

- Research and development for environmental management
- Policy for human resource development for technology and infrastructure for environmental management

With these policies in place, the economic growth would gradually convert into environmentally sustainable development.

16.6.2.1 Policy for Restoration, Conservation, and Management of the Environment

While designing the environmental management policy for each of the five sectors, the focus would be on conserving natural resources like air, water, soil, forests, and biodiversity. The restoration of the natural environment and natural ecosystem services would become necessary in the areas where the natural environment is already degraded. The foremost activities needed would be:

- Monitoring of environmental quality standards for air and water
- Assurance of environmental quality standards for air and water
- Efficient management of water: water availability and depletion monitoring; water consumption and wastage reduction

- Forest cover and urban forest plantation for:
  - air pollution and dust control
  - water quality and soil erosion control
  - protection of habitat for biodiversity of flora and fauna
- Sea plastic pollution preservation activities on river banks and for coastal areas

The policy agenda:

- Environmental impact assessment (EIAs) and environmental management plans (EMPs) would be implemented as a necessary policy tool
- EIAs and EMPs would ensure that there is no hidden cost of development to the natural ecosystems

Energy Sector: The policy agenda for energy sector infrastructure construction would be to conduct EIAs and prepare EMPs. Every ecosystem undergoes some process of displacement with development of the built infrastructure. There are environmental impacts of three types of human activities in relation to the energy sector infrastructure construction:

- Clearing of site
- Energy infrastructure construction and gridline connection
- Operation of the energy infrastructure

EIAs and EMPs would focus on monitoring and assurance of the environmental quality standards for air and water in the surrounding areas. EMPs would also focus on competing demands for water availability and water efficiency for water consumption by humans and the water requirements of the surrounding ecosystem.

As a future energy sector focus, the global and regional interest in reaching net-zero carbon emissions by 2050 would bring in to focus four possibilities:

- Reduction in emissions
- Replace emission-producing fuels with renewable energy resources
- Carbon capture and storage by emission producers
- Sequestration in natural and plantation forests

A combination of change in the energy mix and in energy consumption behavior would be reinforced with carbon capture and storage technology. Applying technology to change captured carbon into a reusable product would add to this process. The traditional system of plantation forests for sequestration would provide a safe ecosystem. With this, the air quality monitoring system would ensure that air quality standards are maintained for a healthy lifestyle.

The Five Planning Sectors: For urban centers, during the site clearing, construction, and operation phases, it would be imperative that monitoring of environmental quality standards for air and water be observed. For this, local and imported equipment would be needed. With the monitoring of environmental quality standards, the equipment and infrastructure would become necessary for assurance and maintenance of environmental quality standards for air and water. Along with water quality, water supply efficiency would be necessary. In all urban planning, water availability and conservation efficiency are to be planned well in advance.

Urban forest parks and forested areas near dense urban locations would provide forest cover and urban forest plantation. Forest plantations, increasing tree density for natural forests, and protecting park areas would be an ongoing process. This would provide the transport and industry sectors with sequestration forests, and this would be included in the urban green planning process. Agriculture sector land use planning would include plantation of forests of productive trees, natural forest density increase, and plantation for sequestration in wilderness and open grounds.

From 2024 to 2060, each decade would focus on protection of marine life. While designing the environmental management policy for conservation and restoration of the marine ecosystem, ocean plastic pollution reduction activities on river banks and for coastal areas would get priority. During 2024 to 2045, the main task would be to clean the plastic that pollutes the ocean ecosystem. From 2046 to 2060, ecosystem restoration and preservation of the restored ecosystem would be part of the policy planning process.

A policy for restoration, conservation, and management of the environment for the aerospace technologies sector would mean traveling in orbit and outer space and clearing up the satellite debris. Decades from now, building biospheres with artificial ecosystems, sustainable and livable, on ocean surfaces on planet Earth, and far from the planet Earth, on orbital space stations, moon mining colonies, planet Mars, and other planets and their satellites, will influence humanity's behaviour. A strict regime would ensure that space stations and colonies on planets and their satellite moons have minimal environmental impact. This includes creating self-sustaining water, air, and waste management systems.

16.6.2.2 Investment Policy for Environmental Management

The funding provided by the energy sector, the renewable energy industry and fossil fuel industry, would be utilized for two tasks for environmental management policy: restoring the degraded environment by returning environmental services to their optimal condition and maintaining environment and natural resources in a pristine condition and protecting environmental services. This funding would ensure that human activity continues without ecosystem collapse. A suggestion for the Qudrati Tawanai Taraqiati Bank was launched (Mallick May 2021) for financing renewable energy for a future energy resource base in Pakistan.

Every sector of the economy that includes the energy sector is concerned about the health of the ecosystem. This concern would eventually be converted into funding for specific abatement tasks (tasks specific to reducing environmental degradation). As each economic activity has some environmental repercussion, EMPs would specify abatement tasks with estimated abatement costs (financial cost of fixing environmental problems). Each industry in the industrial sector would fund sustainable entrepreneurship as a role model for environmental management within the industry.

Corporate sector participation along with the public sector would ensure that abatement costs are met. Public and private participation would fund the activities assigned for environmental management in the EMPs and identified as necessary in the EIAs. At the other end of these processes and actual abatement costs would be better environmental conditions, with full ecosystem services.

16.6.2.3   Infrastructure Policy for Environmental Management

Human economic activities add the built environment to natural ecosystems. This hampers some ecosystem services like quality of air and natural flow of water. Abatement activities for remedies of environmental degradation need planning and funding and new infrastructure or modifications to the existing infrastructure. Artificial wetlands are one of the most popular ecosystem restoration examples. These wetlands are built at new sites or on sites where there were wetlands that were drained due to water needs in nearby urban localities or agricultural land. The other less common ecosystem restoration is linking wilderness and natural forest by constructing animal-accessible paths under roads with traffic or over motorways. These linking passages are built to connect forest and plantation reserves to a river, lake, or seashore. These provide linkage, mobility, and connection for flora and fauna biodiversity.

Ecosystem restoration of the terrain after mining activities have been completed provides park rangelands or artificial lakes. The area has previously experienced heavy machinery and vehicular activities. To restore air quality, plantation forests are needed in parks built on mine sites. Dams and barrages on the rivers disturb the free movement of marine life. Ecologists and the engineers design and build fish and other marine life passages at dams and barrage sites.

Floating plastic in the oceans is an anthropogenic problem created in the last few decades. Ecosystem restoration requires special planning and equipment. It is a macro problem with micro solutions, and everything takes time; removal of plastic from the sea and ocean has already begun. What is difficult to reverse is microplastics in the food chain. After decades of scientific activities, by 2060, it is hoped that appropriate equipment will be designed to clear all water bodies of plastic.

16.6.2.4   Research and Development Policy for Environmental Management

Special interest groups are needed for initiating ecosystem restorations locally. This process needs support from the public sector and the corporate sector. Once the public sector provides political and legal support to the ecosystem restoration movement, the process becomes mainstream. The abatement costs then become justifiable. Due to consumer demand pressures, the corporate and industrial sectors participate actively to restore the ecosystem. R&D funding from the public and corporate sector would ensure that consumer demand for environmental management efficiency is met. Housing estates, suburbs, industrial estates, and transport systems have their own ecosystem footprints. Ecosystem restoration requires systematic planning, funding, and technology development. R&D for environmental management needs to take place at research institutions that are directly linked with the industrial processes that impact the environment and need ecosystem restoration. For example, the fossil fuel drilling and refinery industries and the renewable energy infrastructure installation industry need to be open to these research institutions.

16.6.2.5   Policy for Human Resource Development for Technology and Infrastructure for Environmental Management

The environmental management sector could open up employment for scientists and technicians. The monitoring of environmental quality is a scientific process based on equipment and apparatus. Teams of dedicated trained staff would take these responsibilities at the provincial and regional levels. The restoration of the degraded environment and the process of maintaining a pristine environment could be technology- and labor-intensive activities. A combination of scientific research institutions and localized civil society organizations could identify short- and long-term environmental management activities.

16.7 CONCLUSION

The idea is to find a set of policies that would turn economic growth into environmentally sustainable development. The focus is on the energy mix. Goals and objectives are specified. Five sectors and multiple industries consume the provided energy in the year 2060. The high export potential of some of these five sectors is given as a basis for expansion in the energy sector accordingly. Section 16.6 explains the energy and environment policies for these five sectors. Nine policy items, four policies for energy and five for the environment, provide the motivation for conversion of economic growth into environmentally sustainable development.

For the year 2060, the assumption is that Pakistan would become a net-export–based industrialized economy, providing 400,000 MW of electricity to its citizens to actively participate in productivity progress. The energy mix would include fossil fuels and renewable energy resources.

An innovative set of policies for the energy sector would result in environmentally sustainable development by the year 2060. Environmentally sustainable energy growth prospects are important for understanding the challenges that developing countries face at the beginning of the century. Innovation in energy technology in the 21st century would impact environmental management policies. It is easy to preach an increase of renewable energy in the energy mix of a country, but in actuality, changing direction away from fossil fuels and constructing expensive infrastructure is a difficult deliberate choice. Consumer demand is the only persistent pressure that forces policymakers to opt for renewable energy. Once the direction is changed towards renewable energy and constructing expensive infrastructure, a door opens to a new option.

The direction set by the financing of energy and environmental management policies would provide a clearer picture of an environmentally sustainable outlook. A system would develop where the journey from shared goals to rational and sound policies would manifest. Policy makers are responsible for designing policies, experimentation with technical options for policy suggestions, and implementation of the approved policies.

REFERENCES

Chinese Government. (June 1, 2022). Tracking China's transition to sustainable energy: 14th Five-year plan for renewable energy development. *China Energy Portal*. Beijing, China. https://chinaenergyportal.org/en/14th-five-year-plan-for-renewable-energy-development/

Clean Energy Ideas. (July 23, 2019). What is tidal power? *Tidal Energy Explained*. clean-energy-ideas.com. www.clean-energy-ideas.com/hydro/tidal-power/what-is-tidal-power-tidal-energy-explained/

Corbett, J. (June 15, 2021). *Share of Fossil Fuels in Global Energy Mix 'Has Not Moved by an Inch' in a Decade*. Jessica Corbett is a staff writer for Common Dreams. Common Dreams, Inc. Founded 1997. Registered 501(c3) Non-Profit. www.commondreams.org/news/2021/06/15/share-fossil-fuels-global-energy-mix-has-not-moved-inch-decade

DAWN. (2019). *Pakistan's Population May Double in Next 30 Years.* Accessed December 25, 2019. www.dawn.com/news/1524251

Demirtas, O. (2013). Evaluating the best renewable energy technology for sustainable energy planning. *International Journal of Energy Economics and Policy.* Vol. 3, Special Issue, pp. 23–33. ISSN: 2146-4553. www.econjournals.com. Selected Papers from "International Conference on Energy Economics and Policy, 16–18 May 2013, Nevsehir, Turkey". https://dergipark.org.tr/en/download/article-file/361259

DW. (July 1, 2021). BUSINESS: How much power will Germany need for its energy revolution? *Author Insa Wrede.* Deutsche Welle. Germany. www.dw.com/en/how-much-power-will-germany-need-for-its-energy-revolution/a-58116209

Enerdata. (2022). *Turkey: Key Forecasts.* Accessed Wednesday, June 8, 2022. https://eneroutlook.enerdata.net/country-snapshot/turkey-energy-forecast.html

Mallick, S. (2021*). *Link between Innovation in the Circular Economy Design and Efficiency Improvement in the Renewable Energy Technologies.* Paper visualizes renewable energy for circular economy design for the year 2060. Submitted to: Social Science Research Network (SSRN) at: https://papers.ssrn.com/sol3/papers.cfm?abstract_id=3900035

Mallick, S. (May 2021). *Socially & Environmentally Responsible PSDP: A Way Forward to Sustainable Growth. I Was Guest Contributor for PPAF (Pakistan Poverty Alleviation Fund) Development Dialogue on: Rethinking Public Investment for Pro-Poor Growth.* PPAF, Islamabad, Pakistan.

Mallick, S. (September 2021). Sustainable circular economy design in 2050 for water and food security using renewable energy. In *Circular Economy and Sustainability: Volume 2: Environmental Engineering.* Edited by: Alexandros Stefanakis and Ioannis Nikolaou, 509–521. Elsevier-Science. Direct. 2022. www.sciencedirect.com/book/9780128216644/circular-economy-and-sustainability#book-description

Mallick, S. (2018). *Energy Mix—Change of Focus for the Energy Sector of Pakistan.* Conference Paper Visualizes the Pakistan Energy Sector in the Year 2047. Conference Paper Published in the: Proceedings of the International Conference on Renewable, Applied and New Energy Technologies, ICRANET-2018, 19–22 November 2018, Air University, Islamabad, Pakistan.

Mallick, S., Sinden, J.A. and Thampapillai, D.J. (2000). The relationship between environmentally sustainable income, employment and wages in Australia. *Australian Economic Papers*, Vol. 39, no. 2, p. 231 (14).

Martinez-Alier, J. (2021). The circularity gap and the growth of world movements for environmental justice. *Academia Letters*, Article 334. https://doi.org/10.20935/AL334.

Metalgrass. (August 8, 2021). Perovskite Info: Researchers design smart windows with perovskite PV cells. Source: *Nature Communications PV-Magazine.* Posted: by Roni Peleg. www.perovskite-info.com/researchers-design-smart-windows-perovskite-pv-cells

New Scientist. (October 9, 2009). *ITER: How It Works.* By Valerie Jamieson, Roger Highfield, Neil Calder and Robert Arnoux. www.newscientist.com/article/dn17950-iter-how-it-works/

NTDC. (May 2021). *Indicative Generation Capacity Expansion Plan (IGCEP) 2021–30.* Driven by the Future. Power System Planning. Load Forecast and Generation Planning. National Transmission and Dispatch Company, Pakistan.

Oyewo, T., Olarewaju, O.M., Cloete, M.B., and Adenuga, O.T. (2021). Environmental costs estimation and mathematical model of marginal social cost: A case study of coal power plants. *Environmental Economics*, Vol. 12, no. #1, pp. 90–102.

USA Department of Energy. (2022). *Annual Technology Baseline: Coal. National Renewable Energy Laboratory.* Accessed May 17, 2022. https://web.archive.org/web/20200429032024/https://atb.nrel.gov/electricity/2017/index.html?t=cc

U.S. Energy Information Administration. (2022). Cost and performance characteristics of new generating technologies. *Annual Energy Outlook 2022.* www.eia.gov/outlooks/aeo/assumptions/pdf/table_8.2.pdf

17 Global Warming and Climate Change
Projections and Implications

Taddeo Rusoke

17.1 INTRODUCTION

Greenhouse gas (GHG) emissions from human activities have continued to increase from 2010 to 2019 in comparison to other decades (IPCC, 2022). These GHGs from anthropogenic sources, minus removals by anthropogenic sinks, per the United Nations Framework Convention on Climate change, include CO_2 from fossil fuel combustion and industrial processes (CO_2-FFI); net CO_2 emissions from land use, land-use change, and forestry (CO_2-LULUCF); methane (CH_4); nitrous oxide (NO); and fluorinated gases (F-gases), including hydrofluorocarbons (HFCs), perfluorocarbons (PFCs), and sulfur hexafluoride (SF_6), as well as nitrogen trifluoride (NF_3) per IPCC (2022). On average, global per capita net anthropogenic GHG emissions increased from 7.7 to 7.8 tCO-eq, ranging from 2.6 tCO-eq to 19 tCO-eq across regions, as shown in Table 17.1 (IPCC, 2022). These are the major contributors to global warming and climate change.

TheUnited Nations (2022) defines climate change as long-term shifts in temperatures and weather patterns. These shifts may be natural, such as through variations in the solar cycle. But since the 1800s, human activities have been the main driver of

TABLE 17.1
Net Greenhouse Gas Emission Increase from 1990 to 2019

	2019 Emissions (GtCO2-eq)	1990–2019 Increase (GtCO2-eq)	Emissions in 2019 Relative to 1990 (%)
CO_2-FFI	38 ± 3	15	167
CO_2-LULUCF	6.6 ± 4.6	1.6	133
CH_4	11 ± 3.2	2.4	129
N_2O	2.7 ± 1.6	0.65	133
F-gases	1.4 ± 0.41	0.97	354
Total	59 ± 6.6	21	154

Source: IPCC (2022). Summary for Policymakers. In: *Climate Change 2022: Mitigation of Climate Change. Contribution of Working Group III to the Sixth Assessment Report of the Intergovernmental Panel on Climate Change.* Cambridge, UK and New York, NY, USA: Cambridge University Press.

DOI: 10.1201/9781032715438-17

climate change, primarily due to burning fossil fuels like coal, oil, and gas. Global warming is the gradual increase in the overall temperature of the earth's atmosphere generally attributed to the greenhouse effect caused by increased levels of carbon dioxide, CFCs, and other pollutants (UN, 2022).

Globally, the known implications of climate change are increased heat, drought, and insect outbreaks; increasing wildlife fires; declining water supplies, reduced agricultural yields; health impacts in cities due to heat and flooding; and erosion in coastal and riparian regions. In Africa, the main long-term impacts of climate change are changing rainfall patterns, which are affecting agricultural production systems and consequently making communities across Africa vulnerable to food insecurity; reducing water security; shrinking fish resources in lakes and rivers due to increasing temperatures and unsustainable harvest of fishery resources; and rising sea levels likely to affect coastal ecosystems (Tadesse, 2010).

Climate change is also likely to create insecurity in Africa as climate and environmental disasters threaten human security. It is most likely to influence human and wildlife migration, especially in countries north of the Sahara, which has limited water access options. It should be noted that constraints on water availability are a growing area of concern in Africa in the advent of harnessing water resources to produce energy. Therefore, if no appropriate measures or action is taken, available evidence shows that the net damage costs associated with climate change are likely to increase significantly over time, affecting human and animal wellbeing (IPCC, 2017).

The observable effects of a changing climate are changes in animal migration and life cycles, changing rain and snow patterns, damaged coral, rising sea levels, changes in plant life cycles, warming oceans, higher temperatures and more heat waves, more droughts and wildfires, thawing permafrost, and less snow and ice, as illustrated in Figure 17.1. It's projected that by 2025, if urgent climate action is not taken, the global temperature is expected to rise by 2° of warming (Neukom et al.,

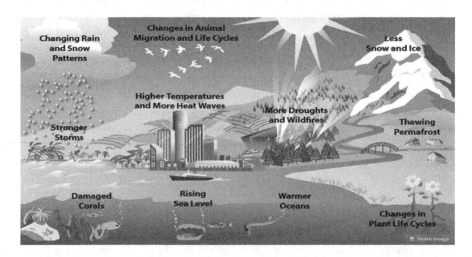

FIGURE 17.1 Observable indicators of climate change and global warming.

Source: NCA4, 2018.

2019). Projections of changing climate are also shown in Figure 17.2. Reducing this temperature by a quarter by 2030 and achieving net-zero emissions by 2070 is urgent (Ng et al., 2022).

17.1.1 THE NEXUS BETWEEN GLOBAL WARMING AND CLIMATE CHANGE

Although the two terms (global warming and climate change) are used interchangeably, global warming is just one aspect of climate change. "Global warming" refers to the rise in global temperatures due mainly to the increasing concentrations of greenhouse gases in the atmosphere. "Climate change" refers to the increasing changes in the measures of climate over a long period—including precipitation, temperature, and wind patterns (USGS, 2021a). Some of the signs of climate change include rising temperature—temperatures are rising worldwide due to greenhouse gases trapping more heat in the atmosphere; prolonged drought spells—droughts are becoming longer and more extreme around the world; tropical storms—tropical storms are becoming more severe due to warmer ocean water temperatures; and less snowpack—as temperatures rise, there is less snowpack in mountain ranges such as Rwenzori, Kilimanjaro, and Mount Kenya in Uganda, Tanzania, and Kenya, respectively, whereas in artic and polar areas, and the snow melts faster. Overall, glaciers are melting faster, and sea ice in the Arctic Ocean around the North Pole is melting faster with warmer temperatures. The melting of permafrost releases methane, a powerful greenhouse gas, into the atmosphere, accelerating further climate change. Increasing sea levels are threatening the lives and livelihoods of coastal communities and estuarine ecosystems (USGS, 2021b).

17.1.2 HOW DOES GLOBAL WARMING INFLUENCE CLIMATE CHANGE?

There are three major ways in which global warming will make changes to regional climate: melting or forming ice, changing the hydrological cycle (evaporation and precipitation), and changing currents in the oceans and air flows in the atmosphere. Global warming is a term used for the observed century of the Earth's climate system and its related effects. Scientists note that global warming is caused by increasing concentrations of other human-caused emissions. Within the earth's atmosphere, accumulating greenhouse gases like carbon dioxide, methane, and nitrous oxide and emitting heat radiation are responsible for climate change. Increasing or decreasing amounts of greenhouse gases within the atmosphere act to either hold in or release more of the heat from the sun. Figure 17.1 shows the observable effects of global warming and climate change. The atmosphere is getting hotter, more turbulent, and more unpredictable because of the "boiling and churning" effect caused by the heated atmosphere. With each increase of carbon, methane, or other greenhouse gases in the atmosphere, our local weather and global climate are further agitated, heated, and "boiled." Global warming is gauged by the increase in the average global temperature of the Earth (IPCC, 2014).

Global warming is a term used for the observed century-scale rise in the average temperature and its related effects. Nearly 95% of global warming is caused by increasing concentrations of greenhouse gases within the earth's atmosphere; the accumulation of greenhouse gases like water vapor and ozone contributes to global warming. Increasing or decreasing amounts of greenhouse gases within the atmosphere act to either hold in or release more of the heat from the sun. Observable effects of global warming include climate change, which has had catastrophic implications on the environment such as shifting of animal breeding ranges and premature gestation for plants (NASA, 2021).

Our atmosphere is getting hotter, more turbulent, and more unpredictable because of the "boiling and churning" effect caused by the heat-trapping greenhouse gases within the upper layers of our atmosphere. For instance, the planet was warmer by 1.2° Celsius from January to October 2020 compared to the pre-industrial era average recorded between 1850 and 1900 (Sangomla, 2020). With each increase of carbon, methane, or other greenhouse gas levels in the atmosphere, our local weather and global climate are further agitated, heated, and "boiled." Global warming is gauged by the increase in the average global temperature of the Earth. Along with our currently increasing average global temperature, some parts of the Earth may get colder while other parts get warmer—hence the idea of average global temperature. The average global temperature in 2020 was estimated at 1.27° Celsius above the global average temperature of 0.02° Celsius since 1850 (Rohde, 2021).

It is projected that the emergence of a moderate La Niña reduced global mean temperatures in the earlier half of 2020 and kept 2021 cooler. La Niña is the emergence of a large area of relatively cool water in the western equatorial Pacific. This has an immediate cooling effect on the Pacific and can disrupt weather patterns for an extended period (Rohde, 2021)

Greenhouse gas–caused atmospheric heating and agitation also increase the unpredictability of the weather and climate and dramatically increase the severity, scale, and frequency of storms, droughts, wildfires, and extreme temperatures. These adverse effects have far-reaching implications for livelihoods. Global warming can

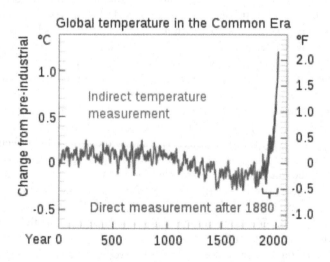

FIGURE 17.2 Global surface temperature reconstruction over the last millennia using proxy data from tree rings, corals, and ice cores in blue (Neukom et al., 2019).

reach levels of irreversibility (Harvey, 2015), and increasing levels of global warming can eventually reach an extinction level where humanity and all life on earth will end.

Irreversible global warming is referred to as a continuum of increasing temperatures that causes the global climate to rapidly change until those higher temperatures become irreversible on practical human time scales (Rusoke, 2017). The eventual temperature range associated with triggering and marking the beginning of the irreversible global warming processes is an increase in average global temperature of 2.2°–4° Celsius (4°–7.2° Fahrenheit) above preindustrial levels. Extinction-level global warming can further be defined as temperatures exceeding preindustrial levels by 5–6° Celsius (9–10.8° Fahrenheit), the extinction of all planetary life, or the eventual loss of our atmosphere. Changes in temperature increase physiological intolerances, harming species and driving them to extinction (Cahill et al., 2013).

There is a new way of thinking regarding global warming, climate change, and its implications on the planet. How about if the atmosphere is lost to increased greenhouse gas concentrations, climate change is affecting several species of animals and could impair their reproductive ability, for instance, male dragonflies have been observed to shed their black colorant pigment off their wings to cope with increasing temperature. This adaptation mechanism is likely to result in less observation by their female counterparts for mating and consequently will affect their production (Moore, 2021). These could become the irreversible implications of extinction-level global warming across the globe.

Carbon dioxide is currently the most important greenhouse gas related to global warming. Carbon dioxide can remain in the atmosphere for between 300 and 1000 years (Buis, 2019). As we continue to create greenhouse gases, we are creating a serious global warming crisis that will last far longer than we ever thought possible. The need to stop the use of fossil fuels is urgent, and switching to green energy is urgent. The next time you fill your car tank, reflect upon this: "The climatic impacts of releasing fossil fuel CO_2 to the atmosphere will last longer than Stonehenge. Longer than time capsules, longer than nuclear waste, far longer than the age of human civilization so far" (Inman, 2008).

Tracking carbon dioxide concentrations in the atmosphere is imperative to understand how it influences global warming and climate change. Important to note is that atmospheric carbon from fossil fuel burning is the main human-caused factor in the increasing carbon dioxide concentrations escalating global warming that we are experiencing now. The current level of carbon in our atmosphere is tracked using what is called the Keeling curve, after Charles David Keeling (Oreskes, 2021). The Keeling curve is a measure of the daily record of global atmospheric carbon dioxide concentration in parts per million (ppm). Each year, many measurements are taken at Mauna Loa, Hawaii, to determine the parts per million (ppm) of carbon in the atmosphere at that time. At the beginning of the Industrial Revolution around 1880, before we began fossil fuel burning, our atmospheric carbon ppm level was at about 270 (EPA, 2017). According to Keeling, if human-made emissions were to drop to zero, this could take many years to detect against the natural carbon cycling (Jones, 2017).

Earth's climate is changing. Continuous climate destabilization can quickly move us from one fairly stable state of dynamic balance and equilibrium into a new transitional state of instability and greater unpredictability. Eventually, the global climate will settle at a new, but different, stable state of dynamic equilibrium and balance, but it will be at a new level and range (a dynamic equilibrium is not static or unchanging; it varies within a range of some climate quality, e.g., average temperature, average humidity). A transitional state of escalating global climate instability, this state is characterized by greater unpredictability, which lasts until the global climate eventually finds a new and different stable state of dynamic equilibrium and balance at some different level of temperature and other climate qualities from what it has held for hundreds or thousands of years. This is likely to drive human extinction in many years to come (Turchin, 2010).

The three degrees defined here help individuals and organizations better understand the relative boundary ranges and levels of threat that are occurring or will occur based on measured increases in global warming (Cook, 2020). The temperature, carbon ppm, and loss or cost levels described for each degree of climate destabilization are not hard and rigid boundaries but boundary ranges designed to help you think about a set of related consequence intensities closely associated with that degree of climate destabilization. The temperature, carbon, cost, and loss boundary levels below may be modified by future research. The three degrees and definitions for climate destabilization are: 1. Catastrophic climate destabilization is associated with a measurement of carbon 400–450 ppm. At the estimated current 1.2° Celsius (2.2° Fahrenheit) temperature increase, we are already in the beginning stages of catastrophic climate destabilization. The eventual temperature range associated with catastrophic climate destabilization will be an increase in average global temperature of about 2.7° Celsius (4.9° Fahrenheit). The lack of urgency to act and make an impact could result in climate fatalism (Mayer and Smith, 2018).

When global warming–caused storms, floods, seasonal disruption, wildfires, and droughts begin to cost a nation $30 to 100 billion per incident to repair, we will have reached the level of catastrophic climate destabilization. We are already in this phase of climate destabilization (Appropedia, 2010). Hurricane Sandy in New York cost the United States between $50 and 60 billion to repair. Irreversible climate destabilization is associated with measurements beginning around carbon 425 ppm and going up to about carbon 550–600 ppm (Wollersheim, 2015).

The eventual temperature range associated with triggering irreversible climate destabilization is an increase in average global temperature of 2.2°–2.7° Celsius (4°–4.9° Fahrenheit) to 4° Celsius (7.2° Fahrenheit). Irreversible climate destabilization occurs when we have moved away from the relatively stable dynamic equilibrium of temperature and other key weather conditions, which we have experienced during the hundreds of thousands of years of our previous cyclical Ice Ages. Once a new dynamic equilibrium finally stabilizes the climate in these carbon ppm ranges, we will have crossed from catastrophic climate destabilization into irreversible climate destabilization. Irreversible climate destabilization is a new average global temperature range and a set of destabilizing climate consequences we most likely will never recover from—or that could take hundreds or even thousands of years to correct or re-balance. Irreversible climate destabilization will eventually cost the nations of the world hundreds of trillions of dollars. There is a likelihood that there will be an increase in rainfall in high latitudes and a decrease in rainfall in most subtropical land regions (Figure 17.2).

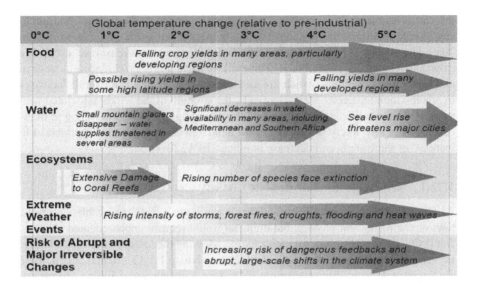

FIGURE 17.3   Global temperature change (relative to the pre-Industrial era).

(Source: IPCC 2007.)

Extinction-level climate destabilization. Extinction-level climate destabilization as defined here is begins around the measurement of carbon parts per million in the atmosphere in the range of 600 ppm or more. The eventual temperature range associated with extinction-level climate destabilization is an increase in average global temperature of 5 ° to 6 ° C (9 ° to 10.8 ° Fahrenheit). Extinction-level climate destabilization is also defined as the eventual extinction of approximately up to half or more of the species on earth and most, if not all, of humanity. This occurs when the climate destabilizes to a level where the human species and/or other critical human support species can no longer successfully exist. Extinction-level climate destabilization has occurred several times previously during Earth's evolution. Extinction-level climate destabilization will cost the nations of the world hundreds of trillions of dollars and potentially billions of lives—maybe the survival of the human species itself. There is a possibility that extinction-level climate destabilization may never correct or re-balance itself to some new equilibrium level. If the climate were able to correct or re-balance itself from this level of destabilization, it could take hundreds, thousands, or even hundreds of thousands of years.

17.2   KEY GLOBAL PROJECTIONS OF CLIMATE CHANGE

There is an expected decrease in global annual precipitation through the end of the 21st century (Tabari, 2021). Across Europe, projections of annual precipitation show a general decrease except over the central regions where annual precipitation is increasing as a result of winter. Consequent dry days are projected to be higher across Europe due to a significant increase during spring, summer, and autumn (Lorenzo and Alvarez, 2020). Though global average annual precipitation through the end of the century is expected to increase, changes in the amount and intensity of precipitation will vary significantly by region across the globe from 2021 to 2050. The intensity of precipitation events will likely increase on average, escalating tropical cyclones (Thomas, 2021). This will be particularly pronounced in tropical and high-latitude regions, which are also expected to experience overall increases in precipitation (epa.gov, 2021). Generally, the strength of the winds associated with tropical storms is likely to increase, and the amount of precipitation falling in tropical storms is also likely to increase, with annual average precipitation projected to increase in some areas and decrease in others (EPA, 2017).

17.2.1 KEY PROJECTIONS OF CLIMATE CHANGE ACROSS AFRICA

As Skea (2021) observed, limiting global warming to 1.5° C is possible within the laws of chemistry and physics, but doing so would require unprecedented changes. But IPCC, the world's foremost authority for assessing the science of climate change, says it is still possible to limit global temperature rise to 1.5° C—if, and only if, there are "rapid and far-reaching transitions in land, energy, industry, buildings, transport, and cities across cities".

For sub-Saharan Africa, which has experienced more frequent and more intense climate extremes over the past decades, the ramifications of the world's warming by more than 1.5° C would be profound. It has been projected that at 1.5° C, less rain would be experienced in southern Africa, especially around the Zambezi in Zambia and the Limpopo basins and parts of the Western Cape in South Africa. Hotter days are expected to increase across west and central Africa. The western part of southern Africa is set to become drier, with increasing drought frequency and more heat waves toward the end of the 21st century. But at 2° C, southern Africa is projected to face a decrease in precipitation of about 20% and increases in the number of consecutive dry days in Namibia, Botswana, northern Zimbabwe, and southern Zambia. This will cause reductions in the volume of the Zambezi basin projected at 5% to 10% (Shepard, 2018). If the global mean temperature reaches 2° C of global warming, it will cause significant changes in the occurrence and intensity of temperature extremes in all sub-Saharan regions.

With increasing temperatures, we will see eastern Africa at higher risk of intensified droughts, unprecedented floods in west and southern Africa, and increasing ocean acidity around Africa's southern coast. African projections are projected to be higher than the global mean temperature increase; regions in Africa within 15° of the equator are projected to experience an increase in hot nights as well as longer and more frequent heat waves. The odds are long but not impossible, says the IPCC. And the benefits of limiting climate change to 1.5° C are enormous, with the report detailing the difference in the consequences between a 1.5° C increase and a 2° C increase. Every bit of additional warming adds greater risks for Africa in the form of greater droughts, more heat waves, and more potential crop failures (Shepard, 2018).

As of March 2020, climate change projections in the Sahel region, especially the western Sahel region, the temperatures are rising 1.5 times faster than the rest of the world. More intense and violent thunderstorms above normal rainfall and experiences of the strongest drying, with a significant increase in the maximum length of dry spells, have been recorded (Mayans, 2020). The IPCC expects central Africa to see a decrease in the length of wet spells and a slight increase in heavy rainfall (IPCC, 2017).

West Africa has been identified as a climate-change hotspot (Muller et al., 2014), with climate change likely to lessen crop yields and production, with resultant impacts on food security. Few, if any, regions in the world have been affected as much as the Sahel, a region experiencing rapid population growth; with an estimated 1,094,907 persons and 49.3% females, the population growth rate in the Sahel region is 2.8 to 3.7% per year. This population is expected to double in two decades amid shrinking environmental capital, including land and water resources (Wuehler, 2015).

Across the cattle corridor in Uganda and among farmers elsewhere where pastoralism is the mainstay, climate change has pitted crop farmers against pastoralists with several multiplier threats across the divide. The UN secretary-general advisor for the Sahel says the Sahel region is particularly vulnerable to climate change, with 300 million people affected. Drought, desertification, and scarcity of resources have led to heightened conflicts between crop farmers and cattle herders. The situation has been heightened by weak governance, which has led to social breakdowns (World Bank, 2020). After being ranked as the world's largest inland water body, with an open water area of 25,000 km² in the early 1960s (Binh et al., 2020), the shrinking of Lake Chad due to climate change is leading to economic marginalization and providing a breeding ground for recruitment by terrorist groups as social values and moral authority evaporate (Shepard, 2018).

17.3   KEY GLOBAL IMPLICATIONS OF CLIMATE CHANGE

Globally the predicted long-term effects of climate change are a decrease in sea ice and an increase in permafrost thawing, an increase in heat waves and heavy precipitation, and decreased water resources in semi-arid regions (USGS, 2021a). IPCC regional impacts of global climate change are as follows: in North America: Decreasing snowpack in the western mountains (Mortezapour et al., 2022); climate change models are projecting major changes in northern sea ice, including a possible reduction in summer ice extent by 60% for a doubling of carbon dioxide and possibly a complete disappearance of summer ice by 2100. Early ice breakup or complete loss of ice would have a profound effect on northern lifestyles (Vogel and Bullock, 2021).

In southern Africa, there is an increase in yields of rain-fed agriculture from 5 to 20% (Olabanji et al., 2021) and increased frequency, intensity, and duration of heat waves and floods (Tolulope et al., 2022) in South Africa's provinces such as Durban and Kwazulu Natal that killed 443 people in KwaZulu-Natal and left over 40,000 missing, 4,000 houses destroyed, and more than 8,000 houses damaged, mostly across Durban City and its neighborhood (ECHO, 2022).

In Latin America: Gradual replacement of tropical forest by savannah in eastern Amazonia; risk of significant biodiversity loss through species extinction in many tropical areas; significant changes in water availability for human consumption, agriculture, and energy generation. Land-use change for agribusiness expansion together with climate change in the transition zone between eastern Amazonia and the adjacent Cerrado may have induced a worsening of severe drought conditions over the last decade (Marengo et al., 2022).

In Europe: Increased risk of inland flash floods, more frequent coastal flooding and increased erosion from storms and sea-level rise, a glacial retreat in mountainous areas, reduced snow cover and winter tourism, extensive species losses, and reductions of crop productivity in southern Europe (Thieken et al., 2022).

In Asia: Freshwater availability is projected to decrease in central, South, East, and Southeast Asia by the 2050s; coastal areas will be at risk due to increased flooding; the death rate from disease associated with floods and droughts is expected to rise in some regions. Many countries in Asia face an existential threat of displacement as a result of a 2-meter sea-level rise. In a bid to secure coastal lands communities are being encouraged to plant *Rhizophora* mangrove seedlings in estuaries in Bali, Indonesia, to counter erosion (UNDP, 2019).

17.4 IMPLICATIONS OF GREENHOUSE GAS CONCENTRATIONS: THE AFRICAN PERSPECTIVES

Across Africa, fragile ecosystems continue to be affected by drought, hot extremes, heat waves, and heavy rainfall events that will likely continue to become more frequent. There will be an increased incidence of extremely high sea levels, and tropical cyclones will become more intense (higher peak wind speeds, more heavy precipitation). These could increase the number of weather-related disasters and affect agricultural production. For instance, some countries in eastern Africa faced reductions in yields from rain-fed agriculture by up to 50% by 2010 (Collier et al., 2008).

As mentioned earlier, changes in global temperatures have had catastrophic effects on the planet. Climate change scenarios in Africa are not much different from those in other continents. Increasing greenhouse gas concentrations will affect humans, animals such as livestock, and those in the wild alongside their habitats. By 2019 concentrations of major greenhouse gases continued to increase, globally mole fractions of CO_2 exceeded 410 parts per million (ppm) and are expected to exceed 414 ppm in 2021 (WMO, 2021). Future temperature changes could make both aquatic and terrestrial life difficult, Future ice and snowpack melt on glaciated mountains such as Rwenzori in Uganda, Kilimanjaro in Tanzania, and Mount Kenya could continue receding, yet these are key climate change indicators. According to Patoway (2014), it was reported that the glaciers on Mount Kilimanjaro were fast disappearing, and the increasing loss of glaciers has implications for increasing sea level.

Global warming is causing global mean sea-level change, glaciers are melting, adding water to the oceans, and the volume of the ocean is expanding as the water warms. For example, on Rwenzori Mountain in Uganda, the ice of snow sheets is uncertain due to global warming. In 1906, 43 glaciers were located across six peaks within the mountain ranges, and these sheets covered about 7.5 km², almost half of the total area of Africa's glaciers at the time. As of 2005, the Rwenzori glacier had reduced to 1.5 km² (Russell, 2009).

Future precipitation and storm events as a result of climate variability are likely to affect farming systems across Africa. As farmers in Africa are unable to fully cushion themselves from future unpredictable precipitation and storm events, especially in arid and semi-arid environments (Karienye and Macharia, 2021), it's incumbent upon the current governments to develop appropriate technologies and early warning systems against extreme weather conditions experienced as a result of climate change in Africa.

Oceans across Africa, especially the Atlantic Ocean, could increase the level of acidity as a result of anthropogenic CO_2 emissions. Such a scenario is likely to continue impacting survival, growth, development, and physiology in marine invertebrates. For instance, increased acidification in the Atlantic Ocean's sea urchins resulted in their slower somatic and gonadal growth. This reflects a shift in energy budgets linked to additional costs of pHe and pHi (Dupont and Thorndyke, 2013).

17.5 GLOBAL CLIMATE CHANGE: KEY IMPACTS FOR AFRICA

Continued emissions of greenhouse gases will lead to further climate changes. Future changes are expected to include a warmer atmosphere, a warmer and more acidic ocean, higher sea levels, and larger changes in precipitation patterns. The extent of future climate change depends on what we do now to reduce greenhouse gas emissions. The more we emit, the larger future changes will be. Greenhouse gas concentrations in the atmosphere will continue to increase unless the billions of tons of our annual emissions decrease substantially. Increased concentrations are expected to increase Earth's average temperature, influence the patterns and amounts of precipitation, reduce ice and snow cover, and raise the sea level. Increase in the acidity of the oceans; increase in the frequency, intensity, and/or duration of extreme events; shifting ecosystem characteristics; and increase in threats to human health as a result of human-wildlife interactions and exposure to vectors due to climate change are inevitable. Such changes will impact our food supply, water resources, infrastructure, ecosystems, and even our own health.

It has been projected that across Africa, 380 million people could face water stress by 2050 (Dunne, 2020). Water stress is a measure of the ratio between water use and availability. In comparison to projections for 2050, between 75 and 250 million people are projected to be exposed to increased water stress; this is a threefold projection from 2010. In regard to the number of people who could be distressed by water stress in Africa, efforts must be geared at keeping the global temperature below 2° C. People living in the Middle East and north Africa are prone to water stress. Water stress in Africa is likely to be exacerbated by future population growth and the likelihood of increasing future emissions, which could peak in 2060 (Dunne, 2020).

17.5.1 FISHERIES

Regarding the fisheries sector, there is a noticeable decline in stocks of fish in large lakes due to rising water temperatures, a scenario that has adversely affected local food supply. This could have implications for nutrition and food security among fishing communities across Africa.

According to Magda (2017), climate change has led to rising sea temperatures, making fish stocks migrate toward colder waters away from equatorial latitudes. This is resulting in shrinking individual fish sizes, especially in east Africa where ocean warming has degraded parts of the coral reef, which is a critical habitat as a fish breeding ground. In comparison to global statistics, climate change is predicted to reduce fish catch by 7.7%, and this could translate into reduced revenue from fishing by 10.4% by 2050 (Lam et al., 2016). In sub-Saharan Africa, more than 12 million people are engaged in the fisheries sector, and this creates remarkable employment multiplier effects where one fisherman indirectly or directly employs more than three people in ten in most West African countries. Both inland and marine fisheries are in peril as a result of climate change as temperatures keep increasing (Muringai R, 2021). The need to understand the scale and extent of climate change's effects on the fisheries sector should not be underestimated.

The World Bank report (2019), "Climate Change and Marine Fisheries in Africa, Assessing Vulnerability and Strengthening Adaptation Capacity", assesses the ecological impacts of climate change to project future changes in maximum catch potential (MCP) for the main species within the exclusive economic zones (EEZs) of African nations. With specific reference to changes in catch potential under the impacts of climate change, the projected changes in MCP vary greatly geographically, with substantial differences between African countries (World Bank, 2019). Based on multispecies size spectrum ecological modeling, also known as Mizer (MSSEM), in Table 17.2, the projections show that by the end of the century, the largest decrease (40% or more) will likely occur in tropical African countries, including Ghana, São Tomé and Príncipe, Liberia, and Côte d'Ivoire, but over the longer term, potential catches are also projected to decrease substantially (20% or more) in the temperate northeast and southeast Atlantic. In higher-latitude regions, by contrast, catch potential is projected to increase or at least decrease much less, as expected in temperate regions (e.g., Senegal, The Gambia, Cabo Verde) and shown in Table 17.2. MSSEN is an R package to run dynamic multi-species size-spectrum models of fish communities. The package has been developed to model marine ecosystems that are subject to fishing (Scott et al., 2014).

17.5.2 Health

Concerning health in Africa, climate change is expected to adversely affect human health in a myriad of ways, such as increasing vulnerability to malnutrition, exacerbating poor air quality, polluting drinking water points, and changing the geographical range of infectious disease pathogens and vectors. The projections and likely increase of vector-borne diseases and human displacement due to climate change are inevitable.

There is an expected increase in cases of malaria, especially in southern African countries and the east African highlands. Flood- and drought-related illnesses will increase. Unprecedented flooding is likely to cause an increase in more cases of cholera, malaria, and Rift Valley fever outbreaks across Africa. Effects of the outbreak of such diseases are likely to impact already strained healthcare systems. Most countries further at risk of climate change effects on health are those that are already plagued by civil war and ethnic tensions such as the Democratic Republic of Congo, Cameroon, Central Africa Republic, Ethiopia, Mozambique, Somalia, Nigeria, Mali, Sudan, and South Sudan (Etobbe, 2017).

TABLE 17.2

Percentage Changes in Maximum Catch Potential (MCP) under Low and High Greenhouse Gas Emission Scenarios, by 2050 and 2100 (Multispecies Size Spectrum Ecological Modeling)

Exclusive Economic Zone	Low GHG Emission Scenario (Representative Concentration Pathway 2.6)		High GHG Emission Scenario (Representative Concentration Pathway 8.5)	
	2050	2100	2050	2100
Angola	−5.10	−3.40	−11.12	−34.43
Benin	−17.57	−15.54	−16.93	−33.02
Cameroon	−8.64	−4.76	−12.29	−22.87
Cabo Verde	−10.73	−5.93	−19.33	−36.15
Comoros	−12.38	−10.90	−14.31	−26.05
Congo Dem. Rep	−5.83	−4.30	−9.61	−19.73
Congo Republic	−7.51	−6.82	−11.41	−23.92
Côte d'Ivoire	−22.73	−18.00	−20.46	−35.31
Equatorial Guinea	−10.63	−6.65	−12.38	−28.37
Gabon	−6.28	−4.56	−7.86	−18.72
The Gambia	−18.43	−10.39	−17.64	−35.14
Ghana	−22.66	−15.15	−20.34	−38.36
Guinea	−20.07	−15.88	−15.66	−29.72
Guinea-Bissau	−24.69	−18.29	−17.37	−32.30
Kenya	−18.78	−11.76	−19.93	−34.92
Liberia	−20.96	−20.14	−19.71	−32.04
Madagascar	−6.16	−4.88	−10.57	−18.86
Mauritania	−2.52	−4.79	−8.57	−16.87
Mauritius	−11.59	−12.37	−13.12	−23.09
Mayotte (France)	−9.49	−8.43	−11.71	−21.86
Morocco	2.64	−2.68	5.05	−8.32
Mozambique	−7.14	−4.84	−10.74	−20.37
Namibia	−2.17	−2.33	−3.64	−11.22
Nigeria	−10.81	−9.42	−11.14	−24.20
Réunion (France)	−7.62	−9.14	−12.38	−21.45
São Tomé and Príncipe	−11.24	−10.59	−13.45	−29.05
Senegal	−16.76	−9.04	−18.98	−36.15
Seychelles	−19.92	−14.86	−21.29	−33.51
Sierra Leone	−22.44	−19.40	−18.70	−31.45
Somalia	−15.46	−11.01	−19.06	−36.53
South Africa	−2.13	−1.84	−2.29	−3.83
Tanzania	−17.44	−12.40	−18.22	−32.24
Togo	−17.97	−15.57	−17.13	−34.72

Source: Data from The World Bank, "Climate Change and Marine Fisheries in Africa, Assessing Vulnerability and Strengthening Adaptation Capacity. Projected Changes in Catch Potential under the Impacts of Climate Change" (2019) P. 17–19. World Bank 2019, © World Bank.

Increasing sea level in oceans across Africa is likely to escalate further degradation of coral reefs, affecting health, fisheries, and tourism. Increasing coastal flooding could put at risk the livelihoods of over 70 million people, with an estimated cost of adaptation to sea-level rise in coastal areas at 5–10% of their GDP. Such countries include Benin, Ivory Coast, Senegal, and Togo. About 56% of these countries' coastal lines are eroding, and the situation is expected to worsen by 2080 (Mafaranga, 2020).

In 2018, the World Health Organization estimated the effect of climate change on human health. Climate change affects social and environmental determinants of health such as clean air, safe drinking water, sufficient food, and secure shelter. Though the study further highlighted that continued progress in health and growth, climate change was projected to cause an additional 250,000 deaths per year between 2030 and 2050 from malnutrition, malaria, diarrhea, and heat stress (WHO, 2018). The IPCC AR5 Synthesis report projected with high confidence that climate change will amplify existing risks and create new risks for natural and human systems. The risks are unevenly distributed and generally greater for disadvantaged people and communities in countries at all levels of development (IPCC, 2014).

A sickly population as a result of climate change affects economic development, which is an important component of possible adaptation to climate change. Economic growth on its own, however, is not sufficient to insulate the African population from disease and injury due to climate change. Future vulnerability to climate change will depend not only on the extent of social and economic change but also on how the benefits and costs of change are distributed in society (WHO, 2018).

17.5.3 AGRICULTURE

A report by the Stockholm International Water Institute for Africa 2020 notes that millions of rain-fed farms are scattered throughout Africa, and the majority of Africans are largely dependent on rain-fed agriculture. The Sahel is regularly hit by droughts and floods, with enormous consequences to people's food security. As a result of armed conflict, violence, and military operations, some 4.9 million people were displaced by 2020 (SIWI, 2020). Yields from rain-fed agriculture could be reduced by up to 50% in some regions by 2020; agricultural production, including access to food, may be severely compromised. Some countries such as Uganda which are 80% dependent on rain-fed agriculture, with most farmers not receiving reliable weather reports, forecasts, and other climate services to plan their agricultural activities, climate change is likely to affect agricultural production (GEF, 2018). In Malawi, erratic rainfall led to the production of only 2.5 million tons of maize production in 2016, and the country relied on maize importation to feed 34% of its population in the same year (Kalibata, 2016).

17.6   AFRICA'S VULNERABILITY TO CLIMATE CHANGE

In the wake of the COVID-19 pandemic, women in Africa are more vulnerable to climate change in their cities and communities. Vulnerability is brought about by

COVID-19 lockdown and travel bans to communities to collect data on climate change, which would help shape mitigation and adaptation measures. With limited income as a result of lost jobs due to the COVID-19 pandemic, it is most likely that all efforts to access energy would be directed to the use of fuel wood use in Africa as families grapple to pay costs associated with green energy such as the use of liquefied petroleum gas and hydro-electric power.

Vulnerability to climate change in Africa will be exacerbated by unplanned rapid urbanization, weak adaptive capacity, and overdependence on nature. These factors have also been highlighted by the Intergovernmental Panel on Climate Change (IPCC, 2017) to increase the African population's vulnerability to climate change. Dependence on ecosystem goods for livelihoods and less developed agricultural production systems are projected to increase levels of poverty, especially among fishing communities and subsistence farmers in sub-Saharan Africa.

The Africa Development Bank noted that despite Africa's low contribution to GHGs, it remains the most vulnerable continent in the globe to climate change impacts under all climate scenarios above 1.5° C (AFDB, 2020) and is particularly vulnerable to impacts of climate variability and climate change because of its multiple stresses and low adaptive capacity to modern technology. Climate change is already threatening all aspects of the development agenda in Africa. Regarding income poverty and hunger, Africa as a continent is the poorest and hardest hit by the effects of climate change, For instance, of the 1.216 billion people in Africa, 433 million were estimated to live in extreme poverty by 2018. This was a sharp increase from 284 million people in 1990 (Schoch and Lakner, 2020). Climate change effects can destroy up to 95% of some countries' infrastructures in Africa. A case example was the tropical cyclone Idai in Mozambique, which destroyed 95% of all infrastructure in Beira, the fourth-largest city in the country (AFDB, 2020). Climate change represents a major threat to achieving all sustainable development goals in Africa.

17.7 ACCUMULATION OF GREENHOUSE GASES EFFECT ON CLIMATE TODAY, TOMORROW, AND INTO THE FUTURE

Future climate change scenarios and projections will depend on factors such as the way anthropogenic activities facilitate either the increase or decrease of greenhouse gas concentrations, natural influences on climate such as changes in the sun's intensity and natural processes within the climate system, changes in ocean circulation patterns, and recurring volcanic activities. For instance, the ash from the recent Nyiragongo volcanic eruption is expected to trap toxic gases in the atmosphere such as carbon dioxide and fluorine. The trapping of these greenhouse gases can affect crops or lead to animal and human illness or death (Katoto, 2021). The use of computer models of the climate system to better understand these issues and project future climate changes is imperative. Models such as the Forecasting Coastal Evolution (ForCE) developed by Plymouth have potential to advance coastal evolution science, allowing adaptation at the shoreline to be predicted over longer periods (Davidson, 2021).

17.8 COPING WITH GLOBAL WARMING AND CLIMATE CHANGE

Adaptation to climate change is the main option to cope with it. Adaptation to climate change will require taking appropriate action to mitigate the effects of climate change and adjusting to both the current and predicted effects of climate change. Adaptation to challenges grows with the magnitude and rate of climate change (NASA, Responding to Climate Change, 2020). The UNFCCC defines adaptation to climate change as processes through which societies make themselves better able to cope with an uncertain future. Adapting to climate change entails taking the right measures to reduce the negative effects of climate change by making the appropriate adjustments and changes (UN, 1992). The Intergovernmental Panel on Climate Change defines adaptation as an adjustment in natural or human systems in response to actual or expected climatic stimuli or their effects, which moderates harm or exploits beneficial opportunities for humankind. Adaptation involves learning to manage new risks and strengthening resilience in the face of changing climate (UNFCC, 2011).

17.9 WAY FORWARD

Recognizing the increasing threat of climate change, many countries came together in 2015 to adopt the historic Paris Agreement, committing themselves to limit climate change to well below 2° C. Since COP 21, 196 countries have formally joined the agreement, including almost every African nation. This can be regarded as a formidable effort to scale down global warming. The agreement entered into force in November 2016. Though the United States of America opted out of the Paris Agreement under President Donald Trump on June 1, 2017, on January 20, 2021, the USA officially rejoined the Paris Agreement (Blinken, 2021), and the pact has made progress towards its goal of preventing the average global temperature from increasing by 2° C above the preindustrial era. Many countries have lived up to the promises of curbing emissions. Some countries such as China have even promised to achieve carbon neutrality by 2060 (Normile, 2020). Such promises and achievements underscore the role of international pacts in mitigating greenhouse gas emissions that influence global warming and climate change.

In December 2018, countries met in Katowice, Poland, for the Conference of the Parties to the United Nations Framework Convention on Climate Change—known as COP24—to finalize the rules for implementation of the agreement's work program. As part of the Paris Agreement, countries made national commitments to take steps to reduce emissions and build resilience. The treaty also called for increased financial support from developed countries to assist the climate action efforts of developing countries. But even at the time that the Paris Agreement was adopted, it was recognized that the commitments on the table would not be enough. Even if the countries did everything they promised, global temperatures would rise by 3° C this century. Mitigation implications of climate change on our planet will require the continuous collection of historical, real-time, and forecast data, where historical data elucidate climate statistics, set a context for current data, and allow variability and the occurrence of extremes to be quantified in real-time data, with current climate observations to enable informed decision making and actions.

It's known that many greenhouse gases will stay in the atmosphere for long periods. As a result, even if emissions stopped increasing, atmospheric greenhouse gas concentrations would continue to increase and remain elevated for hundreds of years. The onus is on humans to stabilize concentrations of greenhouse gases, which requires a dramatic reduction in current greenhouse gas emissions. Otherwise, surface temperatures will continue to warm.

Past and present-day greenhouse gas emissions will affect climate far into the future. For instance, there is a likelihood of the inability of oceans which store heat will take longer to fully respond to higher greenhouse gas concentrations (epa.gov, 2021). The ocean's response to higher greenhouse gas concentrations and higher temperatures will continue to impact climate over the next several decades to hundreds of years (He and Ding, 2021). Because it is difficult to project far-off future emissions and other human factors that influence climate, there is a need for climate scientists to use a range of scenarios using various assumptions about future economic, social, technological, and environmental implications of projected climatic conditions. This could go a long way to mitigate the adverse effects climate change could have on the environment.

Climate change poses challenges to fundamental elements of understanding social and economic policies affecting prosperity, growth, equity, and sustainable development (Mearns and Norton, 2010). Whereas developing and least developed countries are producing less when it comes to carbon emissions compared to highly developed and middle-income countries, the principle of equity dimension needs to be applied in dealing with the effects of climate change. Since developed countries emit more carbon dioxide, these countries should continue to provide both financial and technical assistance aimed at remedying climate change impacts.

Adaptation is essential to manage climate change impacts and maximize development outcomes. Development is key to adaptation, as it enhances resilience through reducing vulnerabilities and increasing adaptive capacity to climate change. The European Commission adopted a new EU strategy on adaptation to climate change on February 24, 2021, aiming at building climate resilience by 2050. The EU strategy aims to cut emissions by at least 55% by 2030, and Europe aims to become the world's first climate-neutral continent by 2050 creating a climate-resilient society by adapting to climate effects, too (European Commission, 2021). Adaptation will require economy-wide planning and regional cooperation as well as local engagement. The role of leadership and coordination is essential in mitigating climate change effects, and heads of government, finance, and economic ministries should be seen to actively engage all stakeholders through appropriate funding mechanisms to reverse climate change impacts. Since the impacts of climate change affect farming livelihoods by damaging harvests and lowering crop yields, thereby increasing poverty and food insecurity for farmers across the globe, especially in the Asia-Pacific and Africa (Asia-Pacific Economic Cooperation, 2015), climate funds should target smallholder farmers as a way of building resilience so that smallholder farmers can participate in climate-smart farming, grow green super rice and other produce, and participate in mechanisms that aim at sustainable agriculture, production, and consumption.

17.10 CONCLUSION

Global warming and climate change remain an existential threat to livelihoods and life on earth. Increasing global warming accelerates climate change, and the visible indicators are manifested in receding glaciers, rising sea levels, changes in animal migration and life cycles, shrinking animal ranges, changes in plant lifecycles, strong storms and damaged corals, warmer oceans, and increasing temperatures. Climate change is expected to hit the African continent harder irrespective of the continent's contribution to greenhouse gas emissions. If climate change is not curtailed, extinction-level climate destabilization is projected, and this could lead to the eventual extinction of approximately up to half or more of the species on earth and most, if not all, of humanity. Across Africa, the Intergovernmental Panel on Climate Change expects central Africa to see a decrease in the length of wet spells and intense droughts across east Africa. Southern Africa is poised to experience water shortages in inland water bodies such as the Zambezi basin.

Regarding the fisheries sector, there is a decline in fish stocks in major water bodies inside and surrounding Africa due to increasing temperatures. Climate change is expected to exacerbate malnutrition, diarrhea, and cases of malaria prevalence due to flooding. Yields from rain-fed agriculture are expected to reduce by 50%. Climate change is expected to increase the cost of adaptation. It's imperative to adapt to the changing climate, increase community resilience to climate change, and build resilient economies under climate change adversity.

REFERENCES

AFDB. (2020). *Climate Change in Africa*. Retrieved from afdb.org: www.afdb.org/en/cop25/climate-change-africa/

Appropedia. (2010, July 7). *Global Climate Destabilization*. Retrieved from Appropedia.com: https://lknd.in/dpDrDWv

Asia-Pacific Economic Cooperation. (2015, May 6). Retrieved from Helping Farmers Adapt to Climate Change: https://apec.org

Blinken, A. J. (2021, February 19). *The United States Officially Rejoins the Paris Agreement*. Retrieved from state.gov: www.state.gov/the-united-states-officially-rejoins-the-paris-agreement/

Buis, A. (2019, October 9). *Global Climate Change*. Retrieved from https://climate.nasa.gov/news/2915/the-atmosphere-getting-a-handle-on-carbon-dioxide/6

Cahill A. E., Aiello-Lammens, M. E., Fisher-Reid, M. C., Hua, X., Karanewsky, C. J., Yeong Ryu, H., Sbeglia, G. C., Spagnolo, F., Waldron, J. B., Warsi, O. and Wiens, J. J. (2013). How does climate change cause extinction? In *Proceedings of the Royal Society B: Biological Sciences* (p. 1). London: The Royal Society Publishing.

Cho, R. (2019, December 27). *10 Climate Change Impacts That Will Affect Us All*. Retrieved from news.climate.columbia.edu: https://news.climate.columbia.edu/2019/12/27/climate-change-impacts-everyone/

Collier, P., Conway, G., and Venables, T. (2008). Climate change and Africa. *Oxford Review of Economic Policy*, 24(2), 337–358.

Cook, K. H. (2020, September 23). *thedailytexan.com*. Retrieved from The climate is destabilizing. What do we do?: https://thedailytexan.com/2020/09/23/the-climate-is-destabilizing-what-do-we-do/

Davidson, M. (2021). Forecasting coastal evolution on time-scales of days to decades. *Coastal Engineering*, 168, 103928.

Dunne, D. (2020, June 5). *World Economic Forum*. Retrieved from 380 million people in Africa could face water 'stress' experts warn: www.weforum.org/agenda/2020/06/world-population-water-stress-2050-climate-change

Dupont, S., and Thorndyke, M. (2013). Direct impacts of near-future ocean acidification on sea urchins. *Climate Change Perspectives from the Atlantic: Past, Present and Future*, 461–485.

ECHO. (2022, April 11). *South Africa: Floods and Landslides*. Retrieved from Reliefweb: https://reliefweb.int/disaster/fl-2022-000201-zaf

EPA. (2017, January 19). *Climate Change Science*. Retrieved from Future of Climate Change: www.19january2017snapshot.epa.gov

epa.gov. (2021, May 12). *Climate Change Indicators*. Retrieved from Land Loss Along the Atlantic Coast: https://lknd.in/dKGhF49

Etobbe, D. (2017, March 14). 7 civil wars in Africa we must never forget. *Politics and Society*, 17.

European Commission. (2021, February 24). *EU Climate Action and the European Green Deal*. Retrieved from ec.europa.eu: https://ec.europa.eu/clima/policies/eu-climate-action/

GEF. (2018, June 18). *Power to the Farmers: Climate Information and Early Warnings to Save Lives and Build Resilience in Uganda*. Retrieved from thegef.org: www.thegef.org/news/power-farmers-climate-information-and-early-warnings-save-lives-and-build-resilience-in-uganda/

Harvey, C. (2015, February 10). *Businessinsider.com*. Retrieved from These are the impacts of climate change we will never be able to fix: www.businessinsider.com/irreversible-climate-change-consequences-2015?r=US&IR=T

He, B., and Ding, K. (2021). Localize the impact of global greenhouse gases emissions under an uncertain future: A case study in Western Cape, South Africa. *Earth*, 111–123.

Inman, M. (2008). Carbon is forever. *Nature Climate Change*, 1, 156–158.

IPCC. (2014). Climate Change 2014 Synthesis Report: Summary for Policymakers. Geneva, Switzerland: IPCC.

IPCC. (2017). *Climate Science Special Report: Fourth National Climate Assessment on Effects of Climate Change*. Washington, DC: California Institute of Technology.

IPCC. (2022). *Summary for Policymakers. In: Climate Change 2022: Mitigation of Climate Change. Contribution of Working Group III to the Sixth Assessment Report of the Inter-governmental Panel on Climate Change*. Cambridge and New York: Cambridge University Press.

Jones, N. (2017, January 26). *e360.yale.edu*. Retrieved from how the world passed a Carbon threshold and why it matters: https://e360.yale.edu/features/how-the-world-passed-a-carbon-threshold-400ppm-and-why-it-matters

Kalibata, A. (2016, November 17). *Africa's Farmers Are Among the Most Hurt by Climate Change*. Retrieved from news.trust.org: https://news.trust.org/item/20161117104624-jxar7/

Karienye, D., and Macharia, J. (2021). Adaptive capacity to mitigate climate variability and food insecurity of rural communities along River Tana Basin, Kenya. In *Africa Handbook of Climate Change Adaptation* (pp. 49–60). Nairobi: Springer.

Katoto, D. M. (2021, June 4). *reliefweb.int*. Retrieved from the eruption of Mount Nyiragongo: its health effects will be felt for a long time: www.reliefweb.int/report/democratic-republic-congo/eruption-mount-nyiragongo-its-health-effects-will-be-felt-long-time

Knutson, R., and Thomas, M. V.-L. (2021, March 26). *ScienceBrief.org*. Retrieved from intensity of tropical cyclones is probably increasing due to climate change: www.uea.ac.uk/news/-/article/intensity-of-tropical-cyclones-is-probably-increasing-due-to-climate-change

Lam, V. W. Y., Cheung, W. W. L., and Sumalia, U. R. (2016). Projected change in global fisheries revenue under climate change. *Scientific Reports*, 6, 1–4.

Lorenzo, M. N., and Alvarez, I. (2020). Climate change patterns in precipitation over Spain using CORDEX projections for 2021–2050. *Science of the Total Environment*, 723–726.

Mafaranga. (2020, November 13). *Eos.org*. Retrieved from Sea Level Rise May Erode Development in Africa: https://lnkd.in/dsdUKRp

Magda, L. (2017, November 11). *blogs.worldbank.org*. Retrieved from Climate Impacts on African Fisheries: The Imperative to Understand and Act: https://lknd.in/dmichy3

Marengo, J. A., Jimenez, J. C., and Espinoza, J. C. (2022). Increased climate pressure on the agricultural frontier in the Eastern Amazonia–Cerrado transition zone. *Scientific Reports*, 12, 457.

Mayans, J. (2020, March 17). *The Sahel in the Midst of Climate Change*. Retrieved from reliefweb.int: https://reliefweb.int/report/chad-sahel-climate-change/

Mayer, A. and Smith, E. K. (2018). Unstoppable climate change? The influence of fatalistic beliefs about climate change on behavioral change and willingness to pay cross-nationally. *Climate Policy*, 511–523.

Mearns, R., and Norton, A. (2010). Social dimensions of climate change: Equity and vulnerability in a warming world. *Open Knowledge Repository*. World Bank Group.

Moore, M. P., et al. (2021). Sex-Specific ornament evolution is a consistent feature of climatic adaptation across space and time in dragonflies. *Proceedings of the National Academy of Sciences of the United States of America*, 118.

Mortezapour, M., Menounos, B., and Jackson, P.L. (2022). Future snow changes over the Columbia Mountains, Canada, using a Distributed Snow Model. *Climate Change*, 172, 6. https://doi.org/10.1007/s10584-022-03360-9.

Muller, C., Bondeau, A., Waha, K., and Heinke, J. (2014). Hotspots of climate change impacts in sub-Saharan Africa and implications for adaptation and development. *Global Change Biology*, 2505–2517.

Muringai, R., et al. (2021). Climate change and variability impacts on Sub-Saharan African fisheries. *Reviews in Fisheries Science & Aquaculture*, 1.

NASA. (2020). *Responding to Climate Change*. climate.nasa.gov.

NASA. (2021, July 6). *climate.nasa.gov*. Retrieved from The Effects of Climate Change: https://lknd.in/dqc3mHd

Neukom, R., Steiger, N., Gómez-Navarro, J., and Wang, J. (2019). No evidence for globally coherent warm and cold periods over the preindustrial Common Era. *Nature*, 571(7766), 550–554.

Ng, E., Xue, Y., and Choi, M. (2022, April 5). *Climate Change: Greenhouse Gas Emissions Must Peak by 2025, Deep Cuts Are Needed by End of the Decade to Avoid Catastrophic Impact*. Retrieved from South China Morning Post: www.scmp.com/business/article/3173092/climate-change-greenhouse-gas-emissions-must-peak-2025-deep-cuts-needed

Normile, D. (2020, September 29). *Can China, the World's Biggest Coal Consumer, Become Carbon Neutral by 2060?* Retrieved from sciencemag.org: www.sciencemag.org/news/2020/09/can-china-worlds-bigger-coal-consumer-become-carbon-neutral-2060

Olabanji, M. F., Davis, N., Ndarana, T., Kuhudzai, A. G., and Mahlobo, D. (2021). Assessment of smallholder farmers' perception and adaptation response to climate change in the Olifants catchment, South Africa. *Journal of Water and Climate Change*, 7, 3388–3403.

Oreskes, N. (2021, July 07). *Keelingcurve.ucsd.edu*. Retrieved from The Keeling Curve: https://keelingcurve.ucsd.edu

Patoway, K. (2014, September 15). *The Disappearing Glaciers of Africa*. Retrieved from Amazing Planet: www.amusingplanet.com/2014/09/the-disappearing-glaciers-of-africa.html?m=1

Pham-Duc, B., Sylvestre, F., and Cretaux, J.-F. (2020). The Lake Chad hydrology under current climate change. *Scientific Reports*, 97–108.

Rohde, R. (2021, January 14). *berkeleyearth.org*. Retrieved from Global Temperature Report for 2020: http://berkeleyearth.org/global-temperature-report-for-2020

Rusoke, T. (2017, August 13). *Global Warming and Climate Change: A Planet at Crossroads.* Retrieved from researchgate.net: www.researchgate.net/project/Biodiversity-Conserva-tion-Wildlife-and-Forestry-Resources-Planning-Management-in-Africa

Russell, J., et al. (2009). Paleolimnological records of recent glacier recession in the Rwenzori Mountains, Uganda-D.R. Congo. *Journal of Paleolimnology*, 41, 253–271.

Sangomla, A. (2020, December 15). *DownToEarth.* Retrieved from Looking back: How did climate change alter the world this year?: https://downtoearth.org.in/news/climate-change/looking-back-how-did-climate-change-alter-the-world-this-year-74648

Schoch, M. and Lakner, C. (2020). *The Number of Poor People Continues to Rise in Sub-Saharan Africa, Despite a Slow Decline in the Poverty Rate.* Washington, DC: World Bank Data Blog.

Scott, F., Blanchard, J. L., and Andersen, K. H. (2014). Mizer: An R package for multispecies, trait-based and community size spectrum ecological modelling. *Methods in Ecology and Evolution*, 5(10), 1121–1125. Retrieved from: https://sizespectrum.org/mizer/.

Shepard, D. (2018, May 19). *Environment.* Retrieved from Global warming: Severe consequences for Africa: www.un.org/africarenewal/magazine/december-2018-march-2019/global-warming-severe-consequences-africa#:~:text=The%20IPCC%20expects%20Central%20Africa,resultant%20impacts%20on%20food%20security.

SIWI. (2020). *Transforming Investments in African Rainfed Agriculture (TIARA).* Stockholm: Stockholm International Water Institute.

Skea, J. (2021, July 8). *Reports Synthesis Report.* Retrieved from ipcc.ch: webofknowledge. com

Tabari, H. (2021). Extreme value analysis dilemma for climate change impact assessment on global flood and extreme precipitation. *Journal of Hydrology*, 593–594.

Tadesse, D. (2010). *The Impact of Climate Change in Africa.* Addis Ababa: Africa Union Institute for Security Studies.

Thieken, A. H., Samprogna Mohor, G., Kreibich, H., and Müller, M. (2022). Compound inland flood events: Different pathways, different impacts and different coping options. *Natural Hazards and Earth System Sciences*, 22, 165–185.

Tolulope, E. B., Mukalazi, K. A., Gbenga, A. A., Oludapo, O. O., Oluwayemisi, O., and Ropo, I. O. (2022). Rediscovering South Africa: Flood disaster risk management through eco-system-based adaptation. *Environmental and Sustainability Indicators*, 100–175.

Turchin, A. (2010). *Structure of the Global Catastrophe. Risks of Human Extinction in the XXI Century.* Oxford, UK: Lulu Enterprises Incorporated.

UN. (1992). *United Nations Framework Convention on Climate Change.* New York: United Nations.

UN. (2022, June 5). *What Is Climate Change?* Retrieved from United Nations Climate Action: www.un.org/en/climatechange/what-is-climate-change

UNDP. (2019, November 28). *Climate Change in Asia and the Pacific. What's at Stake?* Retrieved from asia-pacific.undp.org: www.asia-pacific.undp.org/content/rbap/en/home/presscenter/articles/2019/climate-change-in-asia-and-pacific.html

UNFCC. (2011). *Fact Sheet: Climate Change Science—the Status of Climate Change Science Today.* New York: UN.

USGS. (2021a, May 19). Retrieved from What is the difference between global warming and climate change?: www.usgs.gov/faqs/what-difference-between-global-warming-and-climate-change-1?qt-news_science_products=0#qt-news_science_products

USGS. (2021b, July 7). *What Are the Long-Term Effects of Climate Change?* Retrieved from usgs.gov: www.usgs.gov/faqs/what-are-the-long-term-effects-of-climate-change-1?

Vogel, B., and Bullock, R. C. L. (2021). Institutions, indigenous peoples, and climate change adaptation in the Canadian Arctic. *Geo Journal*, 86, 2555–2572.

WHO. (2018, February 1). *Climate Change and Health.* Retrieved from who.int: www.who.int/news-room/facts-sheets/

WMO. (2021). *State of the Global Climate 2020: Unpacking the Indicators*. Geneva, Switzerland: World Meteorological Organisation.

Wollersheim, L. (2015). *Climageddon*. JobOne.

World Bank. (2019). *Climate Change and Marine Fisheries in Africa: Assessing Vulnerability and Strengthening Adaptation Capacity*. Washington, DC: World Bank.

World Bank. (2020, September 21). *Where Climate Change Is Reality: Supporting Africa's Sahel Pastoralists to Secure a Resilient Future*. Retrieved from worldbank.org: www.worldbank.org/en/news/immersive-story/2020/09/2021/where-climate-change-is-reality-supporting-africas-sahel-pastoralists-secure-a-resilient-future

Wuehler, S. (2015). The demographic situation in the Sahel: Two times more inhabitants in the year 2015. *Pop Sahel*, 8–10.

18 Pollution Haven or Pollution Halo?

Environmental Effect of Chinese FDI Regarding the One Belt One Road Project

Yahya Algül, Vedat Kaya, Ömer Yalçınkaya

18.1 INTRODUCTION

China, with its high government investments and rich labor sources, has seen rapid growth rates in the last three decades; however, this growth policy has shown marks of running its course, with the decline of the working-age population following its highest rate in 2012, state investment peaking at 49% of GDP, and a contraction gap among China and other competing developed countries (Haggai, 2016). Considering this economic slowdown in the domestic market, China seeks to transition to rebalanced economic growth by exploiting investment opportunities in the international sphere (Loke, 2018).

In these new circumstances, China has announced the Chinese One Belt and One Road (OBOR) Initiative, or Belt and Road Initiative (BRI), which is shortly planned to consist of the Silk Road Economic Belt and the Maritime Silk Road divisions in aiming to further stimulate the ongoing course of development of China by investing in developing countries of the belt (Johnston, 2019). The land-based Silk Road Economic Belt involves the construction of new roads, railway pipelines, and other infrastructure investments to connect inner China to Europe through Central Asia and the Middle East. On the other hand, the Maritime Silk Road Project involves new marine routes, ports, and other maritime investments to benefit from the opportunities that lie ahead through China to South Asia and Europe through the Indian Ocean and north Africa (Enderwick, 2018).

China is spending approximately 150 billion USD annually on the member countries that have signed up to the project to move their excess production capacity globally (Liu and Kim, 2018). Moreover, the BRI comprises more than 65 nations, which accounted for 30% of the global GDP, 50% of worldwide energy consumption, 64% of the world population, 39% of the earth's land, 35% of the worldwide trade, and 54% of worldwide carbon emissions (Du and Zhang, 2018, as cited in An et al., 2021; Fan et al., 2019, as cited in An et al., 2021).

The BRI project initiates many important infrastructure investments to create a new trade route, such as cross-border high-speed railways, freight railways, oil and gas pipelines, and telecommunications, as well as electricity infrastructure (Li and Hilmola, 2019). But this project is not only positioned as an international trade route, the BRI is also expected to greatly change the geographies of Asia and beyond through global geopolitical, economic, and cultural connections (Teo et al., 2020). Those effects can be grouped under a few headings, such as creating channels for international policy dialogues; strengthening the infrastructure connectivity internationally; creating cross-country industrial value chains; generating financial support mechanisms such as the BRICS New Development Bank, AIIB, and Silk Road Fund for countries in need; and finally creating channels for academic, cultural, and talent exchanges (Huang, 2016).

China promotes the BRI from a win-win relationship perspective, since, while other countries receive infrastructure investments, new markets emerge for Chinese excess production capacity, and beyond that, China claims that, with new policy measures, the BRI might be a strong vision for a green, low-carbon embodied, circular and sustainable growth perspective that might mitigate ecological and other socioeconomic problems (Harlan, 2021; Carey and Ladislaw, 2019; Cuiyun and Chazhong, 2020).

However, alternative perspectives exist regarding the BRI, contending that it is not as benign as China asserts and it may exert adverse effects on the countries within the Belt. Some of the concerns regarding the BRI are evaluated from military, political, sociocultural, socioeconomic, economic, environmental, and other perspectives (Fang et al., as cited in Gu and Zhou, 2020). Intensive investments made by private and state-supported Chinese companies and the strong dominance of these companies in Asian and African countries aroused suspicion of a new type of colonialism that may be applied by China (Shichor, 2018; Dunford and Liu, 2019; Li, 2020; el-Shafei and Metawe, 2021). Moreover, the financial policies implemented by China in the context of providing the financial resources necessary for the underdeveloped and developing countries of the BRI to reach their desired growth potential is seen from a predatory lending or debt trap perspective by some researchers and policymakers (Hurley et al., 2019; Rajah et al., 2019; Lai et al., 2020).

Other concerns about the BRI project are related to the energy geopolitics of China. Due to the various problems of the Malacca Strait, which constitutes an important part of China's energy trade route, such as security problems due to territorial disputes and pirate and terrorist assaults, it is argued that with the BRI, China wants to both eliminate the problems caused by the Malacca Strait and to have a say in the world energy corridors against Western countries by reaching energy sources in Asia and Africa (Len, 2015; Blanchard and Flint, 2017; Aydın, 2018; Muhammad and Long, 2020).

In addition to all of these concerns, undoubtedly, one of the most discussed issues in the literature is the impact of the BRI project on the environment. Since, in the last decade, the average growth rate of CO_2 in China and Belt and Road host countries is about 8% and 5%, respectively (An et al., 2021). Moreover, it is estimated that by 2050, assuming the growth of CO_2 continues at the current rate, 66% of global CO_2 emissions will be accounted for by only the BRI countries, excluding China (Ma et al., 2019, as cited in Li et al., 2021). Considering its important effect on global CO_2 emission, some researchers and policymakers are skeptical of the Chinese BRI project. Some researchers argue that substantial foreign direct investment by China may lead to the relocation of polluting industries and associated pollution issues from China to the countries participating in the BRI.

According to the Statistical Bulletin of China's Outward Foreign Direct Investment, while China is the second-biggest recipient of foreign direct investment (FDI), with a share of 10.7% of global FDI, it is also the second biggest foreign direct investor globally, with a share of 14.1% in 2018 (Mahadevan and Sun, 2020). Moreover, China has spent $1.94 trillion globally in the last decade on other countries in the form of FDI, primarily in the energy, infrastructure, and transportation sectors (An et al., 2021). It is estimated that Chinese investments in the countries of the BRI will add up to over $1 trillion in the form of FDI flow (Mahadevan and Sun, 2020).

According to the pollution haven hypothesis (PHH), due to strict regulations in developed countries related to environmental concerns, firms in highly pollutive sectors may transfer their production to other less developed countries with weaker environmental regulations (Taylor, 2005; Levinson and Taylor, 2008). As a result of this theory, there might be a positive relationship between foreign direct investment received and pollution levels in the host countries. Therefore, according to some researchers, there might be a possibility of transferring air pollution, which has already been experienced in China for years, to the countries of the BRI through FDI flows (Cai et al., 2018; Liu and Kim, 2018; An et al., 2021)

According to other competing and more optimistic argument in contrast to the PHH, there might be a negative relationship between foreign direct investment and pollution levels, since developed countries that invest in other host countries generally have more a environmentally friendly and advanced and efficient production structure than less developed host countries (Balsalobre-Lorente et al., 2019; Mert and Caglar, 2020). Therefore, with foreign direct investment, China may move its better and cleaner production technics and technology to the other host countries (Ahmad et al., 2020; Li et al., 2021; Liu et al., 2020). According to pollution halo hypotheses (PAHO), as a result, these host countries may benefit from this situation environmentally.

From this point of view, we intend to empirically analyze the relations between foreign direct investment and air pollution in terms of BRI countries. For this purpose, the effects of FDI from China and other countries are analyzed separately to understand the dynamics of air pollution in the host countries of the BRI. In this inquiry, econometric methods consider the cross-sectional correlation among the units together with a data set from 1995 through 2020. The study conducted on the 57 BRI countries is expected to contribute to the development of the literature on this subject owing to the separation of the variables used to represent FDI from China and other countries, the countries covered, and the modern econometric methods utilized.

In the second section following the introduction, the relevant literature is summarized, and the positioning of this study in the literature is explained. In the third section, the data set utilized in the study is introduced, and its scope is explained. In the fourth section, the empirical findings of the study are discussed. In the fifth and last section, based on the empirical findings of the study, policy implications and suggestions for future studies are offered.

18.2 FOREIGN DIRECT INVESTMENT AND ENVIRONMENTAL POLLUTION

FDI is a type of investment made by a firm in a country other than its base location. FDI's importance and volume have increased rapidly after the liberal policies that were observed globally in the 1980s. While FDI's contribution to the global economy has been accepted widely, its environmental effect on host countries is still dubious. Studies on this subject for many years have yielded very diverse results. This variety in the literature is mainly due to the differences in the geographical areas handled such as city, state, and countrywide studies, as well as the difference in the time spans considered in the studies; the differences in the variables of pollution such as CH_4, CO_2, N_2O, and total greenhouse gas emissions or the rate of emissions per capita; the differences in the variables used to represent FDI; and the difference in econometric models and processes used (Cai et al., 2018).

Beyond that variety of differences in the methodologies and results, two distinct streams of thought about the environmental effect of FDI on the host countries can be observed. According to the first argument based on the PHH, FDI may expand environmental pollution in host countries. It is argued that this is because industrialized countries may desire to relocate their old-fashioned environmentally detrimental sectors, such as metal and energy, to other less developed countries where environmental regulations are generally applied loosely. Some of the studies that have found some support for this hypothesis were conducted by Akbostanci et al., 2007; Ben Kheder and Zugravu-Soilita, 2008; Lan et al., 2012; Neequaye and Oladi, 2015; Shahbaz et al., 2015; Solarin et al., 2017; Wang et al., 2019; Terzi and Pata, 2019; Sadik-Zada and Ferrari, 2020; Khan and Ozturk, 2020; and Singhania and Saini, 2021.

However, according to the counter-argument, the PAHO, companies that are based in developed countries may transfer their more efficient and cleaner production technology to the less-developed countries by investing in them. Therefore, FDI may rehabilitate the production techniques and reduce pollution levels in the host countries. Some of the studies that have found some evidence in favor of the pollution halo hypothesis were conducted by Zhu et al., 2016; Doytch and Narayan, 2016; Wang, 2017; Balsalobre-Lorente et al., 2019; Mert and Caglar, 2020; Repkine and Min, 2020; Kisswani and Zaitouni, 2021; Polloni-Silva et al., 2021; and Chen et al., 2021.

In this study, the effect of China's FDI on air pollution in BRI countries has been investigated both to evaluate two different hypotheses theoretically and to understand the environmental effects of foreign direct investment of China on the host countries within the scope of BRI. The case of BRI is specifically chosen to investigate since a major part of the foreign direct investment in the initiative consists of energy and pollution-intensive sectors. Moreover, as specified earlier, the enormous size of the BRI, which constitutes 30% of global GDP, 50% of worldwide energy consumption, and 64% of the earth's population (Du and Zhang, 2018, as cited in An et al., 2021; Fan et al., 2019, as cited in An et al., 2021), makes it an important and interesting query. For the same reason, it is necessary to take a look at studies that have investigated FDI and pollution studies specifically the case of the Belt and Road Initiative.

For example, Ahmad et al. (2020) conducted a study to investigate the relationship between foreign direct investment, financial development, and CO_2 emissions for the 90 countries of the BRI with data from 1990 to 2017 and found that foreign direct investment reduces CO_2 emissions in the host countries, which supports PHH. Moreover, researchers have also found that beyond FDI, trade openness mitigates air pollution in host countries.

On the other hand, Liu and Kim (2018), by utilizing the panel vector autoregression model and data between 1990 and 2016, concluded that for the 44 countries of the BRI, the pollution havens hypothesis is validated. Similar results that support the PHH for 65 of the BRI countries were found by Muhammad et al. (2020) by utilizing data set ranges from 2000 to 2016.

Cai and colleagues (2018) also investigated the validity of the PHH. But instead of FDI, they investigated greenhouse gases emissions embodied in exports and imports and their effect on pollution levels by utilizing the multiregional input-output model. According to estimations, China might be a pollution haven for 22 developed countries, while 19 of the developing countries may become pollution havens for China, which validates the PHH. However, there was no strong evidence in favor of the PHH for the case of the other 22 countries (Cai et al., 2018).

An et al. (2021) investigated the validity of the environmental Kuznets curve (EKC) hypothesis in the multivariate conception of people connectivity and technological innovation. Moreover, researchers have also examined the environmental effect of Chinese FDI on the 64 BRI countries with a method of moments quantile regression between 2003 and 2018. According to estimation results, FDI indicates the maximum coefficient rates for countries with higher pollution rates, such as Sri Lanka, Bangladesh, and Pakistan. This means that while Chinese FDI has a lower pollution risk in environmentally less polluted countries, for countries with emission rates from medium to high, Chinese FDI presents substantial pollution risks.

Similar to the study of An et al. (2021), Mahadevan and Sun (2020) also predicted that the effect of Chinese FDI may be differentiated according to the host country. According to the estimations of Mahadevan and Sun (2020), Chinese FDI to upper-middle and high-income BRI countries does not affect host countries' pollution level, has a diminishing impact on low-income BRI countries, and has a growing impact on lower- to middle-income BRI countries. Muhammad and Long (2021) also concluded that the effect of FDI for BRI countries differs according to the income level of the host country. According to Muhammad and Long (2021), the PAHO is verified in high-income countries, while the PHH is verified in low-income, lower-middle-income, and upper-middle-income countries.

According to the nonlinear Granger causality test conducted by Gong et al. (2021) for 19 of the BRI countries in the period running from 1979 to 2017, there is mixed evidence about the relation between FDI and CO_2 emissions. The estimations indicate that Malaysia, China, South Africa, Poland, and Greece have nonlinear causality of FDI to CO_2, while for India, nonlinear causality runs from CO_2 to FDI. On the other hand, for Turkey and Brazil, two-way causality has been found between FDI and CO_2, and no nonlinear causal relation has been found for the case of Malta. Therefore, Gong et al. (2021) put forward different conclusions for different countries.

A different study for BRI countries was conducted by Li et al. (2021). The authors in the study, instead of investigating the relation among FDI, CO_2, and other related variables, performed a forecast analysis of CO_2 emissions in the BRI region for different Chinese investment level scenarios. According to the zero Chinese investment case, the BRI region's CO_2 emission rate will increase by 1973 Gt between 2018 and 2100. On the other hand, in the business-as-usual scenario, the CO_2 emission rate is expected to decrease by 44.16 Gt, and this abatement of CO_2 is expected to reach 79.48 Gt in the strengthening Chinese investment scenario for the BRI region and may help to keep the temperature rise below $2°$ C (Li et al., 2021).

While most of the studies regarding the environmental effects of FDI in the BRI framework were conducted from the macroeconomic perspective, Liu et al. (2020) conducted a study from a microeconomic perspective. Liu et al. (2020) investigated 729 Chinese firms' FDI events specifically in the energy sector between 2005 and 2016 in 118 countries. According to estimations of the feasible generalized least and single-equation logit models, if host countries have better natural sources, political domain, energy efficiency, and lower CO_2 intensities, Chinese firms would tend to invest in green projects. On the other hand, if the host country has a highly developed energy structure and more advanced technologies and infrastructure, Chinese firms would not be so eager to invest in green projects. Moreover, the study further discerned a pronounced policy impact of the BRI on the environmental orientation of Chinese outward foreign direct investment (OFDI) in affiliated nations, particularly within developing regions.

As seen from reviewed literature, there is no clear consensus about the relationship between foreign direct investment and pollution level. Moreover, the same situation is valid for the studies that specifically investigated the case of BRI, too. It is hoped that by differentiating the foreign direct investment data between Chinese FDI and other countries' FDI for the BRI countries, and by utilizing specific econometric models that take into account cross-sectional correlation, our study will contribute to the development of the relevant literature.

18.3 DATA SETS AND SCOPE OF THE STUDY

In this section, the data sets used and data conversion/derivation processes are explained together with the scope of the study that aimed to examine the effects of foreign direct investments on carbon dioxide emissions in the Belt and Road countries. However, contrary to most of the studies in the literature, the effect of foreign direct investment in the BRI countries is investigated according to the source country of the FDI. Most of the studies in the literature do not differentiate among source countries of FDI. This departure from the conventional approach is motivated by the recognition that the carbon dioxide effect of foreign direct investment may vary based on the source country's production efficiency and technology, a distinction often overlooked in existing literature. Therefore, while a country's FDI may mitigate the total CO_2 in a host country, another country's FDI may intensify. To avoid this problem, FDI data sets on the host countries are divided into Chinese FDI and other countries' FDI investments.

With this aim, the study is carried out for 57 countries, for which data is accessible during the sample period covering 1995 to 2020. The sample period of the study was determined as 1995 to 2020. The chosen sample period is from 1995 to 2020, as it ensures continuous data accessibility for the variables across the 57 BRI countries. These 57 BRI countries are listed alphabetically as follows; Albania, Armenia, Azerbaijan, Bahrain, Bangladesh, Belarus, Bosnia and Herzegovina, Brunei Darussalam, Bulgaria, Cambodia, Croatia, Cyprus, Czech Republic, Egypt, Estonia, Georgia, Greece, Hungary, India, Indonesia, Iran, Iraq, Israel, Jordan, Kazakhstan, Kuwait, Kyrgyzstan, Latvia, Lebanon, Lithuania, Macedonia, Malaysia, Moldova, Mongolia, Myanmar, Nepal, Oman, Pakistan, Philippines, Poland, Qatar, Romania, Russia, Saudi Arabia, Serbia, Singapore, Slovakia, Slovenia, Sri Lanka, Tajikistan, Thailand, Turkey, Turkmenistan, Ukraine, United Arab Emirates, Uzbekistan, and Vietnam.

The variables and data sources of the variables used in the econometric analyses of the study to examine the effects of foreign direct investments on carbon dioxide emissions in 57 BRI countries are specified in Table 18.1.

The CH-FDI and OT-FDI variables that are listed in Table 18.1 are derived from the net foreign direct investment data for China found in the HFD database, and the

TABLE 18.1

Variables

Variables	Definition of Variables	Unit of Variables	Data Source
CO	Carbon dioxide (CO_2) emissions	Metric tons per capita	The World Bank (WB) (World Development Indicators [WDI],
DP	Gross domestic product	Constant 2010 US dollar per capita	2021)
UP	Urban population	10,000 persons	
EC	Energy consumption	Kg of oil equivalent per capita	The World Bank (World Development Indicators, 2021) International Energy Agency (IEA) (Data and Statistics, 2021)
CH-FDI	China foreign direct investment	Current US dollar net outflow (% of gross domestic product)	The Heritage Foundation Database (HFD) (China Global Investment Tracker, 2021)
OT-FDI	Other foreign direct investment	Current US dollar net inflow (% of gross domestic product)	The World Bank (World Development Indicators, 2021)

TABLE 18.2
Descriptive Statistics of Variables

Variables	Mean	Median	Maximum	Minimum	Std. Dev.	Skewness	Kurtosis
CO	1.319	1.519	3.864	−2.435	1.229	−0.750	3.297
EC	7.446	7.537	10.004	4.912	1.004	−0.007	2.635
DP	8.620	8.634	11.151	5.529	1.267	−0.104	2.222
UP	15.624	15.399	19.993	12.225	1.484	0.467	2.745
CH-FDI	0.051	0.000	5.074	0.000	0.227	12.268	210.434
OT-FDI	5.504	2.930	280.131	−40.329	14.991	11.426	171.487
Observations	1482	1482	1482	1482	1482	1482	1482

current US dollar values for net inflows of foreign direct investment data and gross domestic product data are obtained from the WB database. The derivation and conversion steps of data sets are explained as follows.

The CH-FDI variable was obtained by dividing the levels of Chinese net foreign direct investment in the BRI countries by the gross domestic product of those respective countries. To obtain the OT-FDI variable, first, the Chinese outflow of net foreign direct investment data was extracted from the total net inflows of foreign direct investment data of the relevant BRI countries. Then, to find the OT-FDI data, the net foreign direct investment made by countries other than China to the relevant BRI countries was divided into the gross domestic product of relevant countries' data.

In the econometric analysis of the study, the natural logarithmic values of the CO, EC, DP, and UP variables and the level values of the CH-FDI and OT-FDI variables are used. After these conversion processes, the descriptive statistics of the values of the CO, EC, DP, UP, CH-FDI, and OT-FDI variables for the 1995–2020 period to be used in econometric analyses are presented in Table 18.2.

18.4 METHODOLOGY AND RESULTS

In this study, the effects of FDI on the BRI countries' carbon dioxide emissions are examined by panel data analysis, and it is aimed to econometrically determine the direction and the size of the long-term effects of FDI on carbon dioxide emissions. For this purpose, in this study, by following the panel data analysis methodology, which was designed with the help of the empirical literature and taking into account cross-sectional dependence (CDS), the econometric models were constructed to be estimated separately for the period 1995–2020 are shown in the following equations.[1] In determining the variables and models to be estimated to understand the long-term effects of foreign direct investments originating from China and other countries on carbon dioxide emissions of BRI countries, the studies of Al-Mulali and Tang (2013) Sun et al. (2019), Muhammad et al. (2020), Ahmad et al. (2020), and Muhammad and Long (2021) were taken as a reference point.

Model 1: $CO_{it} = \alpha_{it} + \beta_1 EC_{it} + \beta_2 DP_{it} + \beta_3 UP_{it} + \beta_4 CH - FDI_{it} + \varepsilon_{it}$ (18.1)

Model 2: $CO_{it} = \vartheta_{it} + \partial_1 EC_{it} + \partial_2 DP_{it} + \partial_3 UP_{it} + \partial_4 OT - FDI_{it} + \varepsilon_{it}$ (18.2)

Model 3: $CO_{it} = \varphi_{it} + \delta_1 EC_{it} + \delta_2 DP_{it} + \delta_3 UP_{it} + \delta_4 CH - FDI_{it} + \delta_5 OT - FDI_{it} + \varepsilon_{it}$ (18.3)

From the terms in the equality equations, α, ϑ, and φ represent the constant parameters; β, ∂, and δ the slope parameters; and ε, i, and t the error term, cross-section, and time dimension of the panel, respectively. To prevent the phenomenon of spurious regression in panel data analysis, the stationarity condition of the model variables was examined with unit root tests, and, according to the findings of the unit root tests, other consecutive tests to be used in model predictions should be determined without bias (Tatoğlu, 2013, p. 199). Unit root tests that guide the econometric methodology in panel data analysis and other successive tests are grouped as first and second generation according to whether CDS exists among the panel units. It is assumed that in first-generation tests, all units are affected at the same rate by a certain shock occurring in one of the panel units, while in second-generation tests, each unit is assumed to be affected at different levels. These assumptions require first investigating CDS in variables and models to achieve unbiased results in panel data analyses and determining the unit root and other successive tests that should be used in the analyses (Menyah et al., 2014). In panel data analysis, CDS in variables and models can be investigated by CD-LM tests, taking into account the time (T) and cross-section (N) dimensions of the panel. In the case of $T > N$, Breusch and Pagan's (1980) CD-LM1 test can be used, while in the case of $T < N$, Pesaran's (2004) CD-LM2 test can be used, whereas in all alternative cases between T and N, Pesaran et al.'s (2008) CD-LM$_{adj}$ test can be used. The CD-LM1 and CD-LM2 tests are calculated using Eq. 18.4:

$$CD - LM = \bar{\rho}_{ji} = \frac{\sum_{t=1}^{T} e_{it} e_{jt}}{\left(\sum_{t=1}^{T} e_{it}^2\right)^{1/2} \left(\sum_{t=1}^{T} e_{jt}^2\right)^{1/2}}$$ (18.4)

While the term $\bar{\rho}_{ji}$ in Eq. 18.4 indicates the correlation between the error term series, the term e_{it} denotes the error series obtained from the cross-section units while going to $i = 1, \ldots n$ for t observations using the least squares method. CD-LM1 and CD-LM2 tests can give biased results in cases where the group mean is zero and the unit mean is different from zero. To address these issues, Pesaran et al. (2008) introduced the CD-LM$_{adj}$ test, which incorporates the addition of the mean ($\propto_{Tij}$) and variance (υ_{Tij}) of the cross-sections into the test statistics, advancing previous tests as seen in Eq. 18.5:

$$CD - LM_{Adj} = NLM^{**} = \sqrt{\frac{2T}{N(N-1)}} \left(\sum_{i=j}^{n-1} \sum_{j=i+1}^{n} \frac{(T-K)\rho_{ij} - \mu_{Tij}^{\sigma^2}}{\upsilon_{Tij}} \right)$$ (18.5)

TABLE 18.3
CD-LM TEST RESULTS

C + T	Test Statistics		
Variables/Models	CD-LM2	CD-LM$_{adj}$	L
CO	25.13[a] [0.000]	4.24[a] [0.000]	4
EC	19.34[a] [0.000]	2.17[a] [0.004]	3
DP	**21.92**[a] [0.000]	3.42[a] [0.000]	2
UP	8.34[a] [0.000]	1.77[b] [0.040]	2
CH-FDI	12.72[a] [0.000]	6.30[a] [0.000]	2
OT-FDI	13.91[a] [0.013]	9.28[a] [0.000]	2
Model-1	78.22[a] [0.000]	1.99[b] [0.023]	2
Model-2	70.36[a] [0.000]	1.92[b] [0.028]	2
Model-3	62.20[a] [0.000]	1.73[b] [0.030]	2

Note: [a] and [b] indicate that CDS is present at the 1% or 5% significance level, respectively. The "L" column shows the optimal lag lengths determined with the Schwarz information criterion, and the values in the square brackets ([]) show the probabilities of the test statistics.

Thus, the calculated CD-LM$_{adj}$ test statistic can give more consistent results compared to the CD-LM1 and CD-LM2 test statistics and can be used in all alternative cases of time and cross-section dimensions of the series that make up the panel. In the CD-LM tests, the presence of CDS among the units in the panel is investigated with the basic hypothesis of "no CDS in the variable/model". In CD-LM tests, which are assumed to have a standard normal distribution, it is concluded that if the basic hypothesis is rejected, CDS is present in the variable/model (Pesaran et al., 2008). The CDS results in the form of constant + trend (C + T) performed with CD-LM2 and CD-LM$_{adj}$ tests, considering T and N conditions in the variables and cointegration equations of the models, are reported in Table 18.3.

Upon examination of the findings in Table 18.3, it is seen that the CD-LM test statistics probabilities for which cointegration equations are calculated and the probabilities of the variables in all defined models are less than 0.01 or 0.05, and the basic hypotheses are rejected at the 1% to 5% significance level. These findings indicate that cross-sectional units exhibit dependency in terms of variables and cointegration equations in the panel of BRI countries and point to the necessity of using second-generation panel data analyses that take this dependency into account (Baltagi, 2008).

Therefore, the stationarity of the variables in the defined models is investigated with the cross-sectional augmented Dickey-Fuller (CADF) panel unit root test developed by Pesaran (2007), which takes into account the CDS among the panel units. In this test, first, CADF test statistics are calculated for the cross-sectional units in the panel, and then cross-sectionally augmented Im, Pesaran, and Shin (CIPS) statistics are generated across the panel by using the arithmetic average of the CADF values. The CADF and CIPS test statistics, which can be used in all alternative situations between the T and N dimensions of the panel and give consistent results, are calculated based on the following equations:

$$t(N,T) = \frac{\Delta y_i' \bar{M}_i y_{i-1}}{\bar{\sigma}^2 \left(\Delta y_{i-1}' \bar{M}_i y_{i-1}\right)^{1/2}} \quad \text{and} \quad CIPS = N^{-1} \sum_{i=1}^{n} t(N,T) \quad (18.6)$$

The calculated CADF and CIPS test statistics are compared with the critical table values generated by Monte Carlo simulations, and the hypotheses are tested for stationarity. If the CADF and CIPS test statistics values are greater than the critical table values in absolute value, the basic hypothesis that "the variable has a unit root" is rejected (Pesaran, 2007). The results of the CIPS panel unit root test, which investigates the stationarity of the variables in the models defined in the study in the form of Constant + Trend (C + T), are presented in Table 18.4.

Upon reviewing the findings in Table 18.4, it is seen that all the variables in the defined models are not stationary at the 1% or 5% significance level but become stationary at the first difference. This conclusion is reached because the CIPS statistics calculated at the first differences of the variables are greater than the critical table values at the 0.01 or 0.05 significance level, hence leading to the rejection of the basic hypothesis.

According to the results of the CIPS panel unit root tests, taking the first differences of variables makes them stationary. This differencing operation helps remove the effects of short-term shocks on the variables and potential long-term integrated relationships between model variables. It is possible to find a combination where these variables are stationary, and this situation can be determined by cointegration analysis (Tarı, 2010, p. 415). The existence of long-term cointegration relationships between the variables in the models is investigated with the Durbin-Hausman (DH)

TABLE 18.4
CIPS Panel Unit Root Test Results

		Test Statistics	
C + T		**CIPS**	
Variables	**Level**	**First Difference**	**L**
CO	−2.11	−3.10[a]	4
EC	−1.69	−2.75[b]	3
DP	−2.29	−2.94[a]	2
UP	−2.46	−2.76[b]	2
CH-FDI	−2.40	−3.35[a]	3
OT-FDI	−2.60	−3.80[a]	4
Critical Table Values %1	−2.85		
%5	−2.71		

Note: [a] and [b] indicate that the variables are stationary at the 1% and 5% significance level, respectively. Critical table values indicate values taken from Pesaran's (2007) study according to T and N conditions. For the L column in the table, see Table 18.3.

second-generation panel cointegration test developed by Westerlund (2008). By using the DH panel co-integration test, DH panel (DH_p), and DH group (DH_g) test statistics, long-term relationships between series can be examined simultaneously across the panel and in the cross-section units that make up the panel. In the DH_p test, it is assumed that the autoregressive parameter is the same among the cross-section units in the panel, and in the DH_g test, it is assumed that the autoregressive parameter differs among the cross-section units in the panel. DH_p and DH_g test statistics, which take into account the presence of common factors between series, are calculated as follows:

$$DH_p = \hat{S}_n \left(\tilde{\varnothing} - \tilde{\varnothing} \right)^2 \sum_{i=1}^{n} \sum_{t=2}^{T} \hat{e}_{it-1}^2 \quad and \quad DH_g = \sum_{i=1}^{n} \hat{S}_i \left(\tilde{\varnothing}_i - \tilde{\varnothing}_i \right)^2 \sum_{t=2}^{T} \hat{e}_{it-1}^2 \quad (18.7)$$

where the terms $(\hat{\varnothing}_i^2)$ and $(\hat{\sigma}_i^{\,2})$ denote the long-term and simultaneous variance estimators of $(\varnothing_i^2)$, respectively. In the equation, $(\hat{S}_n = \hat{\varnothing}_n^2 / \left(\sigma_n^2 \right)^2)$ denotes the estimated variances of the overall panel (n) and $(\hat{S}_i = \hat{\varnothing}_i^2 / \hat{\sigma}_i^4)$ denotes the estimated variances of the cross-sectional units (i) constituting the panel.

DH_p and DH_g test statistics calculated based on the equations and the long-term relationships between the series throughout the panel and in the cross-sectional units that make up the panel are investigated with the basic hypothesis that "there is no cointegration relationship between the series". According to the DH_p and DH_g test statistics, if the basic hypotheses are rejected, it is assumed that there is a cointegration relationship throughout the panel and at least in some of the cross-sectional units in the panel. However, according to the DH_p and DH_g test statistics, the rejection of the basic hypotheses is determined by the normal distribution critical table values. If the calculated DH_p and DH_g test statistics are greater than the critical table value (2.33), the basic hypotheses are rejected, and it is concluded that there is a cointegration relationship at the 1% significance level throughout the panel and/or at least some of the cross-section units in the panel (Westerlund, 2008). The results of the DH test statistics investigating the long-term relationships between variables in the form of Constant + Trend (C + T) are listed in Table 18.5.

Findings in Table 18.5 indicate that the basic hypotheses established according to the DH_p and DH_g test statistics for all defined models were rejected at the 1% significance level. This is evident from the fact that the DH_p and DH_g test statistical values calculated for all defined models are greater than the critical value (2.33).

TABLE 18.5

DH PANEL CO-INTEGRATION TEST RESULTS

C + T	Modeler	Model-1	Model-2	Model-3
Test Statistics	DH_g	48.96[a][0.000]	41.84[a] [0.000]	48.40[a] (0.000)
	DH_p	88.70[a] [0.000]	84.39[a] [0.000]	71.12[a] [0.000]

Note: [a] indicates that there is a cointegration relationship at the 1% significance level between the variables in the model. For the square brackets, "[ ]", in the table, see Table 18.2.

This suggests that the variables in the models exhibit cointegration across the entire panel and among the cross-sectional units within the panel.

Since it is determined that variables in models are affected by each other, become stationary in their first difference, and are cointegrated, it becomes necessary to estimate the long-term coefficients with appropriate methods. In this context, the long-term effects of foreign direct investments originating from China and other countries on carbon dioxide emissions in BRI countries can be examined with the panel augmented mean group (AMG) method, which can make estimations by taking all these conditions into account. In the panel AMG method developed by Eberhardt and Bond (2009), the long-term cointegration coefficients for the panel as a whole are calculated by weighting the arithmetic averages of the cointegration coefficients of the cross-sections in the panel. In the panel AMG method, the estimation of the long-run cointegration coefficients for the panel as a whole and the cross-sections in the panel is based on the following equations:

$$y_{it} = \beta_i' x_{it} + u_{it} \qquad u_{it} = \alpha_i + \lambda_i' f_t + \varepsilon_{it} \qquad (18.8)$$

$$x_{mit} = \pi_{mi} + \delta_{mi}' g_{mt} + \rho_{1mi} f_{1mt+...} + \rho_{nmi} f_{nmt} + v_{mit} \qquad (18.9)$$

$$f_t = \varphi' f_{t-1} + \epsilon_t \qquad g_t = \varrho' g_{t-1} + \omega_t \qquad (18.10)$$

The term x_{it} in the equations represents the vector of observable covariates, the terms f_t and g_t represent unobserved common factors, and the term λ_t represents the factor loadings of the sections in the panel. In this respect, in the panel AMG method, long-term cointegration coefficients are estimated by considering the common factors and dynamic effects in the series. The panel AMG estimator, which can be used even if there is an endogeneity problem arising from the error term, can also produce effective results for unbalanced panel data sets (Eberhardt and Bond, 2009, pp. 1–4). Panel AMG results for the models identified in the study are presented in Table 18.6.

TABLE 18.6
Panel AMG Test Results

Models	Model-1		Model-2		Model-3	
Variables	Coefficients	SE	Coefficients	SE	Coefficients	SE
EC	0.6233[a]	0.0663[0.000]	0.6198[a]	0.0685[0.000]	0.6212[a]	0.0691[0.000]
DP	0.3071[a]	0.0943[0.001]	0.2679[a]	0.0938[0.004]	0.2764[a]	0.0930[0.003]
UP	0.6336[b]	0.3253[0.045]	0.7073[b]	0.3531[0.045]	0.6958[b]	0.3498[0.043]
CH-FDI	−0.1465[b]	0.0617[0.018]	–	–	−0.1517[b]	0.0643[0.018]
OT-FDI	–	–	0.0071[b]	0.0032[0.028]	0.0076[b]	0.0022[0.031]
Constant (C)	−15.2606[a]	5.2044[0.003]	−16.1365[b]	5.1905[0.002]	−15.9738[a]	5.2916[0.003]
WaldChi²(4)	170.75[0.000]		140.80[0.000]		165.12 [0.000]	

Note: [a] and [b] indicate that the t-statistics of the coefficients are significant at the 1% and 5% significance level, respectively. SE indicates the standard error of the coefficients, and the values in square brackets, "[ ]", indicate probabilities.

The findings in Table 18.6 indicate that the coefficients calculated for the constant term and independent variables in all defined models are of similar size and significance level. This shows that, at first glance, the defined models are well organized and produce coherent findings. Moreover, checking the independent variables for the main determinants of carbon dioxide emissions, it is seen that the coefficients of the EC, DP, and UP variables are all positive and statistically significant in all models without exception, which is in accordance with the expectations. These results indicate that enhancement in energy consumption, economic growth, and urban population in 57 BRI countries during the sample period lead to air pollution by increasing carbon dioxide emissions.

On the other hand, checking the foreign direct investment variables, which constitute the core of the study, the coefficients of OT-FDI and CH-FDI variables in all models are statistically significant positive in the range of (0.0071/0.0076) and negative in the range of (−0.1465/−0.1517), respectively. These findings reveal that a 1% rise in inward foreign direct investment for 57 BRI countries originating from countries other than China increases carbon dioxide emissions in host countries by approximately 0.0071–0.0076% and consequently increases air pollution. Therefore, this part of the empirical findings has some support for the PHH.

On the other hand, the empirical findings also indicate that a 1% rise in foreign direct investment from China to 57 of the BRI countries reduces carbon dioxide emissions by about −0.1465/−0.1517% and consequently reduces air pollution in the host countries. This means that, against the FDI from other countries, Chinese FDI to BRI countries shows some support in favor of the PAHO.

18.5 CONCLUSION AND POLICY IMPLICATIONS

For the last few decades, the relations between FDI and air pollution have been investigated in terms of the level of economic development, environmental regulations, and other dynamics. These investigations have yielded two very distinct opposing hypotheses. The first of these is the pollution haven hypothesis, which states that foreign direct investments increase air pollution, and the second is the pollution halo hypothesis, which states that foreign direct investments reduce air pollution.

From this point of view, this study aimed to empirically examine which of the two hypotheses can explain the relationship between foreign direct investment and air pollution in BRI countries. For this purpose, the effects of foreign direct investments of China and other countries on air pollution in BRI countries are analyzed separately with econometric methods that take into account the cross-sectional correlation between the units that make up the panel and the data set for the period 1995–2020. In the estimated models, it was determined that foreign direct investments originating from countries other than China increased air pollution in the 57 BRI countries during the review period. These empirical findings are consistent with the studies of Akbostanci et al., 2007; Ben Kheder and Zugravu-Soilita, 2008; Lan et al., 2012; Neequaye and Oladi, 2015; Shahbaz et al., 2015; Solarin et al., 2017; Wang et al., 2019; Terzi and Pata, 2019; Sadik-Zada and Ferrari, 2020; Khan and Ozturk, 2020; and Singhania and Saini, 2021 and indicate that the relation between FDI and air pollution for BRI countries can be described with the pollution haven hypothesis.

On the other hand, according to empirical findings, foreign direct investments from China to 57 of the BRI countries decrease air pollution in host countries. Therefore, Chinese FDI to the BRI countries indicates a pollution halo effect, which is compatible with the following studies in the literature; Zhu et al., 2016; Doytch and Narayan, 2016; Wang, 2017; Balsalobre-Lorente et al., 2019; Mert and Caglar, 2020; Repkine and Min, 2020; Kisswani and Zaitouni, 2021; Polloni-Silva et al., 2021; and Chen et al., 2021.

The results indicate that Chinese FDI is more environmentally friendly in comparison to other countries' FDI in the host countries of BRI. This is explicit in some parts, since China is already one of the biggest investors in the global renewable energy sector. Moreover, in 2020, with the efforts to green the BRI project, Chinese renewable energy investments in other countries constituted the majority of its energy sector FDI (Ren21, 2021).

FDI is widely acknowledged as an important source of development and growth for underdeveloped and developing countries. This is because less developed countries often lack the necessary capital and technological expertise required to produce goods and services with higher value-added. Therefore, FDI is generally welcomed without question both in theory and policy. However, while unquestioned FDI investment may bring about growth opportunities, its cost may be higher in the long term due to pollution and other risks. As found in the estimation results, FDI from countries other than China seems that it might bring a pollution risk to the countries of BRI.

Therefore, to minimize the pollution risks that may arise due to incoming FDI, countries should implement dynamic and differentiated FDI policies and preventive measures. For example, a prior license mechanism for FDI flows might be beneficial to analyze the potential impact of foreign projects on the environment and pollution levels beforehand. With these sorts of differentiative mechanisms, environmentally pollutive projects may be discouraged and beneficial ones rewarded with various financial mechanisms not only before the project but also after the investment phase environmental effects of FDI projects may be observed so that it is ensured that foreign firms do not move to less costly and pollutive production technics and practices.

While attracting FDI projects in the less pollutive sectors is important, another important point is the transmission of know-how in the renewable energy sector and other production technologies regarding the environment. Technology transfer in cleaner production technics with spillover effects in the host country may strengthen sustainable development efforts.

For future studies, it is considered that if the required data sets are achieved, the empirical analysis of the relationship between foreign direct investments and air pollution in BRI countries on a sectoral basis will contribute to the development of the literature on this subject.

NOTE

1. Stata 17.0 and Gauss 18.0 econometrics programs are used in the estimation of the models defined in the study and in the empirical analysis of model variables.

REFERENCES

Ahmad, M., Jiang, P., Majeed, A., & Raza, M. Y. (2020). Does financial development and foreign direct investment improve environmental quality? Evidence from belt and road countries. *Environmental Science and Pollution Research*, 27(19), 23586–23601.

Akbostanci, E., Tunc, G. I., & Türüt-Aşik, S. (2007). Pollution haven hypothesis and the role of dirty industries in Turkey's exports. *Environment and Development Economics*, 12(2), 297–322.

Al-Mulali, U., & Tang, C. F. (2013). Investigating the validity of pollution haven hypothesis in the gulf cooperation council (GCC) countries. *Energy Policy*, 60, 813–819.

An, H., Razzaq, A., Haseeb, M., & Mihardjo, L. W. (2021). The role of technology innovation and people's connectivity in testing environmental Kuznets curve and pollution heaven hypotheses across the Belt and Road host countries: New evidence from Method of Moments Quantile Regression. *Environmental Science and Pollution Research*, 28(5), 5254–5270.

Aydın, F. (2018). *Political Economy of Belt and Road Project and Turkestan*, Gazi University Graduate School of Social Sciences Thesis.

Balsalobre-Lorente, D., Gokmenoglu, K. K., Taspinar, N., & Cantos-Cantos, J. M. (2019). An approach to the pollution haven and pollution halo hypotheses in MINT countries. *Environmental Science and Pollution Research*, 26(22), 23010–23026.

Baltagi, B. H. (2008). *Econometric Analysis of Panel Data*, 4th Edition. John Wiley & Sons.

Ben Kheder, S., & Zugravu-Soilita, N. (2008). The pollution haven hypothesis: A geographic economy model in a comparative study. *Journal of Cleaner Production*, 227, 724–738.

Blanchard, J.-M. F., & Flint, C. (2017) The geopolitics of China's maritime silk road initiative. *Geopolitics*, 22(2), 223–245. https://doi.org/10.1080/14650045.2017.1291503

Breusch, T. S., & Pagan, A. R. (1980). The Lagrange multiplier test and its applications to model specification in econometrics. *The Review of Economic Studies*, 47(1), 239–253.

Cai, X., Che, X., Zhu, B., Zhao, J., & Xie, R. (2018). Will developing countries become pollution havens for developed countries? An empirical investigation in the Belt and Road. *Journal of Cleaner Production*, 198, 624–632.

Carey, L., & Ladislaw, S. (2019). *Chinese Multilateralism and the Promise of a Green Belt and Road*. Center for Strategic and International Studies (CSIS).

Chen, Z., Paudel, K. P., & Zheng, R. (2021). Pollution halo or pollution haven: Assessing the role of foreign direct investment on energy conservation and emission reduction. *Journal of Environmental Planning and Management*. https://doi.org/10.1080/09640 568.2021.1882965

Cuiyun, C., & Chazhong, G. (2020). Green development assessment for countries along the belt and road. *Journal of environmental management*, 263, 110344.

Doytch, N., & Narayan, S. (2016). Does FDI influence renewable energy consumption? An analysis of sectoral FDI impact on renewable and non-renewable industrial energy consumption. *Energy Economics*, 54, 291–301.

Du, J., & Zhang, Y. (2018). Does one belt one road initiative promote Chinese overseas direct investment? *China Economic Review*, 47, 189–205.

Dunford, M., & Liu, W. (2019). Chinese perspectives on the Belt and Road Initiative. *Cambridge Journal of Regions, Economy and Society*, 12(1), 145–167.

Eberhardt, M., & Bond, S. (2009). "Cross-section dependence in nonstationary panel models: A novel estimator". *Munich Personal RePEc Archive, MPRA Paper No: 17692*. https://mpra.ub.uni-muenchen.de/17692/

El-Shafei, A. W., & Metawe, M. (2021). China drive toward Africa between arguments of neo-colonialism and mutual-beneficial relationship: Egypt as a case study. *Review of Economics and Political Science*, 7(2), 137–152.

Enderwick, P. (2018). The economic growth and development effects of China's One Belt, One Road Initiative. *Strategic Change, 27*(5), 447–454.

Fan, J. L., Da, Y. B., Wan, S. L., Zhang, M., Cao, Z., Wang, Y., & Zhang, X. (2019). Determinants of carbon emissions in 'Belt and Road Initiative' countries: A production technology perspective. *Applied Energy, 239*, 268–279.

Gong, M., Zhen, S., & Liu, H. (2021). Research on the nonlinear dynamic relationship between FDI and CO2 emissions in the "One Belt, One Road" countries. *Environmental Science and Pollution Research, 28*(22), 27942–27953.

Gu, A., & Zhou, X. (2020). Emission reduction effects of the green energy investment projects of China in belt and road initiative countries. *Ecosystem Health and Sustainability, 6*(1), 1747947. https://doi.org/10.1080/20964129.2020.1747947

Haggai, K. (2016). One Belt One Road strategy in China and economic development in the concerning countries. *World Journal of Social Sciences and Humanities, 2*(1), 10–14.

Harlan, T. (2021). Green development or greenwashing? A political ecology perspective on China's green Belt and Road. *Eurasian Geography and Economics, 62*(2), 202–226.

Huang, Y. (2016). Understanding China's Belt & Road initiative: Motivation, framework and assessment. *China Economic Review, 40*, 314–321.

Hurley, J., Morris, S., & Portelance, G. (2019). Examining the debt implications of the Belt and Road Initiative from a policy perspective. *Journal of Infrastructure, Policy and Development, 3*(1), 139–175.

Johnston, L. A. (2019). The Belt and Road Initiative: What is in it for China? *Asia & the Pacific Policy Studies, 6*(1), 40–58.

Khan, M. A., & Ozturk, I. (2020). Examining foreign direct investment and environmental pollution linkage in Asia. *Environmental Science and Pollution Research, 27*(7), 7244–7255.

Kisswani, K. M., & Zaitouni, M. (2021). Does FDI affect environmental degradation? Examining pollution haven and pollution halo hypotheses using ARDL modelling. *Journal of the Asia Pacific Economy.* https://doi.org/10.1080/13547860.2021.1949086

Lai, K. P., Lin, S., & Sidaway, J. D. (2020). Financing the Belt and Road Initiative (BRI): Research agendas beyond the "debt-trap" discourse. *Eurasian Geography and Economics, 61*(2), 109–124.

Lan, J., Kakinaka, M., & Huang, X. (2012). Foreign direct investment, human capital and environmental pollution in China. *Environmental and Resource Economics, 51*(2), 255–275.

Len, C. (2015). China's 21st Century Maritime silk road initiative, energy security and SLOC access. *Maritime Affairs: Journal of the National Maritime Foundation of India, 11*(1), 1–18. https://doi.org/10.1080/09733159.2015.1025535

Levinson, A., & Taylor, M. S. (2008). Unmasking the pollution haven effect. *International Economic Review, 49*(1), 223–254.

Li, M. (2020). The belt and road initiative: Geo-economics and Indo-Pacific security competition. *International Affairs, 96*(1), 169–187.

Li, W., & Hilmola, O. P. (2019). One belt and one road: Literature analysis. *Transport and Telecommunication, 20*(3), 260–268.

Li, X., Liu, C., Wang, F., Ge, Q., & Hao, Z. (2021). The effect of Chinese investment on reducing CO2 emission for the Belt and Road countries. *Journal of Cleaner Production, 288*, 125125.

Liu, H., & Kim, H. (2018). Ecological footprint, foreign direct investment, and gross domestic production: Evidence of belt & road initiative countries. *Sustainability, 10*(10), 3527.

Liu, H., Wang, Y., Jiang, J., & Wu, P. (2020). How green is the "Belt and road initiative"?– Evidence from Chinese OFDI in the energy sector. *Energy Policy, 145*, 111709.

Loke, B. (2018). China's economic slowdown: Implications for Beijing's institutional power and global governance role. *The Pacific Review, 31*(5), 673–691.

Ma, J., Zadek, S., Sun, T. Y., Zhu, S. Q., Cheng, L., Eis, J., . . . & Stumhofer, T. (2019). *Decarbonizing the Belt and Road: A Green Finance Roadmap.* Tsinghua University Center for Finance and Development, Vivid Economics and the Climateworks Foundation. www.climateworks.org/wp-content/uploads/2019/09/Decarbonizing-the-Belt-and-Road_report_final_lo-res.pdf.

Mahadevan, R., & Sun, Y. (2020). Effects of foreign direct investment on carbon emissions: Evidence from China and its Belt and Road countries. *Journal of Environmental Management, 276,* 111321.

Menyah, K., Nazlioglu, S., & Wolde-Rufael, Y. (2014). Financial development, trade openness and economic growth in African countries: New insights from a panel causality approach. *Economic Modelling, 37,* 386–394.

Mert, M., & Caglar, A. E. (2020). Testing pollution haven and pollution halo hypotheses for Turkey: A new perspective. *Environmental Science and Pollution Research, 27*(26), 32933–32943.

Muhammad, S., & Long, X. (2020). China's seaborne oil import and shipping emissions: The prospect of belt and road initiative. *Marine pollution bulletin, 158,* 111422.

Muhammad, S., & Long, X. (2021). Rule of law and CO2 emissions: A comparative analysis across 65 belt and road initiative (BRI) countries. *Journal of Cleaner Production, 279,* 1–12.

Muhammad, S., Long, X., Salman, M., & Dauda, L. (2020). Effect of urbanization and international trade on CO2 emissions across 65 belt and road initiative countries. *Energy, 196,* 1–15.

Neequaye, N. A., & Oladi, R. (2015). Environment, growth, and FDI revisited. *International Review of Economics & Finance, 39,* 47–56.

Pesaran, M. H. (2004). "General diagnostic tests for cross section dependence in panels". Available at SSRN 572504. https://doi.org/10.1007/s00181-020-01875-7

Pesaran, M. H. (2007). A simple panel unit root test in the presence of cross-section dependence. *Journal of Applied Econometrics, 22*(2), 265–312.

Pesaran, M. H., Ullah, A., & Yamagata, T. (2008). A bias-adjusted LM test of error cross-section independence. *The Econometrics Journal, 11*(1), 105–127.

Polloni-Silva, E., Ferraz, D., Camioto, F. D. C., Rebelatto, D. A. D. N., & Moralles, H. F. (2021). Environmental Kuznets curve and the pollution-halo/haven hypotheses: An investigation in Brazilian Municipalities. *Sustainability, 13*(8), 4114.

Rajah, R., Dayant, A., & Pryke, J. (2019). *Ocean of Debt? Belt and Road and Debt Diplomacy in the Pacific.* https://www.lowyinstitute.org/publications/ocean-debt-belt-road-debt-diplomacy-pacific

Ren21. (2021). *Renewables 2021 Global Status Report.* REN21 Secretariat. ISBN 978-3-948393-03-8.

Repkine, A., & Min, D. (2020). Foreign-funded enterprises and pollution halo hypothesis: A spatial econometric analysis of thirty Chinese Regions. *Sustainability, 12*(12), 5048.

Sadik-Zada, E. R., & Ferrari, M. (2020). Environmental policy stringency, technical progress and pollution haven hypothesis. *Sustainability, 12*(9), 3880.

Shahbaz, M., Loganathan, N., Zeshan, M., & Zaman, K. (2015). Does renewable energy consumption add in economic growth? An application of auto-regressive distributed lag model in Pakistan. *Renewable and Sustainable Energy Reviews, 44,* 576–585.

Shichor, Y. (2018). China's belt and road initiative revisited: Challenges and ways forward. *China Quarterly of International Strategic Studies, 4*(1), 39–53.

Singhania, M., & Saini, N. (2021). Demystifying pollution haven hypothesis: Role of FDI. *Journal of Business Research, 123,* 516–528.

Solarin, S. A., Al-Mulali, U., Musah, I., & Ozturk, I. (2017). Investigating the pollution haven hypothesis in Ghana: An empirical investigation. *Energy, 124,* 706–719.

Sun, H., Attuquaye Clottey, S., Geng, Y., Fang, K., & Clifford Kofi Amissah, J. (2019). Trade openness and carbon emissions: Evidence from belt and road countries. *Sustainability*, *11*(9), 2682, 1–20.

Tarı, R. (2010). *Ekonometri*. Umuttepe Yayınları.

Tatoğlu, F. Y. (2013). *İleri Panel Veri Analizi-Stata Uygulamalı*, 2. Second Edition, Beta Publishing house, ISBN10: 6052425946, İstanbul Turkey.

Taylor, M. S. (2005). Unbundling the pollution haven hypothesis. *Advances in Economic Analysis & Policy*, *4*(2).

Teo, H. C., Campos-Arceiz, A., Li, B. V., Wu, M., & Lechner, A. M. (2020). Building a green Belt and Road: A systematic review and comparative assessment of the Chinese and English-language literature. *PLoS ONE*, *15*(9), e0239009. https://doi.org/10.1371/journal.pone.0239009

Terzi, H., & Pata, U. K. (2019). Is the pollution haven hypothesis (PHH) valid for Turkey? *Panoeconomicus*, *67*(1), 93–109.

Wang, H., Dong, C., & Liu, Y. (2019). Beijing direct investment to its neighbors: A pollution haven or pollution halo effect? *Journal of Cleaner Production*, *239*, 118062.

Wang, S. (2017). Impact of FDI on energy efficiency: An analysis of the regional discrepancies in China. *Natural Hazards*, *85*(2), 1209–1222.

Westerlund, J. (2008). Panel cointegration tests of the fisher effect. *Journal of Applied Econometrics*, *23*(2), 193–233.

Zhu, H., Duan, L., Guo, Y., & Yu, K. (2016). The effects of FDI, economic growth and energy consumption on carbon emissions in ASEAN-5: Evidence from panel quantile regression. *Economic Modelling*, *58*, 237–248.

19 Improving the Resilience of Seaports to the Effects of Climate Change
Port Resilience Index as a Tool for Decision-Making

Estefanía Couñago-Blanco, Fernando León-Mateos,
Antonio Sartal, Lucas López-Manuel

19.1 INTRODUCTION

All the evidence indicates that the effects of climate change will unleash significant disruptions in our socio-economic systems over the coming decades, thereby posing new challenges for individuals and their communities across the globe (Howard-Grenville et al., 2014). Coastal communities are especially vulnerable owing to their direct exposure to climate change events, such as rising sea levels and extreme meteorological phenomena. Hence, there is a need to increase their resilience (Greenan et al., 2019).

To address this problem and respond to increasing maritime trade, port authorities have decided to increase investments in capacity and protection (Xia and Lindsey, 2021), because ports are particularly susceptible to the effects of climate change given their location. In global supply chains, ports play a key role as nodal transport infrastructures. Maritime transport accounts for approximately 80% to 90% total freight transport. However, research usually considers these infrastructures isolated elements rather than key facilities for the supply chain, its logistics system, and urban structures (Kontogianni et al., 2019; Panahi et al., 2020). Furthermore, these studies do not cover port systems in depth in resilience assessment or port connectivity analysis and most often focus only on large ports (Kontogianni et al., 2019).

In turn, operational managers also often take decisions individually, disregarding other stakeholders indirectly involved in the development of port operations. Thus, adaptation actions frequently correspond to partial solutions (McIntosh and Becker, 2017) without the support of these stakeholders because they do not identify with or feel represented in them (Panahi et al., 2020). The development of strategies and the implementation of practical measures remain considerable challenges for decision-makers despite the increasing need to implement adaptation measures (Panahi et al., 2020).

In a previous study (León-Mateos et al., 2021), we developed a port resilience index (PRI) to precisely address these shortcomings. Based on five dimensions of

DOI: 10.1201/9781032715438-19

resilience, governance, infrastructure and facilities, operational environment, risk management, and society, we determined that the operational resilience of Punta Langosteira (A Coruña) in 2019 was 51.80%. In the current study, we assessed the operational resilience of the port in 2021, showing the applicability of the PRI to facilitate decisions, both by managers and policymakers, with regard to increasing the resilience of ports to the effects of climate change.

The PRI proved key to increasing port resilience by highlighting the areas where investments were most needed after incorporating the perspectives of all relevant stakeholders. In only two years, the port's operational resilience reached 64%, significantly reducing the gap with the ideal benchmark of 80%. It is also worth noting that there has been a significant increase in operational resilience (more than 30%) with regard to the risk management dimension in only two years. Back in 2019, the performance of this dimension was as low as 36.95%. Without the PRI, it would have been difficult to identify the low performance of this dimension and carry out such improvements.

In the next section, we specify the concept of port resilience and how it can be measured. Next, we explain the methodology behind the PRI. Subsequently, we discuss the current state of the port of A Coruña and compare it with the situation in 2019. Finally, our results and conclusions highlight the validity the proposed indicator as a tool for measuring and improving port resilience, but also the comparison of the current situation with the baseline situation allows us to determine the effectiveness of the measures carried out.

19.2 PORT RESILIENCE AND ITS MEASUREMENT

19.2.1 IMPLICATIONS OF CLIMATE CHANGE ON INFRASTRUCTURE AND PORT OPERATIONS

TheIPCC defined climate change as

> a change in the state of the climate that can be identified by changes in the mean and/ or the variability of its properties and that persists for an extended period, typically decades or longer. Climate change may be due to natural internal processes or external forces such as modulations of the solar cycles, volcanic eruptions and persistent anthropogenic changes in the composition of the atmosphere or in land use.
>
> (IPCC, 2019a, p. 808)

Climate change is one of the main challenges facing society in our time (León-Mateos et al., 2021; Sagna et al., 2021). The IPCC definition reflects the existence of two types of factors that favor climate change: those of natural origin and those of anthropogenic origin (IPCC, 2018). We have abundant empirical evidence that since the beginning of the first Industrial Revolution, the latter type of factors have been altering the natural cycles of climate change (i.e. Şahin et al., 2019; Yoro and Daramola, 2020). According to available data, there is an exponentially increasing trend in the presence of greenhouse gases (GHGs) in the atmosphere (IPCC, 2018; Yoro and Daramola, 2020). This increase in GHGs is causing global warming, which results in a series of climate modifications that are already evident today and, according to forecasts, will become much more so in the future (IPCC, 2018; Lenderink et al., 2021).

This situation generates a range of effects on

lives; livelihoods; health and well-being; ecosystems and species; economic, social and cultural assets; services (including ecosystem services); and infrastructure, the risks of which derive from the interactions of climate-related hazards, exposure and vulnerability.

(IPCC, 2019a, p. 815)

In this context, the sixth IPCC report (2022) describes exposure as the presence of people; livelihoods; species or ecosystems; environmental functions, services and resources; infrastructure; or economic, social, or cultural assets in places and environments that could be adversely affected. Vulnerability, on the other hand, is understood as the susceptibility to being negatively affected. Some transport infrastructures that may be most affected by the effects of climate change are those related to maritime transport (Becker et al, 2013; EUROCONTROL, 2008). In this regard, it should be noted that ports are complex systems with numerous interacting components, so multiple factors can affect them (Asariotis et al., 2018), among which climate-related factors stand out (Economic Commission for Europe, 2013, 2019a). Seaports, given their location in coastal areas and estuaries, are susceptible to damage from sea level rise (SLR), increasing extreme weather events (i.e. surge storms, wave heights, heavy precipitation, sea-surface temperature, etc.), and their intensity, which could affect port protection infrastructure (Camus et al., 2019; Ng et al., 2019; Yang and Ge, 2020). In addition, the development of these activities is largely subject to wave characteristics, particularly the entry and exit of ships, but also ordinary port operations (Rusu and Soares, 2011), which in this context affect risk levels and economic productivity (Sierra et al., 2017).

This issue becomes particularly relevant if we consider that ports are gateways to various continents and form a key infrastructure in global supply chains (Christodoulou and Demirel, 2018). In the European Union alone, nearly three quarters of foreign trade and almost one fifth of domestic trade circulates by sea (EUROSTAT, 2022). Globally, this figure is 80%–90% of the volume of international trade in goods (León-Mateos et al., 2021). Despite the current scenario's uncertainty, an issue that implies moderation of the initially predicted growth, between 2022 and 2026, an annual rate of increase in world maritime trade volumes of +2.4% is expected (UNCTAD, 2021).

Although many existing port infrastructures incorporate in their design the capacity to cope with various hazards, the increased frequency and intensity of climatic events will accelerate their degradation (Christodoulou and Demirel, 2018) because many port infrastructures have been designed for current climatic conditions, so the potential effects of climate change are likely to affect them (Sánchez-Arcilla et al., 2016). Authors such as Christodoulou and Demirel (2018) have pointed out that if the IPCC's (2013) projections for SLR are met, 64% of seaports in the European Union could be flooded.

Leaving aside possible indirect consequences for ports (i.e. possible changes in population distribution and patterns of production, consumption and trade (Economic Commission for Europe, 2019b) and modification of coastal populations' socioeconomic activities (IPCC, 2019b)), it is estimated that climate change will have two main types of direct effects on ports: i) effects on the infrastructure itself and ii) effects on port operations. In terms of the infrastructure, issues such as rising sea levels, storm surges, and extreme precipitation can lead to flooding, which can compromise the port's structural integrity (IPCC, 2019b), affecting cargo terminals, inland roads, and areas of interconnection with other transport systems (USDOT, 2012). Flooding can also cause damage to the various electrical supply and tele-communications equipment as well as maritime traffic systems (Wright, 2013; Karagiannis et al., 2017). Also, extreme cold and heat waves can affect electronic equipment and deteriorate ports' paved areas (UNCTAD, 2020). In addition, phenomena such as increased salinity and acidification of the sea can endanger parts of the infrastructure, such as dikes, docks, and piers (McEvoy and Mullett, 2013).

In terms of operability, we must take into account that ports are systems integrating various interrelated components, such as ships, cranes, docks, and warehouses, that must work together in sequential processes that allow for the movement of cargo between producers and consumers (Ng and Liu, 2014). This synchronous operation implies that failure in any of the components leads to operational disruption of the port (Davydenko, 2019). Its operability can be compromised by sudden-onset events (i.e., precipitation and wind) and slow-onset events (i.e., temperature or SLR). Processes such as loading and unloading of vessels can be disrupted by extreme winds and precipitation, which impede the work of cranes and operators (McEvoy and Mullett, 2013). Storm surges can cause flooding that can paralyze port activity (Verschuur, 2020). These types of events as well as fog, wind, and extreme precipitation can hinder ship entry and exit tasks to and from the port (Zhao et al., 2021). Flooding, wind, rain, and fog can also disrupt internal transport and storage processes (McEvoy and Mullett, 2013). On the other hand, extreme temperatures (cold or heat waves) can compromise operators' safety and disrupt the port's connections with other transport systems (PIANC, 2020). Figure 19.1 presents a detailed description of the relationships between weather events, port infrastructure, and the various processes carried out therein.

In view of this situation, the need to increase ports' resilience, protect them, and adapt them for potential disruptions seems clear (Becker et al., 2013; Monios and Wilmsmeier, 2020). For this purpose, it is essential to involve stakeholders and policymakers, providing them tools that allow them to evaluate and improve the adaptation of their infrastructures to the new situation (Becker et al., 2018; Yang et al., 2018).

Currently, the effects of climate change are faced through two types of strategies: mitigation and adaptation. Although mitigation is meant to prevent and lessen the effects of climate change through a reduction in carbon emissions (Monios and Wilmsmeier, 2020; Nyong et al., 2007), adaptation strategies focus on acting on these phenomena's consequences. The ultimate goal of adaptation is to protect critical infrastructure through proactive responses (Ng et al., 2019).

Although these strategies reduce the climatic risks port infrastructures face (Jiang et al., 2021), ports commonly experience budgetary restrictions that prevent them from allocating resources to both strategies; therefore, they must prioritize one over the other (Gong et al., 2020). Investments in adaptation are encouraged, as the

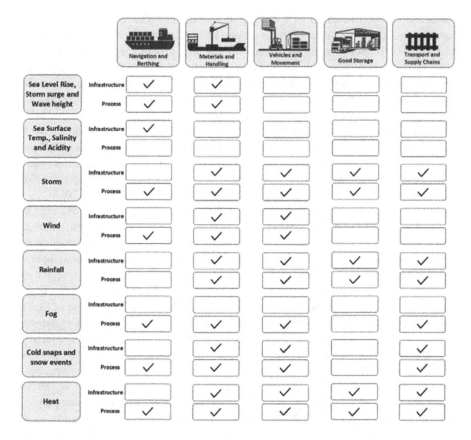

FIGURE 19.1 Relationships between weather events, port infrastructure, and port processes.
Source: Own elaboration based on McEvoy and Mullett (2013)

expected occurrence of adverse events increases, whereas the variation in the probability of occurrence causes the opposite effect. The fact that, in practice, knowledge about climate change and its possible impacts is scarce in terms of adaptation makes it clear that adaptation measures are much more difficult to implement than mitigation measures (Wang and Zhang, 2018).

The importance of mitigation for tackling climate change is unquestionable, but these strategies present a long-term path and may take decades to produce results. Therefore, adaptation measures must be employed to respond effectively to the effects of climate change in less time (Füssel and Klein, 2006). A proactive adaptation to such risks may be a much more profitable strategy than mitigation-based or reactive strategies (Pielke, 2007). Accordingly, Jiang et al. (2021) suggested that in a sector in which multiple complementary stakeholders converge, such as a supply chain, policymakers should focus their efforts on promoting adaptation strategies.

Considering the previous, this chapter focuses on the methodological design of a tool, the PRI, which can facilitate the implementation of strategies for port adaptation to the potential effects of climate change. In the following subsections, we delve into the concept of port resilience and discuss the methods for its measurement.

19.2.2 PORT RESILIENCE DEFINITION

The concept of resilience, which emerged in the field of ecology (Holling, 1973), has gradually spread to a very varied set of disciplines, including supply chains (Pettit et al., 2019; Hosseini et al., 2019). Resilience is used to assess the capacity of a system to maintain its functions in the face of adverse effects caused by external disruption, be it in the short term (Bruneau et al., 2003; Rose, 2004) or in the long term, as in the case of climate change (Djalante and Thomalla, 2011; Aldunce et al., 2015). It is precisely in this field of climate change where the use of the concept of resilience has grown exponentially in recent decades (Woodruff et al., 2018). Several organizations have pointed out the importance of adaptation to the effects of climate change through the resilience of systems. In addition, they have adopted very similar definitions of resilience. The Intergovernmental Panel on Climate Change, for example, defines resilience as

> The ability of a system and its component parts to anticipate, absorb, accommodate, or recover from the effects of a hazardous event in a timely and efficient manner, including through ensuring the preservation, restoration, or improvement of its essential basic structures and functions.
>
> (IPCC, 2012, p. 563)

The Report of the World Conference on Disaster Risk Reduction in Kobe, Hyogo, Japan, defines resilience as

> The capacity of a system, community or society potentially exposed to hazards to adapt, by resisting or changing in order to reach and maintain an acceptable level of functioning and structure. This is determined by the degree to which the social system is capable of organizing itself to increase this capacity for learning from past disasters for better future protection and to improve risk reduction measures.
>
> (UN-ISDR [United Nations International Strategy for Disaster Reduction], 2005, p. 4)

Both the UN-ISDR and the IPCC, in addition to referring to the ability to withstand and recover quickly from interruptions, also include in their definitions the need for systems to prepare for and adapt to changing conditions. This concept has led to the term "operational resilience", highlighting the need to "make the system better able to absorb the impact of an event without losing the capacity to function" (Alderson et al., 2015, p. 562). We adopted this operational resilience approach in our study. This will allow us to evaluate the capacity of a system to adapt to changes in a dynamic and proactive way, with the participation of all stakeholders (Kamalahmadi and Parast, 2016). According to Alderson et al. (2015) this type of capacity allows systems to absorb the impact of adverse events, such as those caused by climate change, without compromising the operational capacity of port infrastructures.

Verschuur et al. (2020) indicated that interruptions in port processes show marked differences depending on the port and the type of event. Furthermore, the dynamics of these interruptions is determined by the different actors involved in minimizing their consequences throughout the supply chain. These authors identified the following main logistics strategies: i) recovering production by increasing cargo once the operation has recovered and ii) replacing the port by fully or partly diverting the cargo to (an)other port(s). The observation and study of these interruptions provide key information to improve risk management and assess the scope of the potential recovery capacity of the port and its network.

However, the strategy of investing in ports to prevent disasters has not been widely adopted, due both to its complexity and the enormous underlying challenges it poses. Considering this limitation, and given the fact that some disturbances are clearly unavoidable, organizations must learn to adapt their operational routines and procedures to increase their operational resilience (Hohenstein et al., 2015; Scholten et al., 2019).

Xiao et al. (2015) suggest that, by making the appropriate investments, it is possible to prevent and mitigate the damage caused by disasters. In turn, Panahi et al. (2020) highlighted that studies on this topic are based on hypothetical cases and that practical solutions should be found as a preliminary step to address other problems such as financial barriers. In addition, they emphasized the need for empirical research focused on designing pragmatic and applicable strategic solutions geared towards adaptation to climate change. Bie et al. (2017) suggested that proactive strategies for vulnerability reduction are much more cost effective and feasible than strategies focused on the recovery of affected systems. This authors concluded that a roadmap to improve the resilience of a system should combine strengthening and operational resilience strategies to optimize investments.

In line with these authors' perspective regarding the need to seek solutions for adaptation to the potential effects of climate change in port operations, we believe that only measurable parameters can be improved. Consequently, the first step to develop a port adaptation strategy is to design and implement a measurement system that enables stakeholders to determine both the initial status of the port in relation to operational resilience to climate change and its evolution with the implementation of measures for its improvement. Establishing a global resilience metric is difficult given the different definitions of resilience found in the academic literature, including the factors that contribute to resilience and how these factors can be evaluated and improved upon (Klein et al., 2003; Grafton et al., 2019).

Considering these limitations, among the existing alternatives to conduct these measurements, we opted for a composite index (see methodology in Section 19.3). In the following subsection, we explain the reasons that led us to this choice.

19.2.3 PORT RESILIENCE MEASUREMENT METHODS

A composite index is made up of the mathematical aggregation of a series of variables, called indicators. These indicators allow summarizing the characteristics of a system, community, or society with respect to the target parameter (Salvati and Carlucci, 2014; McIntosh and Becker, 2017). These types of indicators are commonly used to assess the magnitude of risk exposure derived from various threats, such as the social vulnerability index (SOVI), disaster risk index (DRI), and coastal infrastructure vulnerability index (CIVI). The application of these indices makes it easier to make information accessible to decision makers regardless of their level of expertise (Yoon et al., 2016).

Furthermore, because these indicators help operationalize the observable variables of systems, an evaluation method based on such indicators is useful when the measured concepts are not directly quantifiable in nature (McIntosh and Becker, 2019). This is precisely the case of concepts such as resilience so that, although it is not possible to measure it directly, it can be operationalized "by mapping them to functions of observable variables called indicators" (McIntosh and Becker, 2017, p. 210). However, it should be taken into account that this method also has weak points, because it is possible that some relevant and necessary information for a correct evaluation may go unnoticed (Davis et al., 2015). These weaknesses result from negotiations between stakeholders, who may be biased towards the most powerful interest groups. They may also lead to overconfidence in a single variable (with its shortcomings in explaining complex relationships), while others with the same or even greater importance are overlooked. In addition, the complementarity of mixed variables, such as those related to recovery strategies after an interruption, hinders both the evaluation of the adequacy of strategies for improving port resilience and the supply chain and the determination of their current resilience level (Davis et al., 2015; Verschuur et al., 2020).

In order to alleviate these aspects, we have designed a composite index in which the participation of the different port stakeholders has been incorporated, in line with Greco et al. (2019), who indicated that including stakeholders in the process of preparing composite indicators ensures robustness. In addition, the participation of different stakeholders is associated with the success of composite indicator initiatives (Mickwitz and Melanen, 2009) in overcoming barriers to the development of indicators and implementation processes (Geng et al., 2012). Furthermore, involving the stakeholders in the design of the index promotes their commitment to subsequent tasks aimed at increasing resilience (Mojtahedi and Oo, 2014; Cookey et al., 2016).

Notwithstanding the importance of port infrastructure due to its key role in the supply chain, its contribution to the economy (not only in coastal areas but also globally), and the growing interest in this field, the number of studies focused on determining its interrelation with the possible effects derived from climate change remains limited. Furthermore, many of these studies do not go beyond theory or maintain an incomplete perspective in which only isolated climatological elements are taken into account. However, the way to obtain a more realistic view of the factors that determine the degree of vulnerability/adaptability of the port to climate change is to include as many potential impacts as possible in its assessment (McIntosh and Becker, 2019).

In this sense, we identified contributions in terms of port resilience both at the community level (Mayunga, 2007; Summers et al., 2017) and among coastal communities (Orencio and Fujii, 2013; Smith et al., 2019), but only a few studies have met the need to develop indicators focused on port infrastructure (Laxe et al., 2017). Izaguirre et al. (2021) determined a series of threats to global operations in ports, such as disturbances in storm surge, rainfall, wind, or waves. Monioudi et al. (2018) offered a methodology to determine the minimum thresholds necessary to maintain the operability of port infrastructures. Verschuur et al. (2020) established metrics for projecting the scope of interruptions in port operations and their recovery capacity, whereas Dui et al. (2021) proposed a new method for optimizing the management of port resilience and maritime transport routes. Asadabadi and Miller-Hooks (2020) proposed models for assessing and improving the resilience of port systems; Zhang et al. (2020) measured resilience in network systems using different approaches. In turn, Chen et al. (2020) designed a resilience assessment model based on the cost structure of the supply chain susceptible to interruptions. Dhanak et al. (2021) proposed a qualitative and quantitative predictive resilience assessment tool using a microscopic traffic flow simulation model (VisSiM). Liu and Chen (2021) developed an integrated system that helps decision makers, in situations of uncertainty, to assess resilience under different risk situations. In turn, Kara et al. (2020) developed a mixed-methods approach, consisting of using a cognitive map of interrelationships between climate change and supply chain performance management designed by surveying managers from various industries, and a system dynamics (SD) model to assess the cascading effects of climate change risk on supply chain performance. Mutombo and Ölçer (2017) evaluated the "exposure to climate" risk of port infrastructures using a form composed of a matrix that combines potential extreme weather events with the processes involved in port operations for identifying high-risk scenarios. Last, Izaguirre et al. (2020, 2021) proposed a multilevel method and cluster analysis to study risk components and the risk itself using clustering indicators, respectively. This method provides useful information to stakeholders and policymakers to identify the most vulnerable areas to the risks derived from climate change analyzed and to design appropriate adaptation strategies.

In general, disaster resilience is measured using semi-quantitative approaches (Hosseini et al., 2016) for preparing composite indicators summarizing complex or multidimensional characteristics of a community of infrastructure. Accordingly, some of the main limitations indicated by different authors are summarized as follows.

i. Most of the evaluations carried out are qualitative in nature; hence, they can serve as a guide to define resilient systems but their description is insufficient for policymakers who need an explanation to maximize efficient resource allocation (Cutter et al., 2014).

ii. Quantitative evaluations of complex and highly connected systems often employ inadequate and/or unspecified assumptions of structures and characteristics (Dessavre et al., 2016).

iii. Another weak point noted by both academics and policymakers is that a large part of these indicators represent stakeholders in a partial way. Some authors, including Bryson (2004) and Few et al. (2007), point out that the inclusion of the perspectives of the different stakeholders regarding the development of resilience in general is a very important aspect to consider

iv. Last, another main problem of many resilience indicators is that they are not validated. Previous studies point to the importance of this, but failed to test a method to assess whether quantitative assessment results can truly show a system's preparedness for adverse events (i.e., Mayunga, 2007).

In order to overcome these limitations, we designed a renewed method for the elaboration of a quantitative PRI that includes all the agents involved in port activity.

19.3 DESCRIPTION OF METHODOLOGY EMPLOYED TO DEVISE PRI

The PRI was developed with the participation of all stakeholders towards providing both port managers and policymakers with a tool for facilitating decision-making and implementing adaptation measures (León-Mateos et al., 2021).

With this idea in mind, a composite index with indicators was developed, since this option allows valuable information to be displayed in a parsimonious manner, facilitating its understanding by both professionals and non-professionals (Yoon et al., 2016). The proposed methodology is structured as follows (Figure 19.2).

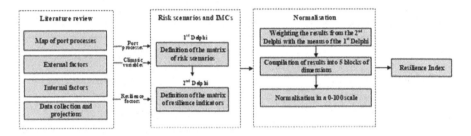

FIGURE 19.2   PRI design.

Source: León-Mateos et al. (2021)

From the perspectives of port stakeholders (both internal and external), we were able to select three sets of parameters for the design of the index: (i) those related to processes for maintaining port operations, (ii) parameters associated with ocean-meteorological issues affecting the previous, and (iii) resilience factors that could moderate the impact of the selected climate and ocean-meteorological elements. Thus, the literature review also allowed us to establish the five dimensions that comprise our resilience index: infrastructure and facilities, governance, operating environment, society, and risk management (Figure 19.3).

After preparing a matrix of risk scenarios consisting of intersections between climatic and ocean-meteorological elements and port processes, in the first Delphi session, we presented this matrix to a group of experts, who assessed, in three Delphi rounds, the different risk scenarios, using a Likert scale ranging from 0 to 3 (where 0 = no risk, 1 = low risk, 2 = medium risk, and 3 = high risk). At the end of the three rounds, the scenarios that passed the following two filters were selected:

i. The average score was equal to or higher than 2, that is, the most important operational risk scenarios for the experts.
ii. Scenarios with a coefficient of variation equal to or less than 0.50 to ensure broad consensus.

The results of the first Delphi allowed us to obtain the different risk scenarios. After developing a new relationship matrix in which we cross the higher risk scenarios with the resilience factors, grouped by the different resilience dimensions, we presented

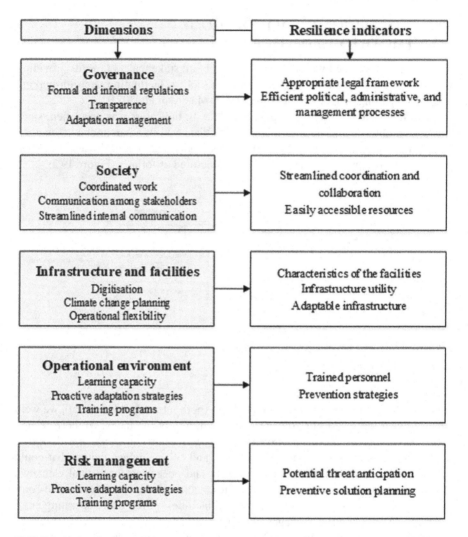

FIGURE 19.3 Conceptual framework of the dimensions of port operational resilience to climate change.

Source: León-Mateos et al. (2021)

this second matrix, in the second Delphi, to a group of experts, who again evaluated this matrix using a Likert scale with the same parameters as those of the first round of consultations for our model. The comparison between risk scenarios versus resilience factors allows us to obtain the "impact moderation coefficients" (IMCs). The IMCs allow us to quantitatively assess the contribution of each dimension of resilience to increased port adaptability.

The IMCs were then weighted according to the importance given by the experts to each risk scenario (mean score from the first Delphi). The mean score of these weighted IMCs was obtained for each resilience factor (matrix columns) in order to determine the impact value of each of the factors (a), as shown in Table 19.1.

We grouped the values obtained according to the resilience dimension to which they belong. We proceeded to add them in order to calculate their individual value (b). Then, we divided the impact values for each resilience factor obtained (a) by the added value of the resilience dimension to which each factor belonged (b). This allowed us to determine the weighted value of each resilience factor (c).

Averaging the resulting values (c), we obtained the impact value for each of the five resilience dimensions (d). The aggregation of these impact values for the five dimensions allowed us to calculate the sum of average of dimensions (e). Finally, we proceeded to divide the impact value of each of the resilience dimensions (d) by the sum of average of dimensions (e). In this way, we obtained the weighted value for each of the resilience dimensions (f).

TABLE 19.1
IMC STANDARDIZATION AND CALCULATION

	Dimension 1				...	Dimension n			
	Resilience factor1	Resilience factor2	Resilience factor 3	Resilience factor4	...	Resilience factor n_1	Resilience factor n_2	Resilience factor n_3	Resilience factor n_4
Weighted scenario 1	$\bar{X}CMI_{11}$	$\bar{X}CMI_{12}$	$\bar{X}CMI_{13}$	$\bar{X}CMI_{14}$	...	$\bar{X}CMI_{1n_1}$	$\bar{X}CMI_{1n_2}$	$\bar{X}CMI_{1n_3}$	$\bar{X}CMI_{1n_4}$
Weighted scenario 2	$\bar{X}CMI_{21}$	$\bar{X}CMI_{22}$	$\bar{X}CMI_{23}$	$\bar{X}CMI_{24}$	...	$\bar{X}CMI_{2n_1}$	$\bar{X}CMI_{2n_2}$	$\bar{X}CMI_{2n_3}$	$\bar{X}CMI_{2n_4}$
Weighted scenario 3	$\bar{X}CMI_{31}$	$\bar{X}CMI_{32}$	$\bar{X}CMI_{33}$	$\bar{X}CMI_{34}$	...	$\bar{X}CMI_{3n_1}$	$\bar{X}CMI_{3n_2}$	$\bar{X}CMI_{3n_3}$	$\bar{X}CMI_{3n_4}$
Weighted scenario 4	$\bar{X}CMI_{41}$	$\bar{X}CMI_{42}$	$\bar{X}CMI_{43}$	$\bar{X}CMI_{44}$	...	$\bar{X}CMI_{4n_1}$	$\bar{X}CMI_{4n_2}$	$\bar{X}CMI_{4n_3}$	$\bar{X}CMI_{4n_4}$
Weighted scenario 5	$\bar{X}CMI_{51}$	$\bar{X}CMI_{52}$	$\bar{X}CMI_{53}$	$\bar{X}CMI_{54}$	...	$\bar{X}CMI_{5n_1}$	$\bar{X}CMI_{5n_2}$	$\bar{X}CMI_{5n_3}$	$\bar{X}CMI_{5n_4}$
Weighted scenario 6	$\bar{X}CMI_{61}$	$\bar{X}CMI_{62}$	$\bar{X}CMI_{63}$	$\bar{X}CMI_{64}$	...	$\bar{X}CMI_{6n_1}$	$\bar{X}CMI_{6n_2}$	$\bar{X}CMI_{6n_3}$	$\bar{X}CMI_{6n_4}$
...	...	...	...	...	...	...	...	...	...
Weighted scenario n	$\bar{X}CMI_{n_1}$	$\bar{X}CMI_{n_2}$	$\bar{X}CMI_{n_3}$	$\bar{X}CMI_{n_4}$	...	$\bar{X}CMI_{nn_1}$	$\bar{X}CMI_{nn_2}$	$\bar{X}CMI_{nn_3}$	$\bar{X}CMI_{nn_4}$

(a) Mean	$\bar{X}_{FR1}$	$\bar{X}_{FR2}$	$\bar{X}_{FR3}$	$\bar{X}_{FR4}$	...	$\bar{X}FR_{n_1}$	$\bar{X}FR_{n_2}$	$\bar{X}FR_{n_3}$	$\bar{X}FR_{n_4}$
(b) Sum of dimensions	$S_1 = \sum \bar{X}_{FR1}..\bar{X}_{FR4}$				...	$S_n = \sum \bar{X}_{FRn_1}..\bar{X}_{FRn_4}$			
(c) Weighting of factors	$\bar{X}_{FR1}/S_1$	$\bar{X}_{FR2}/S_1$	$\bar{X}_{FR3}/S_1$	$\bar{X}_{FR4}/S_1$	...	$\bar{X}_{FRn_1}/S_n$	$\bar{X}_{FRn_2}/S_n$	$\bar{X}_{FRn_3}/S_n$	$\bar{X}_{FRn_4}/S_n$

(d) Average score of each dimension	$\bar{X}_{B1} = \bar{X}(\bar{X}_{FR1}.\bar{X}_{FR4})$	...	$\bar{X}_{Bn} = \bar{X}(\bar{X}FR_{n_1}..\bar{X}FR_{n_4})$
(f) Weighting of dimensions	$P_1 = \bar{X}_{B1}/S_T$	...	$P_n = \bar{X}_{Bn}/S_T$

(e) Sum of averages of dimensions
$S_T = \sum \bar{X}_{B1}..\bar{X}_{Bn}$

After these calculations, the PRI model follows the equation:

$$IRP = \beta_1 Dm_1 + \beta_2 Dm_2 + \beta_3 Dm_3 + \ldots + \beta_n Dm_n \tag{19.1}$$

where $\beta_1, \beta_2, \ldots \beta_n$ = weights determined after weighting and normalizing the results from the second Delphi.

$Dm_1, Dm_2, \ldots, Dm_n$ = values that stakeholders attach to the resilience dimensions according to a self-assessment questionnaire in which they were asked to evaluate port performance based on the IMCs.

19.4 PRI VALIDATION AS A TOOL TO IMPROVE RESILIENCE IN SEAPORTS

19.4.1 Previous Results

The outer port of A Coruña in Punta Langosteira is located at the confluence of the Atlantic Ocean with the Cantabrian Sea (Figure 19.4). This port operates under the "landlord port" model. In such ports, land ownership and, in specific cases, the development of basic infrastructures, is a public responsibility. In contrast, the remaining infrastructures, services, and port operations (especially those related to cargo handling) are managed by private companies through licensing (Notteboom, 2006). Run by the Port Authority of A Coruña (*Autoridad Portuaria de A Coruña*—APAC), this port focuses its activity on solid and liquid bulk cargo. The case study of this port was selected due to it was ideal for our

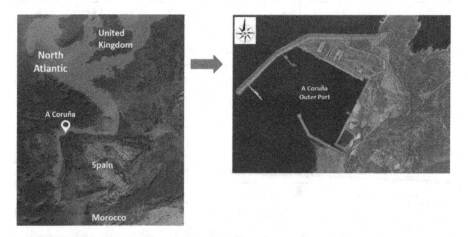

FIGURE 19.4 Location of the outer port in A Coruña.

Source: León-Mateos et al. (2021)

research questions considering its particular characteristics. This port not only conducts comprehensive climatic and ocean-meteorological data collection but also has unique characteristics, which are critical for studying the effects of climate change:

i. This port is part of the main European transport arteries because it is included, as a nodal point, in the Trans-European Transport Network (TEN-T).
ii. It is an outer port, with inherently adverse conditions and lacking natural protection against oceanic and meteorological effects.
iii. It is located on the European western coast, which, according to European Union predictions, will be a tipping point for climate change; this region faces not only an increased risk of flooding due to rising sea levels but also a possible increase in storm surges (European Environment Agency, 2017). According to Izaguirre et al. (2021), ports in northern Spain are exposed to strong waves and "are impacted primarily by agitation in the navigation zone and by heavy rain".
iv. The infrastructure was recently built (2012) and has a projected useful life of 50 years (Puertos del Estado, 2012). Consequently, this infrastructure will most likely be affected by the potential adverse effects of climate change.

The outer port of A Coruña is therefore a valid case study to assess the negative impact associated with climate change. Thus, once selected in September 2018, we applied the previously described method to measure the operational resilience of the port according to Figure 19.5.

In the first prospective phase, we identified the following elements:

• All port stakeholders, relying on the advice of the APAC.
• The port processes based on the master map of processes of the outer port of A Coruña, selecting those that were key to continuing port operations and most likely to be affected by the selected climatic elements.
• The main resilience factors for port infrastructures after reviewing the literature and heeding the advice of experts from the Marine Technological Center (*Centro Tecnológico del Mar*—CETMAR), the Regional Meteorological Agency of Galicia (Meteogalicia), and the Water and Environmental Engineering Research Group (*Grupo de Enxeñaría da Agua e do Medio Ambiente*—GEAMA) of the University of A Coruña (*Universidade da Coruña*—UDC).

After completing the first phase of identification and selection of variables, following the Delphi method, we asked to different experts to evaluate the matrix formed by the port processes and the selected oceanic-meteorological elements based on their experience. The three rounds of this first Delphi were held in early 2019, with a 73.72% participation rate, identifying a total of 13 medium- and high-risk scenarios.

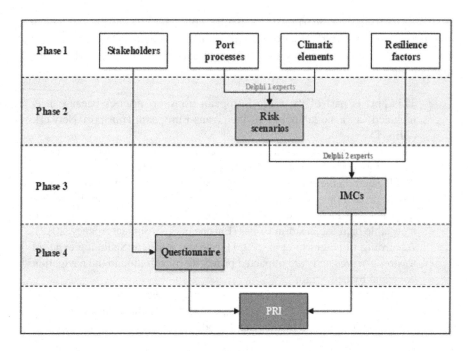

FIGURE 19.5   Preparation phases of the PRI for the outer port of A Coruña.

Source: The authors, based on León-Mateos et al. (2021)

Subsequently, in phase 3, using the Delphi method, we shared a new matrix consisting of risk scenarios and resilience factors with the 22 aforementioned experts. On this occasion, the participation rate was 86.36%. The experts assessed the moderating effect of the different resilience factors identified in phase 1. These assessments enabled us to obtain a matrix of mean scores for the IMCs and to identify the key resilience factors that must be targeted to ensure the operational resilience of the port.

In the last phase, we developed a questionnaire for self-assessment of the performance level in terms of the resilience factors of the port, where the participating stakeholders were asked to evaluate the current situation of the port in terms of the five dimensions of resilience selected (see Table 19.2).

All these calculations allow us to establish the PRI for the outer port of A Coruña (Eq. 19.1) as described in Section 19.3; we obtained the following result of the first measurement performed in 2019:

$$PRI = 19.5 * 65.06\% + 19.3 * 61.94\% + 21.7 * 48.15\% + 20.2 * 47.47\%$$
$$+19.3 * 36.95\% = 51.80\%$$

TABLE 19.2
Estimated Results and Stakeholder Questionnaire Coding

	Weighting	Current state[1]
Governance	19.50%	65.06%
Society	19.30%	61.94%
Infrastructure and facilities	21.70%	48.15%
Operational environment	20.20%	47.47%
Risk management	19.30%	36.95%

(1) Above a maximum performance potential of 100% for each of the dimensions; the values were determined after coding the questionnaire administered to port stakeholders (available on request).

19.4.2 ASSESSMENT OF CURRENT STATE AND COMPARISON BETWEEN INITIAL AND CURRENT SITUATION OF OUTER PORT OF A CORUÑA (SPAIN)

Since the first operational resilience assessment was performed using the PRI in 2019, the APAC has implemented a series of measures that are likely to affect the resilience of the outer port of A Coruña. Among these measures, the following stand out:

- Design of a "Green Port" strategic plan, in agreement with the growing demand of society for an evolution towards green supply chains (Garza-Reyes et al., 2015; López-Manuel et al., 2020), to promote the installation of industries linked to the generation of renewable energies within the port infrastructures. Its objective is to make the Outer Port of A Coruña in 2025 the first self-sustaining Spanish port in terms of energy. In addition, this strategy will serve to alleviate the fall in port traffic triggered by decarbonization, as well as to generate mitigation and adaptation measures to the potential effects of climate change. The first steps that have been carried out are detailed as follows:
  - Agreement with the multinational company Industria de Diseño Textil, S.A. (Inditex), for the installation of three wind turbines within the perimeter of the outer port of A Coruña. In exchange for the transfer of the space, the company finances the new infrastructures that will supply both the port (including the industries established there) and the nearby Inditex facilities.
  - Agreement with the renewable energy company Enerfín to exploit part of the energy generated through the aforementioned agreement with Inditex for the production of green hydrogen.
  - Project, in the study phase, for the installation of an ammonia (NH_3) ecological production plant in the port. According to Bird et al. (2020), the production of green ammonia offers several options (i.e., energy storage, carbon-free fuel) to foster the transition to a net-zero carbon emissions paradigm without generating waste that is harmful to the environment.
- Improvement of prediction models for overtaking and long waves, port operability during agitation events, etc.

- Taking the first steps to using offshore wind energy, thereby advancing the diversification of energy sources towards goods less sensitive to possible alterations resulting from climate change.

In addition, since January 2021, all energy has been entirely supplied by renewable sources, albeit, for now, still derived from the electrical grid. Nevertheless, as mentioned, measures focused on reducing energy dependence continue to be implemented.

To assess how the resilience performance of the outer port of A Coruña evolved after applying these measures, we conducted a second evaluation using the PRI. The method used was the same as that applied in the first measurement to render the results comparable. We used the same variables and weights of the IMCs as in the first evaluation for two reasons: i) according to the literature, climate change has long-term effects. Accordingly, no relevant variations have been identified in either risk scenario (port processes and climatic and oceanic-meteorological factors) or resilience factors because only three years have elapsed since the first measurement, and ii) using the same model would allow us to establish robust comparisons between both evaluations. Therefore, in October 2021, the last phase of the method, consisting of the self-assessment questionnaire of port performance in terms of resilience, was reproduced to determine the impact of introducing adaptation measures on the resilience of the outer port during this period.

TABLE 19.3
Response Rate to the Resilience Assessment Questionnaire by Stakeholders (Grouped by Categories) in the Outer Port of A Coruña

	Secondary Sector	Tertiary Sector	Social Environment	Political and Administrative Environment	Scientific Environment	Total
All stakeholders (number/%	4 (6.6%)	36 (59.0%)	4 (6.6%)	12 (19.6%)	5 (8.2%)	61 (100%)
Emails sent	4	36	4	12	5	61
Responses received (number/%)	4 (7.0%)*	33 (57.9%)*	4 (7.0%)*	11 (19.3%)*	5 (8.8%)*	57 (100%)*

* No significant differences were found between the distribution of stakeholders and the sample of our study.

Note: Although the list of stakeholders of the port of A Coruña includes the primary sector, this sector was not included in the study. This port has two locations: the outer and the inner port, and our study was conducted in the former (with the greatest impact due to climate change), whereas all fishing activities are entirely conducted in the latter.

Table 19.3 shows the percentages of responses. Furthermore, according to the Kruskal-Wallis test there are no significant differences between the subgroups. In addition, participants were asked to rate (Likert scale 1–10) the port's situation in the five defined dimensions of resilience.

The results from this process, after coding the responses to the questionnaire, together with the IMCs calculated in the previous phases, are outlined in Table 19.4.

From these data, the PRI for the outer port of A Coruña could be determined (see Eq. 19.1) with the following result:

$$PRI = 19.50 * 68.49\% + 19.30 * 68.61\% + 21.70 * 58.13\% + 20.20 * 57.52\%$$
$$+ 19.30 * 68.23\% = 64.00\%$$

It should be noted at this point that a yield of 80% is considered optimal, since higher values would be difficult to achieve from an economic point of view. Taking this into account, the IRP of our case is 16 percentage points below the optimal situation.

The PRI also makes it possible to evaluate the relative importance of the different dimensions studied. Accordingly, we found that the dimensions with the highest potential for improving resilience to climate change are the operational environment and infrastructure and equipment. A comparison between these results and those assessed in the first evaluation (Table 19.5) shows that the two aforementioned dimensions, together with risk management (identified as most vulnerable in the first measurement), have experienced the highest degree of improvement, particularly risk management (31.28%).

TABLE 19.4
Estimated Results and Stakeholder Questionnaire Coding

	Weighting	Current State[1]
Governance	19.50%	65.49%
Society	19.30%	68.61%
Infrastructure and facilities	21.70%	58.13%
Operational environment	20.20%	57.52%
Risk management	19.30%	68.23%

(1) Above a maximum performance potential of 100% for each of the dimensions. The values were calculated by coding the responses to the questionnaire administered to the port stakeholders (available on request).

TABLE 19.5
Comparison of Estimated Results and Stakeholder Questionnaire Coding

	Weighting	State in 2019[1]	Current State[1]
Governance	19.50%	65.06%	68.49%
Society	19.30%	61.94%	68.61%
Infrastructure and facilities	21.70%	48.15%	58.13%
Operational environment	20.20%	47.47%	57.52%
Risk management	19.30%	36.95%	68.23%

(1) In relation to a maximum performance potential of 100% for each dimension. The values were determined after coding the questionnaire administered to port stakeholders (available on request).

Last, based on the PRI results, after the second evaluation, the level of resilience of the outer port of A Coruña improved by 12.19 percentage points over the measurement performed in 2019. As such, during the approximately two and a half years between the two measurements, the PRI moved 15.3% towards the ideal situation (PRI = 80%).

19.5   CONCLUSIONS

Climate change is one of the main challenges facing humanity today (Howard-Grenville et al., 2014), where ports are particularly critical environments due not only to their high susceptibility to the adverse problems involved (i.e. sea level rise, extreme weather events, etc.) but also because of their key impact in worldwide value chains. It should not be forgotten that ports play a fundamental role as nodal transport infrastructures.

Faced with this situation, port managers have been working for some time on two key issues to solve this problem (Panahi et al., 2020; Xia and Lindsey, 2021): on the one hand, looking for the most appropriate tools with which to determine the resilience of ports to the challenges posed by climate change, and on the other hand, identifying those indicators that make it possible to anticipate possible problems, ensuring the operation of ports in the best conditions. In short, they are making ports as resilient as possible to climate change (Greenan et al., 2019).

It is curious, however, how little research attention has been paid to these infrastructures or to port systems, in general, in terms of resilience assessment (Kontogianni et al., 2019). It is also noteworthy that, on many occasions, planners approach these types of issues by attending to only part of the stakeholders involved in port operations. Consequently, it is very frequent that the adaptation actions undertaken correspond to partial solutions (McIntosh and Becker, 2017) and do not obtain the support of those stakeholders who do not feel identified with the initiatives carried out (Panahi et al., 2020).

Aware of this situation, we propose in this chapter an integrative approach that allows further progress in this field of research. Following the theoretical approaches and methodology of León-Mateos et al. (2021), we use the port resilience index (PRI) developed by these authors to define the current value of resilience of the port of Punta Langosteira (A Coruña), and we analyze its evolution over time based on data published in 2019. Our work allows us, therefore, not only to validate the proposed indicator as a tool for measuring and improving port resilience, but also, the comparison of the current situation with the baseline situation allows us to determine the effectiveness of the measures carried out (i.e., gauging meters, collaboration with stakeholders to adopt measures that contribute to the port's energy self-sufficiency, first steps to become a green port, etc.).

The results are very enlightening. In just two years, the port's operational resilience reached 64%, significantly reducing the gap with the ideal benchmark of 80%. It is also worth noting that there has been a significant increase in operational resilience (over 30%) in the risk management dimension in this same period. Thus, the level of resilience of the outer port of A Coruña shows an improvement of more than 12 percentage points (an increase of 15.3%) with respect to 2019. We can conclude, therefore, that the PRI proved key to increasing port resilience by highlighting the areas with the greatest potential for improvement and the greatest contribution to global resilience after incorporating the perspective of all relevant stakeholders. In view of these results, it seems right to continue to insist on this type of initiatives which, without requiring significant costs, could contribute very positively to raising each of the dimensions analyzed to reach the recommended 80% PRI target value.

REFERENCES

Alderson, D., Brown, G., and Carlyle, W. (2015). Operational models of infrastructure resilience. *Risk Analysis*, 35(4), 562–586.

Aldunce, P., Beilin, R., Howden, M., and Handmer, J. (2015). Resilience for disaster risk management in a changing climate: Practitioners' frames and practices. *Global Environmental Change*, 30, 1–11.

Asadabadi, A., and Miller-Hooks, E. (2020). Maritime port network resiliency and reliability through co-opetition. *Transportation Research Part E: Logistics and Transportation Review*, 137, 101916.

Asariotis, R., Benamara, H., and Naray, V. (2018). Port industry survey on climate change impacts and adaptation. *Research Paper No. 18*. UNCTAD. Retrieved June 2, 2022, from https://unctad.org/system/files/official-document/ser-rp-2017d18_en.pdf

Becker, A., Acciaro, M., Asariotis, R., Carera, E., Cretegny, L., Crist, P., Esteban, M., Mather, A., Messner, S., Naruse, S., Ng, A. K. Y., Rahmstorf, S., Savonis, M., Song, D., Stenek, V., and Velegrakis, A. F. (2013). A note on climate change adaptation for seaports: A challenge for global ports, a challenge for global society. *Climatic Change*, 120(4), 683–695. https://doi.org/10.1007/s10584-013-0843-z

Becker, A., Ng, A. K., McEvoy, D., and Mullett, J. (2018). Implications of climate change for shipping: Ports and supply chains. *Wiley Interdisciplinary Reviews: Climate Change*, 9(2), e508.

Bie, Z., Lin, Y., Li, G., and Li, F. (2017). Battling the extreme: A study on the power system resilience. *Proceedings of the IEEE*, 105(7), 1253–1266.

Bird, F., Clarke, A., Davies, P., and Surkovic, E. (2020). *Ammonia: Zero-Carbon Fertiliser, Fuel and Energy Store*. Policy Briefing. The Royal Society, London. Retrieved November 23, 2021, from https://royalsociety.org/topics-policy/projects/low-carbon-energy-programme/green-ammonia/

Bruneau, M., Chang, S., Eguchi, R., Lee, G., O'Rourke, T., Reinhorn, A., . . . and vonWinterfeldt, D. (2003). A framework to quantitatively assess and enhance seismic resilience of com-munities. *Earthq Spectra*, 19, 733–752.

Bryson, J. (2004). What to do when stakeholders matter: Stakeholder identification and analysis techniques. *Public Management Review*, 6(1), 21–53.

Camus, P., Tomás, A., Díaz-Hernandez, G., Rodríguez, B., Izaguirre, C., and Losada, I. (2019). Probabilistic assessment of port operation downtimes under climate change. *Coastal Engineering*, 147, 12–24.

Chen, L., Dui, H., and Zhang, C. (2020). A resilience measure for supply chain systems considering the interruption with the cyber-physical systems. *Reliability Engineering and System Safety*, 199, 106869.

Christodoulou, A., and Demirel, H. (2018). *Impacts of Climate Change on Transport—A Focus On Airports, Seaports and Inland Waterways*. EUR 28896 EN. Publications Office of the European Union, Luxembourg, ISBN 978-92-79-97039-9. https://doi.org/10.2760/378464, JRC108865

Cookey, P. E., Darnsawasdi, R., and Ratanachai, C. (2016). Performance evaluation of lake basin water governance using composite index. *Ecological Indicators*, 61, 466–482.

Cutter, S., Ash, K., and Emrich, C. (2014). The geographies of community disaster resilience. *Global Environmental Change*, 29, 65–77.

Davis, K. E., Kingsbury, B., and Merry, S. (2015). Introduction: The local-global life of indicators: Law, power, and resistance. In *The Quiet Power of Indicators: Measuring Governance, Corruption, and Rule of Law* (pp. 1–24). Cambridge University Press.

Davydenko, I. Y., and Fransen, R. W. (2019). Conceptual agent based model simulation for the Port Nautical Services. *IFAC-PapersOnLine*, 52(3), 19–24.

Dessavre, D., Ramirez-Marquez, J., and Barker, K. (2016). Multidimensional approach to complex system resilience analysis. *Reliability Engineering and System Safety*, 149, 34–43.

Dhanak, M., Parr, S., Kaisar, E. I., Goulianou, P., Russell, H., and Kristiansson, F. (2021). Resilience assessment tool for port planning. *Environment and Planning B: Urban Analytics and City Science*, 2399808321997824.

Djalante, R., and Thomalla, F. (2011). Community resilience to natural hazards and climate change impacts: A review of definitions and operational frameworks. *Asian Journal of Environment and Disaster Management (AJEDM)*, 3(3).

Dui, H., Zheng, X., and Wu, S. (2021). Resilience analysis of maritime transportation systems based on importance measures. *Reliability Engineering and System Safety*, 209, 107461.

Economic Commission for Europe. (2013). *Climate Change Impacts and Adaptation for International Transport Networks: Expert Group Report*. New York and Geneva.

Economic Commission for Europe. (2019a). *Report of the Group of Experts on Climate Change Impacts and Adaptation for Transport Networks and Nodes*. ECE/TRANS/WP.5/GE.3/36, Geneva, 27 June.

Economic Commission for Europe. Inland Transport Committee. (2019b). Implication for transport from climate variability and change. *Working Party on Transport Trends and Economics. Group of Experts on Climate Change Impacts and Adaptation for Transport Networks and Nodes*. Seventeenth session, Geneva, 24 and 25 April 2019. Item 4 of the provisional agenda. Discussions on the final report of the Group of Experts. Retrieved June 3, 2022 from https://unece.org/fileadmin/DAM/trans/doc/2019/wp5/ECE-TRANS-WP5-GE3-2019-08e.pdf

Er Kara, M., Ghadge, A., and Bititci, U. S. (2020). Modelling the impact of climate change risk on supply chain performance. *International Journal of Production Research*, 1–19.

EUROCONTROL. (2008). *Performance Review Report 2007: An Assessment of Air Traffic Management in Europe during the Calendar Year 2007*. EUROCONTROL PerformanceReview Commission, Brussels.

European Environment Agency. (2017). *Climate Change, Impacts and Vulnerability in Europe an Indicator-Based Report*. EEA Report No 1/2017. Publications Office of the European Union, Luxembourg. Retrieved December 3, 2021, from www.eea.europa.eu/publications/climate-change-impacts-and-vulnerability-2016

EUROSTAT. (2022). *Maritime Freight and Vessels Statistics*. Retrieved June 7, 2022, from https://ec.europa.eu/eurostat/statistics-explained/index.php?title=Maritime_ports_freight_and_passenger_statistics&oldid=218671#Most_EU_maritime_freight_transport_is_with_extra-EU_partners

Few, R., Brown, K., and Tompkins, E. L. (2007). Public participation and climate change adaptation: Avoiding the illusion of inclusion. *Climate Policy*, 7(1), 46–59.

Füssel, H. M., and R. J. Klein. 2006. Climate change vulnerability assessments: An evolution of conceptual thinking. *Climatic Change*, 75(3), 301–329.

Garza-Reyes, J. (2015). Lean and green–a systematic review of the state of the art literature. *Journal of Cleaner Production*, 102, 18–29.

Geng, Y., Fu, J., Sarkis, J., and Xue, B. (2012). Towards a national circular economy indicator system in China: An evaluation and critical analysis. *Journal of Cleaner Production*, 23(1), 216–224.

Gong, L., Xiao, Y. B., Jiang, C., Zheng, S., and Fu, X. (2020). Seaport investments in capacity and natural disaster prevention. *Transportation Research Part D: Transport and Environment*, 85, 102367.

Grafton, R. Q., Doyen, L., Béné, C., Borgomeo, E., Brooks, K., Chu, L., . . . and Williams, J. (2019). Realizing resilience for decision-making. *Nature Sustainability*, 2(10), 907–913.

Greco, S., Ishizaka, A., Tasiou, M., and Torrisi, G. (2019). On the methodological framework of composite indices: A review of the issues of weighting, aggregation, and robustness. *Social Indicators Research*, 141(1), 61–94.

Greenan, B. J. W., Shackell, N., Ferguson, K., Greyson, P., Cogswell, A., Brickman, D., . . . and Saba, V. S. (2019). Climate change vulnerability of American lobster fishing communities in Atlantic Canada. *Frontiers in Marine Science*, 6, 579.

Hohenstein N.-O., Feisel, E., Hartmann, E., et al. (2015) Research on the phenomenon of supply chain resilience: A systematic review and paths for further investigation. *International Journal of Physical Distribution and Logistics Management*, 45, 90–117.

Holling, C. (1973). Resilience and stability of ecological systems. *Annual Review of Ecology and Systematics*, 4(1), 1–23.

Hosseini, S., Barker, K., and Ramirez-Marquez, J. (2016). A review of definitions and measures of system resilience. *Reliability Engineering and System Safety*, 145, 47–61.

Hosseini, S., Ivanov, D., and Dolgui, A. (2019). Review of quantitative methods for supply chain resilience analysis. *Transportation Research Part E: Logistics and Transportation Review*, 125, 285–307.

Howard-Grenville, J., Buckle, S., Hoskins, B., and George, G. (2014). Climate change and management. *Academy of Management Journal*, 57(3), 615–623.

IPCC. (2013). *Climate Change (2013): The Physical Science Basis. Contribution of Working Group I to the Fifth Assessment Report of the Intergovernmental Panel on Climate Change* [Stocker, T. F., D. Qin, G.-K. Plattner, M. Tignor, S. K. Allen, J. Boschung, A. Nauels, Y. Xia, V. Bex and P. M. Midgley (eds.)]. Cambridge and New York: Cambridge University Press.

IPCC. (2018). Annex I: Glossary [Matthews, J. B. R. (ed.)]. In: *Global Warming of 1.5°C. An IPCC Special Report on the Impacts of Global Warming of 1.5°C above Pre-Industrial Levels and Related Global Greenhouse Gas Emission Pathways, in the Context of Strengthening the Global Response to the Threat of Climate Change, Sustainable Development, and Efforts to Eradicate Poverty* [Masson-Delmotte, V., P. Zhai, H.-O. Pörtner, D. Roberts, J. Skea, P. R. Shukla, A. Pirani, W. Moufouma-Okia, C. Péan, R. Pidcock, S. Connors, J. B. R. Matthews, Y. Chen, X. Zhou, M. I. Gomis, E. Lonnoy, T. Maycock, M. Tignor, and T. Waterfield (eds.)]. In Press.

IPCC (2019a). Annex I: Glossary [van Diemen, R. (ed.)]. In: *Climate Change and Land: An IPCC Special Report on Climate Change, Desertification, Land Degradation, Sustainable Land Management, Food Security, and Greenhouse Gas Fluxes In Terrestrial Ecosystems* [P. R. Shukla, J. Skea, E. Calvo Buendia, V. Masson-Delmotte, H.-O. Pörtner, D. C. Roberts, P. Zhai, R. Slade, S. Connors, R. van Diemen, M. Ferrat, E. Haughey, S. Luz, S. Neogi, M. Pathak, J. Petzold, J. Portugal Pereira, P. Vyas, E. Huntley, K. Kissick, M. Belkacemi, J. Malley, (eds.)]. In Press.

IPCC (2019b). *Special Report on the Ocean and Cryosphere in a Changing Climate, IPCC, 2019: IPCC Special Report on the Ocean and Cryosphere in a Changing Climate* [H.-O. Pörtner, D. C. Roberts, V. Masson-Delmotte, et al., (eds.).]. www.ipcc.ch/srocc/

IPCC. (2022). *Climate Change 2022: Impacts, Adaptation and Vulnerability. Summary for Policymakers. Working Group II to the Sixth Assessment Report of the Intergovernmental Panel on Climate Change*. Retrieved June 3, 2022 from www.ipcc.ch/report/ar6/wg2/downloads/report/IPCC_AR6_WGII_FinalDraft_FullReport.pdf

IPCC (Intergovernmental Panel on Climate Change) Field, C., Barros, V., Stocker, T., Quin, D., Dokken, D., . . . and Midgley, P. (2012). *Managing the Risks of Extreme Events and Disasters to Advance Climate Change Adaptation. A Special Report of Working Groups I and II of the Intergovernmental Panel on Climate Change.* Cambridge and New York: Cambridge University Press.

Izaguirre, C., Losada, I. J., Camus, P., González-Lamuño, P., and Stenek, V. (2020). Seaport climate change impact assessment using a multi-level methodology. *Maritime Policy and Management*, 1–14.

Izaguirre, C., Losada, I. J., Camus, P., Vigh, J. L., and Stenek, V. (2021). Climate change risk to global port operations. *Nature Climate Change*, 11(1), 14–20.

Jiang, M., Lu, J., Qu, Z., and Yang, Z. (2021). Port vulnerability assessment from a supply Chain perspective. *Ocean and Coastal Management*, 213, 105851.

Kamalahmadi, M., and Parast, M. M. (2016). A review of the literature on the principles of enterprise and supply chain resilience: Major findings and directions for future research. *International Journal of Production Economics*, 171, 116–133.

Karagiannis, G. M., Turksezer, Z. I., Alfieri, L., Feyen, L., and Krausmann, E. (2017). *Climate Change and Critical Infrastructure–Floods.* EUR-Scientific and Technical Research Reports. Publications Office of the European Union, Luxembourg.

Klein, R. J., Nicholls, R. J., and Thomalla, F. (2003). Resilience to natural hazards: How useful is this concept?. *Global Environmental Change Part B: Environmental Hazards*, 5(1), 35–45.

Kontogianni, A., Damigos, D., Kyrtzoglou, T., Tourkolias, C., and Skourtos, M. (2019). Development of a composite climate change vulnerability index for small craft harbours. *Environmental Hazards*, 18(2), 173–190.

Laxe, F., Bermúdez, F., Palmero, F., and Novo-Corti, I. (2017). Assessment of port sustainability through synthetic indexes. Application to the Spanish case. *Marine Pollution Bulletin*, 119(1), 220–225.

Lenderink, G., de Vries, H., Fowler, H. J., Barbero, R., van Ulft, B., and van Meijgaard, E. (2021). Scaling and responses of extreme hourly precipitation in three climate experiments with a convection-permitting model. *Philosophical Transactions of the Royal Society A*, 379(2195), 20190544.

León-Mateos, F., Sartal, A., López-Manuel, L., and Quintás, M. A. (2021). Adapting our sea ports to the challenges of climate change: Development and validation of a Port Resilience Index. *Marine Policy*, 130, 104573.

Liu, X., and Chen, Z. (2021). An integrated risk and resilience assessment of sea ice disasters on port operation. *Risk Analysis*, 41(9), 1579–1599.

López-Manuel, L., León-Mateos, F., and Sartal, A. (2020). Closing loops, easing strains: Industry 4.0's potential for overcoming challenges of circularity in manufacturing environments. In *Circular Economy for the Management of Operations* (pp. 23–47). CRC Press.

Mayunga, J. (2007). Understanding and applying the concept of community disaster resilience: A capital-based approach. *Summer Academy for Social Vulnerability and Resilience Building*, 1(1), 1–16.

McEvoy, D., and Mullett, J. (2013). Enhancing the resilience of seaports to a changing climate: Research synthesis and implications for policy and practice. *Work Package*, 4.

McIntosh, R. D., and Becker, A. (2017). Seaport climate vulnerability assessment at the multi-port Scale: A review of approaches. In I. Linkov and J. M. Palma-Oliveira (eds.), *Resilience and Risk* (pp. 205–224). Springer.

McIntosh, R. D., and Becker, A. (2019). Expert evaluation of open-data indicators of seaport vulnerability to climate and extreme weather impacts for U.S. North Atlantic ports. *Ocean and Coastal Management*, 180(104911).

Mickwitz, P., and Melanen, M. (2009). The role of co-operation between academia and policy-makers for the development and use of sustainability indicators–a case from the Finnish Kymenlaakso Region. *Journal of Cleaner Production*, 17(12), 1086–1100.

Mojtahedi, S. M., and Oo, B. (2014). Development of an index to measure stakeholder approaches toward disasters in the built environment. *Procedia Economics and Finance*, 18, 95–102.

Monios, J., and Wilmsmeier, G. (2020). Deep adaptation to climate change in the maritime transport sector–a new paradigm for maritime economics?. *Maritime Policy & Management*, 47(7), 853–872.

Monioudi, I. N., Asariotis, R., Becker, A., Bhat, C., Dowding-Gooden, D., Esteban, M., . . . and Phillips, W. (2018). Climate change impacts on critical international transportation assets of Caribbean Small Island Developing States (SIDS): The case of Jamaica and Saint Lucia. *Regional Environmental Change*, 18(8), 2211–2225.

Mutombo, K., and Ölçer, A. (2017). Towards port infrastructure adaptation: A global port climate risk analysis. *WMU Journal of Maritime Affairs*, 16(2), 161–173.

Ng, A., and Liu, J. (2014). *Port-Focal Logistics and Global Supply Chains*. Springer. Palgrave Macmillan. https//doi.org/10.1057/9781137273697

Ng, A., Monios, J., and Zhang, H. (2019). Climate adaptation management and institutional erosion: Insights from a major Canadian port. *Journal of Environmental Planning and Management*, 62(4), 586–610.

Notteboom, T. (2006). Concession agreements as port governance tools. *Research in Transportation Economics*, 17, 437–455.

Nyong, A., Adesina, F., and Elasha, B. O. (2007). The value of indigenous knowledge in climate change mitigation and adaptation strategies in the African Sahel. *Mitigation and Adaptation strategies for global Change*, 12(5), 787–797.

Orencio, P., and Fujii, M. (2013). A localized disaster resilience index to assess coastal communities based on an analytic hierarchy process (AHP). *International Journal of Disaster Risk Reduction*, 3(1), 62–75.

Panahi, R., Ng, A. K., and Pang, J. (2020). Climate change adaptation in the port industry: A complex of lingering research gaps and uncertainties. *Transport Policy*, 95, 10–29.

Pettit, T. J., Croxton, K. L., and Fiksel, J. (2019). The evolution of resilience in supply chain management: A retrospective on ensuring supply chain resilience. *Journal of Business Logistics*, 40(1), 56–65.

PIANC. The World Association for Waterborne Transport Infrastructure. (2020). *PIANC Report N° 178 Environmental Commission*. Climate Change Adaptation Planning for Ports and Inland Waterways.

Pielke Jr, R. A. (2007). Future economic damage from tropical cyclones: Sensitivities to societal and climate changes. *Philosophical Transactions of the Royal Society A: Mathematical, Physical and Engineering Sciences*, 365(1860), 2717–2729.

Puertos del Estado (Ministerio de Fomento-Gobierno de España). (2012). *Recomendaciones Para Obras Marítimas ROM 2.0-11 Tomo II*. Retrieved November 26, 2021, from www.puertos.es/es-es/BibliotecaV2/ROM%202.0-11.pdf

Rose, A. (2004). Defining and measuring economic resilience to disasters. *Disaster Prev Manage*, 13, 307–314.

Rusu, E., and Soares, C. G. (2011). Wave modelling at the entrance of ports. *Ocean Engineering*, 38(17–18), 2089–2109.

Sagna, P., Dipama, J. M., Vissin, E. W., Diomandé, B. I., Diop, C., Chabi, P. A. B., . . . and Yade, M. (2021). Climate change and water resources in West Africa: A case study of Ivory Coast, Benin, Burkina Faso, and Senegal. In *Climate Change and Water Resources in Africa* (pp. 55–86). Springer.

Şahin, Ü. A., Onat, B., and Ayvaz, C. (2019). Climate change and greenhouse gases in Turkey. In *Recycling and Reuse Approaches for Better Sustainability* (pp. 201–214). Springer.

Salvati, L., and Carlucci, M. (2014). A composite index of sustainable development at the local scale: Italy as a case study. *Ecological Indicators*, 43, 162–171.

Sánchez-Arcilla, A., Sierra, J. P., Brown, S., Casas-Prat, M., Nicholls, R. J., Lionello, P., and Conte, D. (2016). A review of potential physical impacts on harbours in the Mediterranean Sea under climate change. *Regional Environmental Change*, 16(8), 2471–2484.

Scholten, K., Scott, P. S., and Fynes, B. (2019). Building routines for non-routine events: Supply chain resilience learning mechanisms and their antecedents. *Supply Chain Management: An International Journal*, 24(3), 430–442.

Sierra, J. P., Genius, A., Lionello, P., Mestres, M., Mösso, C., and Marzo, L. (2017). Modelling the impact of climate change on harbour operability: The Barcelona port case study. *Ocean Engineering*, 141, 64–78.

Smith, L., Harwell, L., Bousquin, J., Buck, K., Harvey, J., and McLaghlin, M. (2019). Using re-scaled resilience screening index results and location quotients for socio-ecological characterizations in US coastal regions. *Frontiers in Environmental Science*, 7, Article 96, 1–16.

Summers, J., Smith, L., Harwell, L., and Buck, K. (2017). Conceptualizing holistic community resilience to climate events: Foundation for a climate resilience screening index. *GeoHealth*, 1(4), 151–164.

UNCTAD. (2020) *Climate Change Impacts and Adaptation for Coastal Transport Infrastructure: A Compilation of Policies and Practices*. United Nations Publications. Retrieved May 25, 2022, from https://unctad.org/system/files/official-document/dtltlb2019d1_en.pdf

UNCTAD. (2021). *Review of Maritime Transport 2021. UNCTAD/RMT/2021*. Retrieved May 25, 2022, from https://unctad.org/system/files/official-document/rmt2021_en_0.pdf.

UN-ISDR (United Nations International Strategy for Disaster Reduction), Currently United Nations Office for Disaster Risk Reduction (UNDRR). (2005). Hyogo framework for action 2005–2015: Building the resilience of nations and communities to disasters. In *Extract from the Final Report of the World Conference on Disaster Reduction (A/CONF. 206/6)* (Vol. 380). The United Nations international Strategy for Disaster Reduction.

USDOT. (2012). *Impacts of Climate Change and Variability on Transportation Systems and Infrastructure: The Gulf Coast Study, Phase II*. A report by the US Department of Transportation, Center for Climate Change and Environmental Forecasting [Choate A, W Jaglom, R Miller, B Rodehorst, P Schultz and C Snow (eds.)]. Department of Transportation, Washington, DC, USA, 470 pp.

Verschuur, J., Koks, E. E., and Hall, J. W. (2020). Port disruptions due to natural disasters: Insights into port and logistics resilience. *Transportation Research Part D: Transport and Environment*, 85, 102393.

Wang, K., and Zhang, A. (2018). Climate change, natural disasters and adaptation investments: Inter-and intra-port competition and cooperation. *Transportation Research Part B: Methodological*, 117, 158–189.

Woodruff, S. C., Meerow, S., Stults, M., and Wilkins, C. (2018). Adaptation to resilience planning: Alternative pathways to prepare for climate change. *Journal of Planning Education and Research*, 0739456X18801057.

Wright, P. (2013). Impacts of climate change on ports and shipping. *MCCIP Science Review*, 263–270.

Xia, W., and Lindsey, R. (2021). Port adaptation to climate change and capacity investments under uncertainty. *Transportation Research Part B: Methodological*, 152, 180–204.

Xiao, Y., Fu, X., Ng, A., and Zhang, A. (2015). Port investments on coastal and marine disasters prevention: Economic modeling and implications. *Transportation Research Part B: Methodological*, 78, 202–221.

Yang, Y. C., and Ge, Y. E. (2020). Adaptation strategies for port infrastructure and facilities under climate change at the Kaohsiung port. *Transport Policy*, 97, 232–244.

Yang, Z., Ng, A. K., Lee, P. T. W., Wang, T., Qu, Z., Rodrigues, V. S., . . . and Lau, Y. Y. (2018). Risk and cost evaluation of port adaptation measures to climate change impacts. *Transportation Research Part D: Transport and Environment*, 61, 444–458.

Yoon, D., Kang, J., and Brody, S. (2016). A measurement of community disaster resilience in Korea. *Journal of Environmental Planning and Management*, 59(3), 436–460.

Yoro, K. O., and Daramola, M. O. (2020). CO2 emission sources, greenhouse gases, and the global warming effect. In *Advances in Carbon Capture* (pp. 3–28). Woodhead Publishing.

Zhang, C., Xu, X., and Dui, H. (2020). Resilience measure of network systems by node and edge indicators. *Reliability Engineering and System Safety*, 202, 107035.

Zhao, K., He, B., Cui, H., Du, P., Han, B., Chen, M., and Wei, Z. (2021, February). Test of meteorological influence on sail based on fuzzy comprehensive evaluation. In *IOP Conference Series: Earth and Environmental Science* (Vol. 638, No. 1, p. 012040). IOP Publishing.

Index

Note: Page numbers in *italics* indicate figures, **bold** indicate tables in the text, and references following "n" refer notes.

Printed in the United States
by Baker & Taylor Publisher Services